Functional Analysis

An Elementary Introduction

Functional Analysis

An Elementary Introduction

Markus Haase

Graduate Studies in Mathematics

Volume 156

American Mathematical Society
Providence, Rhode Island

2010 *Mathematics Subject Classification.* Primary 46-01, 46Cxx, 46N20, 35Jxx, 35Pxx.

For additional information and updates on this book, visit
www.ams.org/bookpages/gsm-156

Library of Congress Cataloging-in-Publication Data

Haase, Markus, 1970–
Functional analysis : an elementary introduction / Markus Haase.
pages cm. — (Graduate studies in mathematics ; volume 156)
Includes bibliographical references and indexes.
ISBN 978-0-8218-9171-1 (alk. paper)
1. Functional analysis—Textbooks. 2. Differential equations, Partial—Textbooks. I. Title.

QA320.H23 2014
515′.7—dc23

2014015166

∞ The paper used in this book is acid-free and falls within the guidelines
established to ensure permanence and durability.
Visit the AMS home page at `http://www.ams.org/`

10 9 8 7 6 5 4 3 2 1 19 18 17 16 15 14

Für Dietlinde Haase (1938–2012)

Contents

Preface

The present book was developed out of my course, "Applied Functional Analysis", given during the years 2007–2012 at Delft University of Technology. It provides an introduction to functional analysis on an elementary level, not presupposing, e.g., background in real analysis like metric spaces or Lebesgue integration theory. The focus lies on notions and methods that are relevant in "applied" contexts. At the same time, it should serve as a stepping stone towards more advanced texts in functional analysis.

The course (and the book) evolved over the years in a process of reflection and revision. During that process I gradually realized that I wanted the students to learn (at least):

- to view functions/sequences as *points in certain spaces*, abstracting from their internal structure;
- to treat *approximations* in a multitude of situations by virtue of the concept of an abstract *distance* (metric/norm) with its diverse instances;
- to use *approximation arguments* in order to establish properties of otherwise unwieldy objects;
- to recognize *orthogonality* and its fundamental role for series representations and distance minimization in Hilbert spaces;
- to reduce differential and integral equations to abstract fixed point or minimization problems and find solutions via approximation methods, recognizing the role of *completeness*;
- to work with *weak derivatives* in order to facilitate the search for solutions of differential equations via Hilbert space methods;

- to use *operators* as a unified tool of producing solutions to a problem with varying initial data;
- to be aware of the important role of *compactness*, in particular for eigenvalue expansions.

In this book, functional analysis is developed to an extent that serves these purposes. The included examples are of an elementary character and might appear — from the point of view of applications — a little artificial. However, with the material presented in the book at hand, students should be prepared for serious real-world applications as well as for more sophisticated theoretical functional analysis.

For the Student and the Teacher. This book can be used for self-study. Its material is divided into "mandatory" and "optional" parts. The latter are indicated by a star in front of the title; see the table of contents. By "optional" I mean that it can be omitted without affecting the "mandatory" parts. However, optional material from a later chapter may refer to optional material from an earlier one. In principle, "optional" does not necessarily mean "more advanced", although it occasionally may be like that. In effect, the optional parts can be viewed as an "honors track" amendment to the mandatory course.

In the optional parts I sometimes leave the details to the reader, something that I have tried to avoid in the mandatory part.

Being interested mainly in "applied mathematics", one may well stop with Chapter 14. Chapters 15 and 16 are more of a theoretical nature and are supposed to be a bridge towards higher functional analysis. (That, however, does not mean that they are irrelevant for applications.)

Integration Theory. A sensitive point in any introductory course on functional analysis is the use of measure-theoretic integration theory. For this book, no prior knowledge of Lebesgue theory is required. However, such ignorance has to be compensated by the will to take some things for granted and to work with some concepts even if they are only partially understood.

Chapter 7 provides the necessary information. For the later chapters one should have a vague understanding of what Lebesgue measure is and how it is connected with the notion of integral, a more thorough understanding of what a null set is and what it means that something is true almost everywhere, and a good working knowledge of the dominated convergence theorem (neglecting any measurability issues).

As unproven facts from integration theory the following results are used:

- The dominated convergence theorem (Theorem 7.16);

- The density of $C[a,b]$ in $L^2(a,b)$ (Theorem 7.24);
- Fubini's theorem (Section 11.1);
- The density of $L^2(X) \otimes L^2(Y)$ in $L^2(X \times Y)$ (Theorem 11.2).

See also my "Advice for the Reader" on page 125.

Exercises. Each chapter comes with three sets of exercises — labelled Exercises A, B and C. Exercises of category A are referred to alongside the text. Some of them are utmost elementary and all of them have a direct connection to the text at the point where they are referred to. They are "simple" as far as their complexity is concerned, and the context mostly gives a hint towards the solution. One could consider these exercises as recreational pauses during a strenuous hike; pauses that allow one to look back on the distance made and to observe a little closer the passed landscape.

Exercises of category B are to deepen the understanding of the main (mandatory) text. Many of them have been posed as homework exercises or exam questions in my course. The exercises of category C either refer to the mandatory parts, but are harder; or they refer to the optional material; or they cover some additonal and more advanced topics.

Synopsis. In the following I describe shortly the contents of the individual chapters.

Chapter 1: Vector spaces of functions, linear independence of monomials, standard inner products, inner product spaces, norm associated with an inner product, polarization identity, parallelogram law, orthogonality, Pythagoras' lemma, orthonormal systems, orthogonal projections onto finite-dimensional subspaces, Gram–Schmidt procedure, the trigonometric system in $C[0,1]$.

Chapter 2: Cauchy–Schwarz inequality, triangle inequality, ℓ^2, normed spaces, ℓ^1, ℓ^∞, bounded linear mappings (operators), operator norm, isometries, point evaluations, left and right shift, multiplication operators and other examples of operators. *Optional:* ℓ^p-spaces for all $1 < p < \infty$ and Hölder's inequality.

Chapter 3: Metric associated with a norm, metric spaces, discrete metric, convergence in metric spaces, uniform vs. pointwise vs. square mean convergence, mean vs. square mean convergence on $C[a,b]$, closure of a subset, dense subsets, c_{00} dense in ℓ^p $(p = 1,2)$ and in c_0, properties of the closure, Weierstrass' theorem (without proof).

Chapter 4: Open and closed sets (definition, examples, properties), continuity in metric spaces, examples (continuity of metric, norm, algebraic operations, inner product), the closure of a subspace is a subspace, continuity is equal to boundedness for linear mappings, (sequential) compactness and its consequences, Bolzano–Weierstrass theorem, stronger and weaker norms, equivalence of norms. *Optional:* separability and general compactness.

Chapter 5: Cauchy sequences, complete metric spaces, Hilbert spaces, ℓ^2 is a Hilbert space, $(\mathrm{C}[a,b], \|\cdot\|_2)$ is not complete, Banach spaces, examples (finite-dimensional spaces, $(\mathcal{B}(\Omega), \|\cdot\|_\infty)$, ℓ^∞, $(\mathrm{C}[a,b], \|\cdot\|_\infty)$, absolutely convergent series in Banach spaces.

Chapter 6 (optional): Banach's contraction principle, local existence and uniqueness of solutions to ODEs, Google's PageRank algorithm, inverse mapping theorem and implicit function theorem from many-variable calculus.

Chapter 7: Lebesgue (outer) measure, measurable sets and functions, Lebesgue integral, L^p for $p \in \{1, 2, \infty\}$, null sets, equality/convergence almost everywhere, dominated convergence theorem, completeness of L^p, Hölder's inequality, $\mathrm{C}[a,b]$ is dense in $\mathrm{L}^p(a,b)$, for $p = 1, 2$. *Optional:* L^p-spaces for general p.

Chapter 8: Best approximations, counterexamples (nonexistence and non-uniqueness), existence and uniqueness of best approximations in closed convex subsets of a Hilbert space, orthogonal projections, orthogonal decomposition, Riesz–Fréchet theorem, orthogonal series and Parseval's identity, abstract Fourier expansions and Bessel's inequality, orthonormal bases.

Chapter 9: Approximation and permanence principles, proof of Weierstrass' theorem, approximation via truncation, density of $\mathrm{C}_c^\infty(\mathbb{R})$ in $\mathrm{L}^p(\mathbb{R})$, classical Fourier series, the trigonometric system is an orthonormal basis of $\mathrm{L}^2(0,1)$, theorem of Riemann–Lebesgue. *Optional:* strong convergence lemma, Fejér's theorem, convolution operators, uniqueness theorem for Fourier series, extension of bounded linear mappings, Plancherel's theorem.

Chapter 10: Weak derivatives, Sobolev spaces $\mathrm{H}^n(a,b)$, fundamental theorem of the calculus for H^1-functions, density of $\mathrm{C}^1[a,b]$ in $\mathrm{H}^1[(a,b)$, variational method for the Poisson problem on (a,b), Poincaré's inequality for an interval. *Optional:* Poisson problem on $\Omega \subseteq \mathbb{R}^d$.

Chapter 11: Integration on product spaces, Fubini's theorem, integral operators, invertibility of operators and well-posedness of equations, Dirichlet Laplacian, Green's function, Hilbert–Schmidt integral operators, strong vs. norm convergence of operators, perturbation and Neumann series, Volterra integral equations.

Chapter 12: Operators of finite rank, compact operators, Hilbert–Schmidt operators are compact, diagonal argument, representing sesquilinear forms by operators, adjoints. *Optional:* Lax–Milgram theorem, Sturm–Liouville problems, abstract Hilbert–Schmidt operators.

Chapter 13: Eigenvalues and approximate eigenvalues, location of the spectrum, self-adjoint operators, numerical range, spectral theorem for compact self-adjoint operators, eigenvalue equation and Fredholm alternative. *Optional:* spectral theory on Banach spaces (in exercises).

Chapter 14: Eigenvalue expansion of the (one-dimensional) Dirichlet Laplacian and a Schrödinger operator, application to the associated parabolic evolution equation. *Optional:* the norm of the integration operator, best constant in the one-dimensional Poincaré inequality.

Chapter 15: Principle of nested balls, Baire's theorem, uniform boundedness principle, Banach–Steinhaus theorem, Dirichlet kernel, Du Bois-Reymond's theorem, open mapping theorem, closed graph theorem, applications, Tietze's theorem.

Chapter 16: Dual space, sublinear functionals, Hahn–Banach theorem for separable spaces, elementary duality theory, dual operators, pairings and dualities, identification of duals for c_0, ℓ^1, and $L^1[a,b]$. *Optional:* Hahn–Banach theorem for general spaces, geometric Hahn–Banach theorem (without proof), reflexivity, weak convergence, dual of ℓ^p and $L^p[a,b]$ for $1 \le p < \infty$, Riesz representation theorem, dual of $C[a,b]$.

History of Functional Analysis. Many mathematical concepts or results are named after mathematicians, contemporary or past. These names are a convenient help for our memory, but should not be mistaken as a claim about who did what first. Certainly, what I call Pythagoras' lemma in this book (Lemma 1.9) was not stated in this form by Pythagoras, and we use the name since the lemma is a generalization and modernization of a well-known theorem from Euclidean geometry that traditionally is associated with Pythagoras.

Although the taxonomy is sometimes unjustified or questionable, it is not arbitrary. There are in fact *real people* behind functional analysis, and what now appears to be a coherent and complete theory needed more than a century to find its contemporary form.

After the main text and before the appendices I have included a short account of that history with special focus on the parts that are treated in the main text. A brief historical account of the real number system is included in Appendix A.5.

What is Missing. Several topics from the classical canon of functional analysis are not covered: continuous functions on compact spaces (Urysohn's lemma, Arzelá–Ascoli, Stone–Weierstrass theorem), locally convex vector spaces, theory of distributions, Banach algebras and Gelfand theory, weak topologies, Riesz' theory of compact operators on general Banach spaces, spectral theory on Banach spaces, unbounded (symmetric or selfadjoint) operators on Hilbert spaces, the general spectral theorem, Sobolev spaces other than H^n on intervals, elliptic differential equations other than in dimension one, operator semigroups.

Further Reading. A book close in spirit to my text is the work [**GGK03**] by Gohberg, Goldberg and Kaashoek. Beyond that, I recommend the excellent works [**Che01**] by Ward Cheney and [**You88**] by Nicholas Young. These two books were a very valuable assistance during the writing.

In the direction of applications, a suitable follow-up to this book are Eberhard Zeidler's two volumes [**Zei95a, Zei95b**].

If one wants to step deeper into functional analysis there are so many possibilities that to mention just a few would do injustice to all the others. The most profound and comprehensive modern treatment that I know, and certainly a recommendation for the future expert, is Peter Lax's *opus magnum* [**Lax02**].

Acknowledgements. I am very grateful to my students at Delft University of Technology. Confronting their questions about the material and their difficulties in coming to terms with it had a profound influence on me while writing this book. In addition, several students helped to eliminate mistakes from the various preliminary versions.

Special thanks (and many compliments) go to Martijn de Jong (Delft) for producing the vast majority of the figures.

I also want to acknowledge the contributions of many colleagues, most prominently the remarks by Jürgen Voigt (Dresden) and Hendrik Vogt (Dresden, Clausthal) on Chapter 15, which grew out of a discussion of Sokal's article [**Sok11**]. From Bernhard Haak (Bordeaux) I learned — apart from many other things — to view the usual proof of Tietze's theorem in functional analytic terms.

I am grateful to my colleagues from Delft for the excellent working atmosphere they create and the love for functional analysis that we share. In particular, I am indebted to Ben de Pagter, who encouraged me all along to write this text and to Jan van Neerven who read parts of the manuscript and whose comments helped much to improve it.

This book was completed during a sabbatical stay at the University of Auckland, and my warmest thanks go to Tom ter Elst for his generous invitation and his warm hospitality, and for his very valuable remarks on some parts of the text.

I am indebted to Luann Cole from the American Mathematical Society for her very thorough copyediting which led to a considerable improvement of the text.

I thank my friends and co-authors Bálint Farkas (Wuppertal) and Bernhard Haak (Bordeaux) for their support and their patience.

Finally, I would like to thank those from whom I myself learned functional analysis, Wolfgang Arendt (Ulm), Rainer Nagel and Ulf Schlotterbeck (Tübingen). Without them not one single line of this text would ever have been written.

Delft and Auckland, April 2014

Markus Haase

Chapter 1

Inner Product Spaces

The main objects of study in functional analysis are *function spaces*, i.e., vector spaces of real or complex-valued functions on certain sets. Although much of the theory can be done in the context of real vector spaces, at certain points it is very convenient to have vector spaces over the complex number field $\mathbb{C}$. So we introduce the generic notation $\mathbb{K}$ to denote either $\mathbb{R}$ or $\mathbb{C}$. Background on linear algebra is collected in Appendix A.7.

We begin by introducing two central examples of function spaces.

The Space $\mathbb{K}^d$. This is the set of all tuples $x = (x_1, \dots, x_d)$ with components $x_1, \dots, x_d \in \mathbb{K}$:

$$\mathbb{K}^d := \{x = (x_1, \dots, x_d) \mid x_1, \dots, x_d \in \mathbb{K}\}.$$

It is a vector space over $\mathbb{K}$ with the obvious (i.e., componentwise) operations:

$$\begin{aligned}(x_1, \dots, x_d) + (y_1, \dots, y_d) &:= (x_1 + y_1, \dots, x_d + y_d),\\ \lambda(x_1, \dots, x_d) &:= (\lambda x_1, \dots, \lambda x_d).\end{aligned}$$

The Space $\mathrm{C}[a, b]$. We let $[a, b]$ be any closed interval of $\mathbb{R}$ of positive length. Let us define

$$\begin{aligned}\mathcal{F}[a, b] &:= \{f \mid f : [a, b] \longrightarrow \mathbb{K}\},\\ \mathrm{C}[a, b] &:= \{f \mid f : [a, b] \longrightarrow \mathbb{K}, \text{ continuous}\}.\end{aligned}$$

If $\mathbb{K} = \mathbb{C}$, then $f : [a, b] \to \mathbb{C}$ can be written as $f = \operatorname{Re} f + \mathrm{i} \operatorname{Im} f$ with $\operatorname{Re} f$, $\operatorname{Im} f$ being real-valued functions; and f is continuous if and only if both $\operatorname{Re} f$, $\operatorname{Im} f$ are continuous.

Let us define the sum and the scalar multiple of functions pointwise, i.e.,

$$(f + g)(t) := f(t) + g(t), \quad (\lambda f)(t) := \lambda f(t)$$

where $f, g : [a, b] \to \mathbb{K}$ are functions, $\lambda \in \mathbb{K}$ and $t \in [a, b]$. This turns the set $\mathcal{F}[a, b]$ into a vector space over $\mathbb{K}$ (see also Appendix A.7).

If $f, g \in \mathrm{C}[a, b]$ and $\lambda \in \mathbb{K}$, then we know from elementary analysis that $f + g, \lambda f \in \mathrm{C}[a, b]$ again. Since $\mathrm{C}[a, b]$ is certainly not the empty set, it is therefore a subspace of $\mathcal{F}[a, b]$, and hence a vector space in its own right.

We use the notation $\mathrm{C}[a, b]$ for the generic case, leaving open whether $\mathbb{K} = \mathbb{C}$ or $\mathbb{K} = \mathbb{R}$. If we want to stress a particular choice of $\mathbb{K}$, we write $\mathrm{C}([a, b]; \mathbb{C})$ or $\mathrm{C}([a, b]; \mathbb{R})$. If we use the notation $\mathrm{C}[a, b]$ in concrete situations, it is always tacitly assumed that we have the more general case $\mathbb{K} = \mathbb{C}$. Similar remarks apply to $\mathcal{F}[a, b]$ and all other function spaces.

There is an analogy between these two examples. Namely, note that each vector $(x_1, \dots, x_d) \in \mathbb{K}^d$ defines a map

$$x : \{1, \dots, d\} \longrightarrow \mathbb{K} \quad \text{by} \quad x(j) := x_j \quad (j = 1, \dots, d).$$

Conversely, each such map x determines exactly one vector $(x(1), \dots, x(d))$. Apart from a set-theoretical point of view, there is no difference between the vector and the corresponding function, and we will henceforth identify them. So we may write

$$\mathbb{K}^d = \mathcal{F}(\{1, \dots, d\}; \mathbb{K}).$$

A short look will convince you that the addition and scalar multiplication in vector notation coincides precisely with the pointwise sum and scalar multiplication of functions.

How far can we push the analogy between $\mathbb{K}^d$ and $\mathrm{C}[a, b]$? Well, the first result is negative:

Theorem 1.1. *The space $\mathbb{K}^d$ has a basis consisting of precisely d vectors, hence is finite-dimensional. The space $\mathrm{C}[a, b]$ is not finite-dimensional. For example, the set of monomials $\{1, t, t^2, \dots\}$ is an infinite linearly independent subset of $\mathrm{C}[a, b]$.*

Proof. The first assertion is known from linear algebra. Let us turn to the second. Let

$$p(t) := a_n t^n + \cdots + a_1 t + a_0$$

be a finite linear combination of monomials, i.e., $a_0, \dots, a_n \in \mathbb{K}$. We suppose that not all coefficients a_j are zero, and we have to show that then p cannot be the zero function.

Now, if $p(c) = 0$, then by long division we can find a polynomial q such that $p(t) = (t - c)q(t)$ and $\deg q < \deg p$. If one applies this repeatedly, one may write

$$p(t) = (t - c_1)\,(t - c_2)\ldots(t - c_k)\,q(t)$$

for some $k \le n$ and some polynomial q that has no zeroes in $[a, b]$. But that means that p can have only finitely many zeroes in $[a, b]$. Since the interval $[a, b]$ has infinitely many points, we are done. (See Exercise 1.1 for an alternative proof.) □ Ex.1.1

1.1. Inner Products

We now come to a positive result. The **standard inner product** of two vectors $x, y \in \mathbb{K}^d$ is defined by

$$\langle x, y\rangle := x_1\overline{y_1} + x_2\overline{y_2} + \cdots + x_d\overline{y_d} = \sum_{j=1}^{d} x_j\overline{y_j}.$$

If $\mathbb{K} = \mathbb{R}$, this is the usual scalar product known from undergraduate courses; for $\mathbb{K} = \mathbb{C}$ this is a natural extension of it.

Analogously, we define the **standard inner product** on $\mathrm{C}[a, b]$ by

$$\langle f, g\rangle := \int_a^b f(t)\overline{g(t)}\,\mathrm{d}t$$

for $f, g \in \mathrm{C}[a, b]$. There is a general notion behind these examples.

Definition 1.2. Let E be a vector space. A mapping

$$E \times E \longrightarrow \mathbb{K}, \quad (f, g) \longmapsto \langle f, g\rangle$$

is called an **inner product** or a **scalar product** if it is *sesquilinear*:

$$\langle \lambda f + \mu g, h\rangle = \lambda \langle f, h\rangle + \mu \langle g, h\rangle,$$
$$\langle h, \lambda f + \mu g\rangle = \overline{\lambda} \langle h, f\rangle + \overline{\mu} \langle h, g\rangle \qquad (f, g, h \in E,\ \lambda, \mu \in \mathbb{K}),$$

symmetric:

$$\langle f, g\rangle = \overline{\langle g, f\rangle} \qquad (f, g \in E),$$

positive:

$$\langle f, f\rangle \ge 0 \qquad (f \in E),$$

and *definite*:

$$\langle f, f\rangle = 0 \implies f = 0 \qquad (f \in E).$$

A vector space E together with an inner product on it is called an **inner product space** or a **pre-Hilbert space**.[1]

[1] David Hilbert (1862–1943), German mathematician.

There are different symbols used to denote inner products; for example,

$$\langle f, g\rangle, \quad (f\,|\,g), \quad \langle f|g\rangle \quad \text{or simply} \quad (f,g).$$

The latter has the disadvantage that it is the same as for the *ordered pair* (f,g). To avoid confusion, we stick to the notation $\langle f, g\rangle$ in this book.

Ex.1.2 The proof that the standard inner product on $\mathrm{C}[a,b]$ is sesquilinear, symmetric and positive is an exercise. The definiteness is more interesting and derives from the following fact.

Lemma 1.3. *Let $f \in \mathrm{C}[a,b]$, $f \ge 0$. If $\int_a^b f(t)\,\mathrm{d}t = 0$, then $f = 0$.*

Proof. To prove the statement, suppose towards a contradiction that $f \neq 0$. Then there is $t_0 \in (a,b)$ where $f(t_0) \neq 0$, i.e., $f(t_0) > 0$. By continuity, there are $\epsilon, \delta > 0$ such that

$$|t - t_0| \le \delta \quad \Rightarrow \quad f(t) \ge \epsilon.$$

But then

$$\int_a^b f(t)\,\mathrm{d}t \ge \int_{t_0-\delta}^{t_0+\delta} f(t)\,\mathrm{d}t \ge 2\delta\epsilon > 0,$$

which contradicts the hypothesis. □

Using this lemma we prove definiteness as follows: Suppose that $f \in \mathrm{C}[a,b]$ is such that $\langle f, f\rangle = 0$. Then

$$0 = \langle f, f\rangle = \int_a^b f(t)\overline{f(t)}\,\mathrm{d}t = \int_a^b |f(t)|^2\,\mathrm{d}t.$$

Since $|f|^2$ is also a continuous function, the previous lemma applies and yields $|f|^2 = 0$, but this is equivalent to $f = 0$.

Note that by "$f = 0$" we actually mean "$f(t) = 0$ for all $t \in [a,b]$", and we use $|f|$ as an abbreviation of the *function* $t \mapsto |f(t)|$.

Inner products endow a vector space with a geometric structure that allows one to measure lengths and angles. If $(E, \langle\cdot,\cdot\rangle)$ is an inner product space, then the **length** of $f \in E$ is given by

$$\|f\| := \sqrt{\langle f, f\rangle}.$$

The following properties are straightforward from the definition:

$$\|f\| \ge 0, \quad \|\lambda f\| = |\lambda|\,\|f\|, \quad \|f\| = 0 \iff f = 0.$$

The mapping $\|\cdot\| : E \to \mathbb{R}$ is called the **norm induced by** or **associated with** the inner product $\langle \cdot, \cdot \rangle$. We will learn more about norms in the next chapter. Ex.1.3

Example 1.4. For the standard inner product on $\mathbb{K}^d$, the associated norm is

$$\|x\|_2 := \sqrt{\langle x, x \rangle} = \Big(\sum_{j=1}^{d} x_j \overline{x_j} \Big)^{1/2} = \Big(\sum_{j=1}^{d} |x_j|^2 \Big)^{1/2}$$

and is called the **2-norm** or **Euclidean**[2] **norm** on $\mathbb{K}^d$.

For the standard inner product on $\mathrm{C}[a, b]$ the associated norm is given by

$$\|f\|_2 := \sqrt{\langle f, f \rangle} = \Big(\int_a^b f(t)\overline{f(t)}\, \mathrm{d}t \Big)^{1/2} = \Big(\int_a^b |f(t)|^2 \, \mathrm{d}t \Big)^{1/2}$$

and is also called the **2-norm**. Ex.1.4

Let us turn to some "geometric properties" of the norm.

Lemma 1.5. *Let $(E, \langle \cdot, \cdot \rangle)$ be an inner product space. Then the following identities hold for all $f, g \in E$:*

a) $\|f + g\|^2 = \|f\|^2 + 2 \operatorname{Re} \langle f, g \rangle + \|g\|^2$,

b) $\|f + g\|^2 - \|f - g\|^2 = 4 \operatorname{Re} \langle f, g \rangle$ **(polarization identity)**, Ex.1.5

c) $\|f + g\|^2 + \|f - g\|^2 = 2 \|f\|^2 + 2 \|g\|^2$ **(parallelogram law)**. Ex.1.6

Proof. The sesquilinearity and symmetry of the inner product yields

$$\begin{aligned} \|f + g\|^2 &= \langle f + g, f + g \rangle = \langle f, f \rangle + \langle f, g \rangle + \langle g, f \rangle + \langle g, g \rangle \\ &= \|f\|^2 + \langle f, g \rangle + \overline{\langle f, g \rangle} + \|g\|^2 = \|f\|^2 + 2 \operatorname{Re} \langle f, g \rangle + \|g\|^2, \end{aligned}$$

since $z + \overline{z} = 2 \operatorname{Re} z$ for every complex number $z \in \mathbb{C}$; this is a). Replacing g by $-g$ yields

$$\|f - g\|^2 = \|f\|^2 - 2 \operatorname{Re} \langle f, g \rangle + \|g\|^2$$

and adding this to a) yields c). Subtracting it leads to b). □

If $\mathbb{K} = \mathbb{R}$, the polarization identity reads

$$\langle f, g \rangle = \frac{1}{4} \left(\|f + g\|^2 - \|f - g\|^2 \right) \qquad (f, g \in E).$$

[2]Euclid (around 280 BC), Greek mathematician.

Consequently, the inner product is completely determined by its associated norm. Geometrically, this means that lengths determine angles. More precisely: if $\langle\cdot,\cdot\rangle_1$ and $\langle\cdot,\cdot\rangle_2$ are two inner products on a real vector space E with associated norms $\|\cdot\|_1, \|\cdot\|_2$ satisfying

$$\|f\|_1 = \|f\|_2 \qquad \text{for all } f \in E,$$

then $\langle f,g\rangle_1 = \langle f,g\rangle_2$ for all $f,g \in E$.

The same statement is true in the case $\mathbb{K} = \mathbb{C}$; to prove this, one can use the extended polarization identity of Exercise 1.5.

1.2. Orthogonality

As in the case of geometry in three-dimensional Euclidean space, one can use the inner product to define angles between vectors. However, in this book we shall need only right angles, so we confine ourselves to these.

Definition 1.6. Let $(E, \langle\cdot,\cdot\rangle)$ be an inner product space. Two elements $f, g \in E$ are called **orthogonal**, written $f \perp g$, if $\langle f,g\rangle = 0$. For a subset $S \subseteq E$ we let

$$S^\perp := \{f \in E \mid f \perp g \text{ for every } g \in S\}$$

and we also write $f \perp S$ in place of $f \in S^\perp$.

By the symmetry of the inner product we have $f \perp g \iff g \perp f$. The definiteness of the inner product translates into the useful fact

$$f \perp E \quad \iff \quad f = 0,$$

or, in short, $E^\perp = \{0\}$.

Example 1.7. In the (standard) inner product space $\mathrm{C}[a,b]$ we denote by $\mathbf{1}$ the function which is constantly equal to 1, i.e., $\mathbf{1}(t) := 1$, $t \in [a,b]$. Then for $f \in \mathrm{C}[a,b]$ one has

$$\langle f, \mathbf{1}\rangle = \int_a^b f(t)\overline{\mathbf{1}(t)}\,\mathrm{d}t = \int_a^b f(t)\,\mathrm{d}t.$$

Hence $\{\mathbf{1}\}^\perp = \{f \in \mathrm{C}[a,b] \mid \int_a^b f(t)\,\mathrm{d}t = 0\}$.

Let us note a useful lemma.

Lemma 1.8. *Let $(E, \langle\cdot,\cdot\rangle)$ be an inner product space, and let $S \subseteq E$ be any subset. Then $S^\perp$ is a linear subspace of E.*

Proof. Clearly $0 \in S^{\perp}$. If $f, g \in S^{\perp}$, $\lambda \in \mathbb{K}$, then

$$\langle \lambda f + g, s \rangle = \lambda \langle f, s \rangle + \langle g, s \rangle = \lambda \cdot 0 + 0 = 0$$

for arbitrary $s \in S$. Consequently, $\lambda f + g \in S^{\perp}$, and this had to be shown. □

The following should seem familiar from elementary geometry.

Lemma 1.9 (Pythagoras[3]). *Let $(E, \langle \cdot, \cdot \rangle)$ be an inner product space with associated norm $\|\cdot\|$. Let $f_1, \dots, f_d \in E$ be pairwise orthogonal, i.e., $f_i \perp f_j$ whenever $i \neq j$. Then*

$$\|f_1 + \cdots + f_d\|^2 = \|f_1\|^2 + \cdots + \|f_d\|^2 .$$

Proof. For $d = 2$ this follows from Lemma 1.5.a); the rest is induction. □ Ex.1.7

Let I be any nonempty index set. A collection of vectors $(e_j)_{j \in I}$ in an inner product space E is called an **orthonormal system** if

$$\langle e_i, e_j \rangle = \delta_{ij} := \begin{cases} 0 & i \neq j, \\ 1 & i = j. \end{cases}$$

Given an orthonormal system $(e_j)_{j \in I}$ in an inner product space E and a vector $f \in E$, we call the scalar

$$\langle f, e_j \rangle$$

the j-th **abstract Fourier coefficient**[4] and the formal(!) series

$$\sum\nolimits_{j \in I} \langle f, e_j \rangle \, e_j$$

the **abstract Fourier series** of f with respect to $(e_j)_{j \in I}$. We shall keep things simple and confine our study to *finite* orthonormal systems for the moment. It is one of the major achievements of functional analysis to make sense of such expressions when I is not finite; cf. Section 8.4 and Appendix F.

Lemma 1.10. *Let $(E, \langle \cdot, \cdot \rangle)$ be an inner product space with associated norm $\|\cdot\|$, and let $e_1, \dots, e_n \in E$ be a finite orthonormal system.*

a) *Let $g = \sum_{j=1}^{n} \lambda_j e_j$ (with $\lambda_1, \dots, \lambda_n \in \mathbb{K}$) be any linear combination of the e_j. Then*

$$\langle g, e_k \rangle = \sum\nolimits_{j=1}^{n} \lambda_j \langle e_j, e_k \rangle = \lambda_k \qquad (k = 1, \dots, n)$$

and

$$\|g\|^2 = \sum\nolimits_{j=1}^{n} |\lambda_j|^2 = \sum\nolimits_{j=1}^{n} |\langle g, e_j \rangle|^2 .$$

[3]Pythagoras (570–510(?) BC), Greek mathematician and religious leader.
[4]Joseph Fourier (1768–1830), French mathematician.

b) *For* $f \in E$ *let* $Pf := \sum_{j=1}^{n} \langle f, e_j \rangle \, e_j$. *Then*

$$f - Pf \perp \operatorname{span}\{e_1, \dots, e_n\} \quad \textit{and} \quad \|Pf\| \le \|f\| .$$

Proof. a) is just sesquilinearity and Pythagoras' lemma. For the proof of b) note that by a) we have $\langle Pf, e_j \rangle = \langle f, e_j \rangle$, i.e.,

$$\langle f - Pf, e_j \rangle = \langle f, e_j \rangle - \langle Pf, e_j \rangle = 0 \quad \text{for all } j = 1, \dots, n.$$

By Lemma 1.8 it follows that $f - Pf \perp \operatorname{span}\{e_j \mid j = 1, \dots, n\} =: F$. In particular, since $Pf \in F$ we have $f - Pf \perp Pf$ and

$$\|f\|^2 = \|(f - Pf) + Pf\|^2 = \|f - Pf\|^2 + \|Pf\|^2 \ge \|Pf\|^2$$

by Pythagoras' lemma. □

Let us abbreviate $F := \operatorname{span}\{e_1, \dots, e_n\}$. The mapping

$$P : E \longrightarrow E, \qquad Pf = \sum_{j=1}^{n} \langle f, e_j \rangle \, e_j$$

is called the **orthogonal projection** onto the subspace F. The mapping P is *linear*, i.e., it satisfies

$$P(f + g) = Pf + Pg, \quad P(\lambda f) = \lambda Pf \qquad (f, g \in E, \lambda \in \mathbb{K}).$$

By Exercise 1.8.b), P does only depend on the subspace F and not on the
Ex.1.8 chosen orthonormal basis of F used in the construction of P.

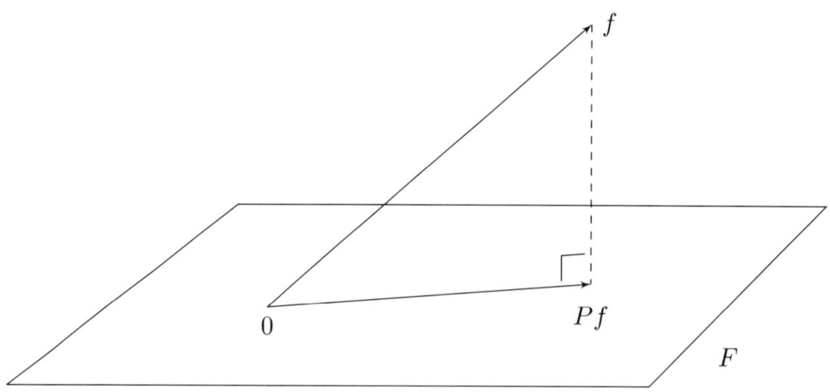

Figure 1. The orthogonal projection onto $F = \operatorname{span}\{e_1, \dots, e_n\}$.

Combining a) and b) of Lemma 1.10 one obtains **Bessel's inequality**[5]

$$\sum_{j=1}^{n} |\langle f, e_j \rangle|^2 = \|Pf\|^2 \le \|f\|^2 \qquad (f \in E). \tag{1.1}$$

[5]Friedrich Wilhelm Bessel (1784–1846), German mathematician and astronomer.

Orthogonal projections are an indispensable tool in Hilbert space theory and its applications. We shall see in Chapter 8 how to construct them in the case that the range space is no longer finite-dimensional.

By Lemma 1.10.a) each orthonormal system is a linearly independent set, i.e., a *basis* for its linear span. Assume for the moment that this span is already the whole space, that is,

$$E = \operatorname{span}\{e_1, \dots, e_n\}.$$

Now consider the (linear!) mapping

$$T : E \longrightarrow \mathbb{K}^n, \qquad Tf := (\langle f, e_1 \rangle, \dots, \langle f, e_n \rangle).$$

By Lemma 1.10.a) T is exactly the **coordinatization mapping** associated with the algebraic basis $\{e_1, \dots, e_n\}$. Hence it is an *algebraic isomorphism*. However, more is true:

$$\langle Tf, Tg \rangle_{\mathbb{K}^n} = \langle f, g \rangle_E \qquad (f, g \in E) \tag{1.2}$$

where $\langle \cdot, \cdot \rangle_{\mathbb{K}^n}$ denotes the standard inner product on $\mathbb{K}^n$. As a consequence, Ex.1.9
one obtains that

$$\|Tf\|_{2,\mathbb{K}^n} = \|f\|_E \qquad \text{for all } f \in E.$$

This means that T maps members of E onto members of $\mathbb{K}^n$ of equal length, and is therefore called a (linear) **isometry**.

The next, probably already well-known result shows that one can always find an orthonormal basis in an inner product space with finite or countable algebraic basis.

Lemma 1.11 (Gram[6]–Schmidt[7]). *Let $N \in \mathbb{N} \cup \{\infty\}$ and let $(f_n)_{1 \le n < N}$ be a linearly independent set of vectors in an inner product space E. Then there is an orthonormal system $(e_n)_{1 \le n < N}$ in E such that*

$$\operatorname{span}\{e_j \mid 0 \le j < n\} = \operatorname{span}\{f_j \mid 0 \le j < n\} \qquad \textit{for all } n \le N.$$

Proof. The construction is recursive. By the linear independence, f_1 cannot be the zero vector, so $e_1 := (1/\|f_1\|) f_1$ has norm one. Let $g_2 := f_2 - \langle f_2, e_1 \rangle e_1$. Then $g_2 \perp e_1$. Since f_1, f_2 are linear independent, $g_2 \neq 0$ and so $e_2 := (1/\|g_2\|) g_2$ is the next unit vector.

[6]Jørgen Pedersen Gram (1850–1916), Danish mathematician.
[7]Erhard Schmidt (1876–1959), German mathematician.

Suppose that we have already constructed pairwise orthogonal unit vectors $e_1, \dots, e_{n-1}$ such that $\operatorname{span}\{e_1, \dots, e_{n-1}\} = \operatorname{span}\{f_1, \dots, f_{n-1}\}$. If $n = N$, we are done. Otherwise let

$$g_n := f_n - \sum\nolimits_{j=1}^{n-1} \langle f_n, e_j \rangle \, e_j .$$

Then $g_n \perp e_j$ for all $1 \le j < n$ (Lemma 1.10). Moreover, by the linear independence of the f_j and the construction of the e_j so far, $g_n \neq 0$. Hence
Ex.1.10 $e_n := (1/\|g_n\|) g_n$ is the next unit vector in the orthonormal system. □

As a corollary we obtain that for each finite-dimensional subspace F of an inner product space E, there exists the orthogonal projection from E onto F. In Chapter 8 we shall be occupied with the extension of this statement to the infinite-dimensional case.

1.3. The Trigonometric System

We now come to an important example of an orthonormal system in the inner product space $E = \mathrm{C}([0,1]; \mathbb{C})$. Consider the functions

$$e_n(t) := \mathrm{e}^{2n\pi i t} \qquad (t \in [0,1],\ n \in \mathbb{Z}),$$

where e is **Euler's constant**[8]. If $n \neq m$, then

$$\langle e_n, e_m \rangle = \int_0^1 e_n(t)\overline{e_m(t)}\, \mathrm{d}t = \int_0^1 \mathrm{e}^{2\pi \mathrm{i}(n-m)t}\, \mathrm{d}t$$

$$= \frac{\mathrm{e}^{2\pi \mathrm{i}(n-m)t}}{2\pi \mathrm{i}(n-m)}\bigg|_0^1 = \frac{1-1}{2\pi \mathrm{i}(n-m)} = 0$$

by the fundamental theorem of calculus. On the other hand,

$$\|e_n\|_2^2 = \int_0^1 \left|\mathrm{e}^{2\pi \mathrm{i} n t}\right|^2 \mathrm{d}t = \int_0^1 \mathbf{1}\, \mathrm{d}t = 1.$$

This shows that $(e_n)_{n\in\mathbb{Z}}$ is an orthonormal system in the complex space $\mathrm{C}([0,1]; \mathbb{C})$, the so-called **trigonometric system**. One can construct from this an orthonormal system in the real space $\mathrm{C}([0,1]; \mathbb{R})$; see Exercise 1.14.

The number

$$\widehat{f}(n) := \langle f, e_n \rangle = \int_0^1 f(t)\overline{e_n(t)}\, \mathrm{d}t = \int_0^1 f(t)\mathrm{e}^{-2\pi \mathrm{i} n t}\, \mathrm{d}t$$

is called the n-th **Fourier coefficient** of f. Note that n ranges over the whole set of integers $\mathbb{Z}$. Bessel's inequality in this context reads

$$\text{(1.3)} \qquad \sum\nolimits_{n=-N}^{N} |\widehat{f}(n)|^2 \le \|f\|_2^2 = \int_0^1 |f(t)|^2\, \mathrm{d}t.$$

[8]Leonhard Euler (1707–1783), Swiss mathematician and physicist.

For $f \in \mathrm{C}[0,1]$ the abstract series

$$f \sim \sum_{n=-\infty}^{\infty} \langle f, e_n \rangle \, \mathrm{e}^{2\pi i n t}$$

with respect to the trigonometric system is called its **Fourier series.**

Exercises 1A

Exercise 1.1. Here is a different way of proving Theorem 1.1. Suppose first that 0 is in the interior of $[a,b]$. Then prove the theorem by considering the derivatives $p^{(j)}(0)$ for $j = 0, \dots, n$. In the general case, find $a < c < b$ and use the change of variables $t = s - c$.

Exercise 1.2. Show that $\langle \cdot, \cdot \rangle : \mathrm{C}[a,b] \times \mathrm{C}[a,b] \to \mathbb{K}$ defined above is indeed sesquilinear, positive and symmetric on $\mathrm{C}[a,b]$.

Exercise 1.3. Show that in an inner product space $\|\lambda f\| = |\lambda| \, \|f\|$ for every $f \in E$ and $\lambda \in \mathbb{K}$. Treat complex scalars explicitly!

Exercise 1.4. a) Compute $\|\cdot\|_2$ of the monomials t^n, $n \in \mathbb{N}$, in the inner product space $\mathrm{C}[a,b]$ with standard inner product.

b) Let $E := \mathrm{P}[0,\infty)$ be the space of all polynomials, considered as functions on the half-line $[0,\infty)$. Define $\|p\|$ by

$$\|p\|^2 = \int_0^\infty |p(t)|^2 \, \mathrm{e}^{-t} \, \mathrm{d}t$$

for $p \in E$. Show that $\|p\|$ is a norm associated with an inner product on E. Prove all your claims.

Exercise 1.5. Let $(E, \langle \cdot, \cdot \rangle)$ be an inner product space over $\mathbb{K} = \mathbb{C}$. Prove that for all $f, g \in E$ one has

$$\|f + ig\|^2 - \|f - ig\|^2 = 4 \operatorname{Im} \langle f, g \rangle .$$

Then conclude that the **polarization identity**

$$\langle f, g \rangle = \frac{1}{4} \left(\|f+g\|^2 - \|f-g\|^2 + i \|f + ig\|^2 - i \|f - ig\|^2 \right)$$

holds for all $f, g \in E$.

Exercise 1.6. Make a drawing that helps you to understand why the parallelogram law carries its name.

Exercise 1.7. Work out the induction proof of Pythagoras' lemma.

Exercise 1.8. Let $\{e_1, \dots, e_n\}$ be a finite orthonormal system in an inner product space $(E, \langle \cdot, \cdot \rangle)$, let $F := \operatorname{span}\{e_1, \dots, e_n\}$ and let $P : E \to F$ be the orthogonal projection onto F. Show that the following assertions hold:

a) $PPf = Pf$ for all $f \in E$.

b) If $f, g \in E$ are such that $g \in F$ and $f - g \perp F$, then $g = Pf$.

c) Each $f \in E$ has a *unique* representation as a sum $f = u + v$, where $u \in F$ and $v \in F^\perp$. (In fact, $u = Pf$.)

d) If $f \in E$ is such that $f \perp F^\perp$, then $f \in F$. (Put differently: $(F^\perp)^\perp = F$.)

e) Let $Qf := f - Pf$, $f \in E$. Show that $QQf = Qf$ and $\|Qf\| \le \|f\|$ for all $f \in E$.

Exercise 1.9. Prove the identity (1.2).

Exercise 1.10. Apply the Gram-Schmidt procedure to the polynomials $1, t, t^2$ in the inner product space $\mathrm{C}[-1,1]$ to construct an orthonormal basis of $F = \{p \in \mathrm{P}[-1,1] \mid \deg p \le 2\}$. (Continuing this for $t^3, t^4 \ldots$ yields the sequence of so-called **Legendre polynomials.**[9])

Exercises 1B

Exercise 1.11. Apply the Gram–Schmidt procedure to the monomials $1, t, t^2$ in the inner product space $\mathrm{P}[0,\infty)$ with inner product

$$\langle f, g \rangle := \int_0^\infty f(t)\overline{g(t)} e^{-t}\, dt.$$

Exercise 1.12. Let us call a function $f : [1,\infty) \to \mathbb{K}$ *mildly decreasing* if there is a constant $c = c(f)$ such that $|f(t)| \le ct^{-1}$ for all $t \ge 1$. Let

$$E := \{f : [1,\infty) \longrightarrow \mathbb{K} \mid f \text{ is continuous and mildly decreasing}\}.$$

a) Show that E is a linear subspace of $\mathrm{C}[1,\infty)$.

b) Show that
$$\langle f, g \rangle := \int_1^\infty f(t)\overline{g(t)}\, dt$$
defines an inner product on E.

c) Apply the Gram–Schmidt procedure to the functions t^{-1}, t^{-2}.

Exercise 1.13. Let E be the space of polynomials of degree at most 2. On E define

$$\langle f, g \rangle := f(-1)\overline{g(-1)} + f(0)\overline{g(0)} + f(1)\overline{g(1)} \qquad (f, g \in E).$$

a) Show that this defines an inner product on E.

b) Describe $\{t^2 - 1\}^\perp$.

c) Show that the polynomials $t^2 - 1, t^2 - t$ are orthogonal, and find a nonzero polynomial $p \in E$ that is orthogonal to both of them.

Exercise 1.14. Let $(e_n)_{n\in\mathbb{Z}}$ be any orthonormal system in $\mathrm{C}([a,b];\mathbb{C})$, with standard inner product. Suppose further that $e_{-n} = \overline{e_n}$ for all $n \in \mathbb{Z}$. Show that

$$\{e_0\} \cup \{\sqrt{2}\,\mathrm{Re}\, e_n \mid n \in \mathbb{N}\} \cup \{\sqrt{2}\,\mathrm{Im}\, e_n \mid n \in \mathbb{N}\}$$

is an orthnormal system in the *real* inner product space $\mathrm{C}([a,b];\mathbb{R})$.

[9]Adrien-Marie Legendre (1752–1833), French mathematician.

Exercise 1.15. Each vector space E over $\mathbb{C}$ is also a vector space over $\mathbb{R}$. Show that if $(E, \langle \cdot, \cdot \rangle)$ is a complex inner product space, then

$$\langle f, g \rangle_r := \operatorname{Re} \langle f, g \rangle \qquad (f, g \in E)$$

is a real inner product on E satisfying $\langle \mathrm{i}f, \mathrm{i}g \rangle_r = \langle f, g \rangle_r$ for all $f, g \in E$.

Conversely, show that if $\langle \cdot, \cdot \rangle$ is a real inner product on the $\mathbb{C}$-vector space E such that $\langle \mathrm{i}f, \mathrm{i}g \rangle = \langle f, g \rangle$ for all $f, g \in E$, then

$$\langle f, g \rangle_c := \langle f, g \rangle + \mathrm{i} \langle f, \mathrm{i}g \rangle$$

is the unique complex inner product on E with $\langle \cdot, \cdot \rangle_{cr} = \langle \cdot, \cdot \rangle$.

Exercise 1.16. With the terminology from the previous exercise, let $(E, \langle \cdot, \cdot \rangle)$ be a complex inner product space. Show that $(e_j)_{j \in I}$ is a $\langle \cdot, \cdot \rangle$-orthonormal system in E if and only if $e_j, \mathrm{i}e_j$ $(j \in I)$ is an $\langle \cdot, \cdot \rangle_r$-orthonormal system in E.

Chapter 2

Normed Spaces

In this chapter we examine further properties of the norm $\|f\| := \langle f, f\rangle^{1/2}$ associated with an inner product $\langle\cdot,\cdot\rangle$ on a vector space E. Then we proceed with an abstract definition of a norm on a vector space, and encounter many examples of normed spaces that are *not* inner product spaces.

2.1. The Cauchy–Schwarz Inequality and the Space ℓ^2

The following is a cornerstone in the theory of inner product spaces.

Theorem 2.1 (Cauchy–Schwarz Inequality[1,2]). *Let $(E, \langle\cdot,\cdot\rangle)$ be an inner product space with associated norm $\|f\| := \sqrt{\langle f, f\rangle}$ for $f \in E$. Then*

$$|\langle f, g\rangle| \le \|f\| \, \|g\| \qquad (f, g \in E),$$

with equality if and only if f and g are linearly dependent.

Proof. If $g = 0$, the inequality reduces to the trivial identity $0 = 0$. So suppose that $g \neq 0$ and consider the orthogonal projection

$$P : E \longrightarrow \operatorname{span}\{g\}, \qquad Pf := \frac{\langle f, g\rangle}{\|g\|^2} g$$

of E onto the one-dimensional subspace of E spanned by g; cf. page 8. By Lemma 1.10 we have $f - Pf \perp g$ and hence, by Pythagoras' Lemma 1.9,

$$\|f\|^2 = \|Pf\|^2 + \|f - Pf\|^2 = \frac{|\langle f, g\rangle|^2}{\|g\|^4} \|g\|^2 + \|f - Pf\|^2 \ge \frac{|\langle f, g\rangle|^2}{\|g\|^2}$$

with equality if and only if $f = Pf$, i.e., $f \in \operatorname{span}\{g\}$. □

[1] Hermann Amandus Schwarz (1843–1921), German mathematician.
[2] Augustin-Louis Cauchy (1789–1857), French mathematician.

Example 2.2. For $x, y \in \mathbb{K}^d$ the Cauchy–Schwarz inequality takes the form

$$\left|\sum_{j=1}^{d} x_j \overline{y_j}\right| \le \left(\sum_{j=1}^{d} |x_j|^2\right)^{1/2} \left(\sum_{j=1}^{d} |y_j|^2\right)^{1/2}$$

and for $f, g \in \mathrm{C}[a,b]$ it is

$$\left|\int_a^b f(t)\overline{g(t)}\,\mathrm{d}t\right| \le \left(\int_a^b |f(t)|^2\,\mathrm{d}t\right)^{1/2} \left(\int_a^b |g(t)|^2\,\mathrm{d}t\right)^{1/2}.$$

With the help of the Cauchy–Schwarz inequality we can establish an important fact about the norm in an inner product space.

Corollary 2.3. *The norm $\|\cdot\|$ induced by an inner product $\langle \cdot, \cdot \rangle$ on a vector space E satisfies*

(2.1) $$\|f+g\| \le \|f\| + \|g\| \qquad (f, g \in E).$$

Proof. Let $f, g \in E$. Then, by Cauchy–Schwarz,

$$\begin{aligned}\|f+g\|^2 &= \|f\|^2 + 2\,\mathrm{Re}\,\langle f, g\rangle + \|g\|^2 \le \|f\|^2 + 2\,|\langle f, g\rangle| + \|g\|^2 \\ &\le \|f\|^2 + 2\,\|f\|\,\|g\| + \|g\|^2 = (\|f\| + \|g\|)^2.\end{aligned}$$

Ex.2.1 Taking square roots proves the claim. □

Inequality (2.1) is called the **triangle inequality**. It says that the direct way is always shorter (or at least not longer) than making a detour via a third stop.

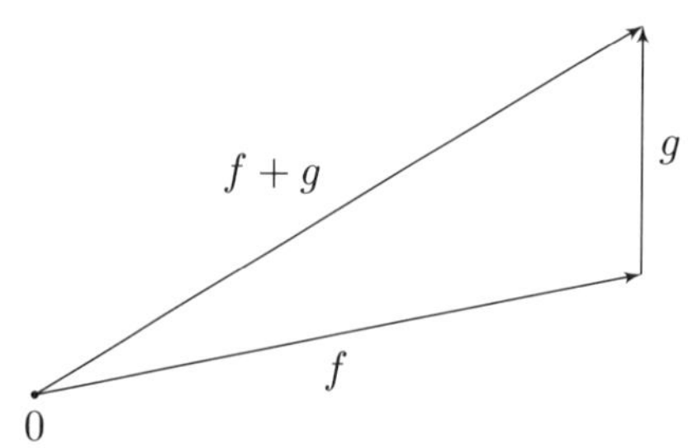

Figure 2. The triangle inequality: $\|f+g\| \le \|f\| + \|g\|$.

The Cauchy–Schwarz inequality helps in constructing new examples of inner product spaces.

Example 2.4. A scalar sequence $(x_n)_{n\in\mathbb{N}}$ is called **square-summable** if $\sum_{n=1}^{\infty} |x_n|^2 < \infty$. We let

$$\ell^2 = \ell^2(\mathbb{N}) := \left\{ (x_n)_{n\in\mathbb{N}} \mid x_n \in \mathbb{C} \text{ for all } n \in \mathbb{N} \text{ and } \sum_{n=1}^{\infty} |x_n|^2 < \infty \right\}$$

be the set of all square-summable scalar sequences. We claim: *The set ℓ^2 is a vector space and*

$$\langle x, y\rangle_{\ell^2} := \sum_{j=1}^{\infty} x_j \overline{y_j} \qquad (x = (x_j)_{j\in\mathbb{N}}, y = (y_j)_{j\in\mathbb{N}} \in \ell^2)$$

is a well-defined inner product on ℓ^2 with associated norm

$$\|x\|_2 := \|x\|_{\ell^2} = \left(\sum_{j=1}^{\infty} |x_j|^2 \right)^{1/2} \qquad (x = (x_j)_{j\in\mathbb{N}} \in \ell^2).$$

This norm is also called the **2-norm**.

Proof. Let $x, y \in \ell^2$. We first prove that the scalar series $\sum_{j=1}^{\infty} x_j \overline{y_j}$ converges absolutely. To this aim, fix $N \in \mathbb{N}$. Then the Cauchy–Schwarz inequality for the (standard) inner product space $\mathbb{K}^N$ yields

$$\begin{aligned} \sum_{j=1}^{N} |x_j \overline{y_j}| = \sum_{j=1}^{N} |x_j|\,|y_j| &\le \left(\sum_{j=1}^{N} |x_j|^2 \right)^{1/2} \left(\sum_{j=1}^{N} |y_j|^2 \right)^{1/2} \\ &\le \left(\sum_{j=1}^{\infty} |x_j|^2 \right)^{1/2} \left(\sum_{j=1}^{\infty} |y_j|^2 \right)^{1/2} =: M. \end{aligned}$$

The right-hand side is a finite number $M < \infty$ (since, by hypothesis, $x, y \in \ell^2$) and is independent of N. Taking the supremum with respect to N yields

$$\sum_{j=1}^{\infty} |x_j \overline{y_j}| \le M < \infty,$$

hence the series converges absolutely. Since every absolutely convergent series in $\mathbb{C}$ converges, our first claim is proved.

To show that ℓ^2 is a vector space, take $x, y \in \ell^2$ and $\lambda \in \mathbb{K}$. Then $\lambda x = (\lambda x_j)_{j\in\mathbb{N}}$ is again square-summable because

$$\sum_{j\ge 1} |\lambda x_j|^2 = \sum_{j\ge 1} |\lambda|^2 |x_j|^2 = |\lambda|^2 \sum_{j\ge 1} |x_j|^2 < \infty.$$

Note that $x + y = (x_j + y_j)_{j\in\mathbb{N}}$. Fixing $N \in \mathbb{N}$ and using the triangle inequality in $\mathbb{K}^N$ we estimate

$$\begin{aligned}\Big(\sum_{j=1}^{N} |x_j + y_j|^2\Big)^{1/2} &\le \Big(\sum_{j=1}^{N} |x_j|^2\Big)^{1/2} + \Big(\sum_{j=1}^{N} |y_j|^2\Big)^{1/2}\\ &\le \Big(\sum_{j=1}^{\infty} |x_j|^2\Big)^{1/2} + \Big(\sum_{j=1}^{\infty} |y_j|^2\Big)^{1/2}\\ &= \|x\|_{\ell^2} + \|y\|_{\ell^2}\,.\end{aligned}$$

The right-hand side is finite and independent of N, so taking the supremum with respect to N yields

$$\sum_{j=1}^{\infty} |x_j + y_j|^2 \le (\|x\|_{\ell^2} + \|y\|_{\ell^2})^2 < \infty.$$

This proves that $x + y \in \ell^2$, thus ℓ^2 is a vector space. The proof that $\langle\cdot,\cdot\rangle_{\ell^2}$
Ex.2.2 is an inner product is left as an exercise. □
Ex.2.3

2.2. Norms

We have seen that inner product spaces allow to assign a length to each of their elements. This length is positive (as lengths should be), scales nicely if one multiplies the vector by a scalar, and obeys the "triangle inequality". Let us put these properties into an abstract definition.

Definition 2.5. Let E be a vector space over the field $\mathbb{K} \in \{\mathbb{R}, \mathbb{C}\}$. A mapping

$$\|\cdot\| : E \longrightarrow \mathbb{R}_+$$

is called a **norm** on E if it has the following properties:

1) $\|f\| = 0 \iff f = 0$ for all $f \in E$ **(definiteness)**,
2) $\|\lambda f\| = |\lambda|\,\|f\|$ for all $f \in E$, $\lambda \in \mathbb{K}$ **(homogeneity)**,
3) $\|f + g\| \le \|f\| + \|g\|$ for all $f, g \in E$ **(triangle inequality)**.

A **normed (linear) space** is a pair $(E, \|\cdot\|)$, where E is a vector space and
Ex.2.4 $\|\cdot\|$ is a norm on it.

We have seen that the natural length function on an inner product space satisfies the axioms of a norm (so we were justified to call it "norm" in the first place). In this rather canonical way, every inner product space is also a normed space. However, there are many vector spaces with norms *not* coming from inner products, and we shall exhibit some of them in the following. The first may be well-known, for example, from undergraduate numerical analysis.

Example 2.6. On $\mathbb{K}^d$ we consider the mappings $\|\cdot\|_1, \|\cdot\|_\infty$ defined by

$$\|x\|_1 := \sum_{j=1}^{d} |x_j|, \quad \|x\|_\infty := \max\{\, |x_j| \mid j = 1, \dots, d\}$$

for $x = (x_1, \dots, x_d) \in \mathbb{K}^d$. The proof that these mappings (called the **1-norm** and the **maximum norm**) are indeed norms, is left as Exercise 2.5. Ex.2.5

The following is an infinite-dimensional analogue.

Example 2.7. On $\mathrm{C}[a, b]$ we consider the mappings $\|\cdot\|_1, \|\cdot\|_\infty$ defined by

$$\|f\|_1 := \int_a^b |f(t)| \,\mathrm{d}t, \quad \|f\|_\infty := \sup\{|f(t)| \mid t \in [a, b]\}$$

for $f \in \mathrm{C}[a, b]$, called **1-norm** and **supremum norm**, respectively. We sketch a proof of the triangle inequality, leaving the other properties of a norm as an exercise. Let $f, g \in \mathrm{C}[a, b]$. Then for each $t \in [a, b]$ one has

$$|(f + g)(t)| = |f(t) + g(t)| \le |f(t)| + |g(t)|.$$

In the first case one simply integrates this inequality, leading to

$$\begin{aligned} \|f + g\|_1 &= \int_a^b |(f + g)(t)| \,\mathrm{d}t \le \int_a^b |f(t)| + |g(t)| \,\mathrm{d}t \\ &= \int_a^b |f(t)| \,\mathrm{d}t + \int_a^b |g(t)| \,\mathrm{d}t = \|f\|_1 + \|g\|_1 . \end{aligned}$$

For the supremum norm note that one can estimate $|f(t)| \le \|f\|_\infty$ and $|g(t)| \le \|g\|_\infty$ for all $t \in [a, b]$, by definition of $\|\cdot\|_\infty$. This leads to

$$|(f + g)(t)| \le |f(t)| + |g(t)| \le \|f\|_\infty + \|g\|_\infty$$

for all $t \in [a, b]$. Taking the supremum over $t \in [a, b]$ we obtain

$$\|f + g\|_\infty = \sup_{t \in [a,b]} |(f + g)(t)| \le \|f\|_\infty + \|g\|_\infty .$$

Ex.2.6

Certainly you recall that a continuous positive function on a compact interval has a maximum, i.e., attains its supremum. So we can write $\|f\|_\infty = \max\{|f(t)| \mid t \in [a, b]\}$. However, in the proofs above this property is never used. With a view towards more general situations below, it is better to use the supremum rather than the maximum.

Suprema and infima play a central role in analysis, and we have collected the relevant definitions in Appendix A.3. It may console you that in this book only suprema and infima over sets of positive real numbers occur.

Examining the supremum norm for $\mathrm{C}[a,b]$ we may realize that in the proof of the norm properties the continuity of the functions actually does not play any role whatsoever. The only property used, was that $\sup\{|f(t)| \mid t \in [a,b]\}$ is a finite number. For continuous functions this is automatically satisfied, since $[a,b]$ is a compact interval. If we leave compact domains, we must include this in the definition of the function space.

Example 2.8. Let Ω be any (nonempty) set. A function $f : \Omega \to \mathbb{K}$ is called **bounded** if there is a finite number $c = c(f) \geq 0$ such that $|f(t)| \leq c$ for all $t \in \Omega$. For a bounded function f, the number

$$\|f\|_\infty := \sup\{|f(t)| \mid t \in \Omega\},$$

called the **supremum norm**, is finite. Let

$$\mathcal{B}(\Omega) := \{f : \Omega \longrightarrow \mathbb{K} \mid f \text{ is bounded}\}$$

be the set of bounded (scalar) functions on Ω. Then $E = \mathcal{B}(\Omega)$ is a linear
Ex.2.7 subspace of $\mathcal{F}(\Omega)$ and $\|\cdot\|_\infty$ is a norm on it (proof as exercise). The norm $\|\cdot\|_\infty$ is usually called the **supremum norm**.

A special instance of $\mathcal{B}(\Omega)$ occurs when $\Omega = \mathbb{N}$. For this we use the symbol

$$\ell^\infty := \mathcal{B}(\mathbb{N}) = \Big\{(x_j)_{j\in\mathbb{N}} \mid x_j \in \mathbb{K} \text{ for all } n \in \mathbb{N} \text{ and } \sup_{j\in\mathbb{N}} |x_j| < \infty\Big\}.$$

Recall that a scalar sequence *is the same* as a scalar function on $\mathbb{N}$!

Finally, we treat a sequence analogue of the 1-norm.

Example 2.9. A scalar sequence $x = (x_j)_{j\in\mathbb{N}}$ is called **absolutely summable** if the series $\sum_{j=1}^\infty x_j$ converges absolutely, i.e., if

$$\|x\|_1 := \sum\nolimits_{j=1}^\infty |x_j| < \infty.$$

We denote by

$$\ell^1 := \Big\{(x_j)_{j\in\mathbb{N}} \mid x_j \in \mathbb{K} \text{ for all } n \in \mathbb{N} \text{ and } \sum\nolimits_{j=1}^\infty |x_j| < \infty\Big\}$$

Ex.2.8 the set of all absolutely summable sequences.

Remark 2.10. We claimed in the beginning that these norms do not come from any inner product on the underlying space. How can one be so sure about that? Well, we know that a norm coming from an inner product satisfies the parallelogram law. A stunning theorem of von Neumann[3] actually states the converse: *A norm on a vector space comes from an inner product if and only if it satisfies the parallelogram law.* See [**Che01**, pp. 65, 66] for

[3]János (John) von Neumann (1903–1957), Austrian-Hungarian mathematician.

a proof. Spaces such as $C[a, b]$ with the supremum norm or the 1-norm do not satisfy the parallelogram law; see Exercise 2.18.

2.3. Bounded Linear Mappings

We shall now learn how normed spaces give rise to new normed spaces. Suppose that E, F are normed spaces. Then we can consider mappings $T : E \to F$ which are **linear**, i.e., which satisfy[4]

$$T(f+g) = Tf + Tg \quad \text{and} \quad T(\lambda f) = \lambda Tf$$

for all $f, g \in E$, $\lambda \in \mathbb{K}$. A linear mapping is also called a (linear) **operator**, and if the target space $F = \mathbb{K}$ is one-dimensional, they are called (linear) **functionals**.

One can add linear mappings and multiply them by scalars by

$$(T+S)f := Tf + Sf, \qquad (\lambda T)f := \lambda(Tf) \qquad (f \in E,\ \lambda \in \mathbb{K})$$

and in this way the set of all linear mappings from E to F becomes a new vector space; see Lemma A.7. Is there a natural norm on that space?

Think for a moment about the finite-dimensional situation. If $E = \mathbb{K}^n$ and $F = \mathbb{K}^m$, then the linear mappings from E to F can be identified with the space of $m \times n$-matrices. As a vector space this is isomorphic to $\mathbb{K}^{m \cdot n}$, and we know already several norms here. But which of them relates naturally to the chosen norms on E and F?

It turns out that if E is infinite-dimensional, then it is impossible to define a norm on the space of *all* linear mappings. However, there is an important *subspace* of linear mappings that allows for a norm.

Definition 2.11. Let $(E, \|\cdot\|_E)$ and $(F, \|\cdot\|_F)$ be normed spaces. A linear operator $T : E \to F$ is called **bounded** if there is a scalar $c \geq 0$ such that

$$\|Tf\|_F \leq c\,\|f\|_E \qquad \text{for all} \quad f \in E.$$

We denote by $\mathcal{L}(E; F)$ the **space of bounded linear operators** from E to F. If $E = F$ we simply write $\mathcal{L}(E)$.

[4]It is very common to write "Tf" instead of "$T(f)$" when T is a linear operator, and use brackets only when it is necessary to avoid misunderstandings.

When considering linear operators $T : E \to F$ we should distinguish the norms on E and on F. We have done this in the above definition explicitly by writing $\|\cdot\|_E$ and $\|\cdot\|_F$, but usually the subscripts are omitted when there is no danger of confusion.

Lemma 2.12. *A linear operator $T : E \to F$ is bounded if and only if its* **operator norm**

$$\|T\| := \|T\|_{\mathcal{L}(E;F)} := \sup_{\|f\|_E \le 1} \|Tf\|_F \tag{2.2}$$

is a finite number. If T is bounded, then

$$\|Tf\|_F \le \|T\| \, \|f\|_E \qquad \text{for all} \quad f \in E. \tag{2.3}$$

Proof. If $\|Tf\| \le c \|f\|$ for all $f \in E$, then obviously $\|T\| \le c$, hence $\|T\|$ is finite. If $\|T\| < \infty$, then for a general $f \in E, f \neq 0$ we write $\lambda := \|f\|$ and compute

$$\|Tf\| = \left\|T(\lambda\lambda^{-1}f)\right\| = \lambda \left\|T(\lambda^{-1}f)\right\| \le \lambda \|T\| = \|T\| \, \|f\|,$$

since $\left\|\lambda^{-1}f\right\| = \lambda^{-1} \|f\| = 1$. This establishes (2.3) for $f \neq 0$. But $\|T(0)\| = \|0\| = 0$, and hence (2.3) also holds for $f = 0$. □

Sometimes one writes $\|T\|_{F \leftarrow E}$ in place of $\|T\|_{\mathcal{L}(E;F)}$, but if there is no danger of confusion, one omits subscripts and simply writes $\|T\|$. The following result shows that the name "operator norm" is justified.

Theorem 2.13. *Let E, F be normed spaces. Then $\mathcal{L}(E;F)$ is a vector space, and the operator norm defined by (2.2) is a norm on $\mathcal{L}(E;F)$.*

Proof. Take $S, T \in \mathcal{L}(E;F)$. Then for all $f \in E$ we have

$$\begin{aligned} \|(T+S)f\| &= \|Tf + Sf\| \le \|Tf\| + \|Sf\| \le \|T\| \, \|f\| + \|S\| \, \|f\| \\ &= (\|T\| + \|S\|) \, \|f\| . \end{aligned}$$

But this means that $S + T$ is bounded and $\|T + S\| \le \|T\| + \|S\|$. If $T \in \mathcal{L}(E;F)$ and $\lambda \in \mathbb{K}$, then

$$\|(\lambda T)f\| = \|\lambda Tf\| = |\lambda| \, \|Tf\| \le |\lambda| \, \|T\| \, \|f\|$$

for all $f \in E$; this means that λT is bounded, with $\|\lambda T\| \le |\lambda| \, \|T\|$. Replacing T by $\lambda^{-1}T$ here yields $\|\lambda T\| = |\lambda| \, \|T\|$. Finally, if $\|T\| = 0$, then $\|Tf\| \le \|T\| \, \|f\| = 0$ and hence $Tf = 0$ for all $f \in E$, i.e., $T = 0$. □

We write ST in place of $S \circ T$ whenever this composition of the operators makes sense; in this case we call ST the **product** of the operators S and T.

Lemma 2.14. *Let E, F, G be normed spaces, and let $T : E \to F$ and $S : F \to G$ be bounded linear operators. Then $ST := S \circ T$ is again a bounded linear operator, and one has*

$$\|ST\|_{G\leftarrow E} \le \|S\|_{G\leftarrow F} \cdot \|T\|_{F\leftarrow E} \,. \tag{2.4}$$

Proof. It is clear that ST is again linear. For $f \in E$ we have

$$\|(ST)f\| = \|S(Tf)\| \le \|S\| \, \|Tf\| \le \|S\| \, \|T\| \, \|f\| = (\|S\| \, \|T\|) \, \|f\| \,.$$

This shows that $ST \in \mathcal{L}(E; G)$ and establishes (2.4). □

Let us conclude this section with an equivalent description of boundedness of linear mappings. This hinges on the following general concept.

Definition 2.15. A subset A of a normed space $(E, \|\cdot\|)$ is called **bounded** if there is $c \ge 0$ such that

$$\|f\| \le c \qquad \text{for all } f \in A.$$

The **closed unit ball** of a normed space E is $B_E := \{f \in E \mid \|f\| \le 1\}$.

The closed unit ball B_E is obviously a bounded set, and a subset $A \subseteq E$ is bounded if and only if there is a constant $c > 0$ such that $A \subseteq cB_E$.

Lemma 2.16. *A linear mapping $T : E \to F$ between normed spaces E and F is bounded if and only if the set $T(B_E)$ is bounded if and only if T maps bounded subsets of E to bounded subsets of F.* Ex.2.9

There is a slight ambiguity in the word "bounded mapping". Analogous to Example 2.8 a mapping $f : \Omega \to E$ from a set Ω into a normed space E is called *bounded* if its image $f(\Omega)$ is a bounded set in E, and a bounded linear operator $T \neq 0$ is *not a bounded mapping* in this sense since $T(E) = \mathrm{ran}(T)$ is a nontrivial linear subspace of F.

On the other hand, note that any linear mapping T is — by scaling — completely determined by its restriction $T|_{B_E}$ to B_E, the unit ball of E. By Lemma 2.16, T is a "bounded linear operator" if and only if $T|_{B_E}$ is a bounded mapping.

2.4. Basic Examples

We shall encounter many bounded linear mappings in this book; cf., in particular, Chapter 11. At this point we look only at a few simple examples.

Example 2.17 (Multiplication by a Scalar)**.** On each normed space E we can consider the operator "multiplication with $\lambda \in \mathbb{K}$"

$$E \longrightarrow E, \qquad f \longmapsto \lambda f. \tag{2.5}$$

For $\lambda = 1$ this operator is simply the **identity** mapping, and we reserve the special symbol I for it. For other values of λ the multiplication operator (2.5) can then be written as $\lambda\mathrm{I}$. If $E \neq \{0\}$, the norm of $\lambda\mathrm{I}$ is

$$\|\lambda\mathrm{I}\| = |\lambda|\,\|\mathrm{I}\| = |\lambda| \cdot 1 = |\lambda|\,.$$

In case $\lambda = 0$ we obtain the **zero operator** $0 \cdot \mathrm{I} = 0$.

Example 2.18 (Isometries)**.** A linear mapping $T : E \to F$ is called an **isometry** if

$$\|Tf\|_F = \|f\|_E \qquad \text{for all} \quad f \in E.$$

An isometry is obviously bounded. It has trivial kernel (only the zero vector is mapped to 0), and hence is injective. If it is also surjective, i.e., if $\operatorname{ran}(T) = F$, we call T an **isometric isomorphism**. In this case, T is invertible and also $T^{-1} : F \to E$ is an isometric isomorphism.

Obviously, on any normed space E the identity mapping $\mathrm{I} \in \mathcal{L}(E)$ is an isometric isomorphism. More general, $\lambda\mathrm{I}$ is an isometric isomorphism for any $\lambda \in \mathbb{K}$ with $|\lambda| = 1$.

Another example of an isometric isomorphism is the *coordinatization mapping*

$$T : E \longrightarrow \mathbb{K}^n, \qquad Tf := \big(\langle f, e_1 \rangle, \dots, \langle f, e_n \rangle \big),$$

where E is a finite-dimensional inner product space with orthonormal *basis*
Ex.2.10 $\{e_1, \dots, e_d\}$ and $\mathbb{K}^d$ is given the standard inner product; see page 9.

Example 2.19. *Any linear mapping $T : \mathbb{K}^d \to F$ is bounded, where F is an arbitrary normed space and on $\mathbb{K}^d$ we consider the standard (Euclidean) norm.*

Proof. Let $e_1, \dots, e_d$ denote the canonical basis of $\mathbb{K}^d$. Then for arbitrary $x = (x_1, \dots, x_d)$ by the triangle inequality (cf. Exercise 2.4)

$$\begin{aligned}\|Tx\|_F &= \Big\|T\big(\sum\nolimits_{j=1}^d x_j e_j\big)\Big\|_F = \Big\|\sum\nolimits_{j=1}^d T(x_j e_j)\Big\|_F \le \sum\nolimits_{j=1}^d \|x_j T e_j\|_F \\ &= \sum\nolimits_{j=1}^d |x_j|\,\|Te_j\|_F \le c\,\|x\|_2\end{aligned}$$

with $c := \big(\sum_{j=1}^d \|Te_j\|_F\big)^{1/2}$, by Cauchy–Schwarz. □

Example 2.20 (Point Evaluations I). Let $E = \ell^p$ with $p = 1, 2, \infty$. The **point evaluation** at $k \in \mathbb{N}$ is

$$\delta_k : \ell^p \longrightarrow \mathbb{K}, \quad x = (x_j)_{j\in\mathbb{N}} \mapsto x_k.$$

So $\delta_k(x)$ picks just the k-th element of the sequence $x = (x_j)_{j\in\mathbb{N}}$. Since sum and scalar multiple of sequences are defined componentwise, δ_k is a linear functional. Moreover,

$$|\delta_k(x)| = |x_k| \le \|x\|_p$$

for each $x \in \ell^p$, and hence δ_k is bounded with $\|\delta_k\| \le 1$. If we consider the k-th **standard unit vector**

$$e_k = (0, 0, \ldots, 0, 1, 0 \ldots) \qquad \text{(the 1 at the } k\text{-th place)},$$

then $\|e_k\|_p = 1$ and $\delta_k(e_k) = 1$. This shows that $\|\delta_k\| = 1$.

Example 2.21 (Point Evaluations II). If $[a, b] \subseteq \mathbb{R}$, then the **point-evaluation** or **Dirac functional** at $x_0 \in [a, b]$,

$$\delta_{x_0} : \mathrm{C}[a, b] \longrightarrow \mathbb{K}, \qquad \delta_{x_0}(f) := f(x_0),$$

is a bounded linear functional on $(\mathrm{C}[a, b], \|\cdot\|_\infty)$ with norm $\|\delta_{x_0}\| = 1$.

Analogously, point evaluations are bounded linear functionals of norm one on the spaces $(\mathcal{B}(\Omega), \|\cdot\|_\infty)$ (Ω an arbitrary nonempty set).

Example 2.22 (Point Evaluations III). In contrast to the previous example, point evaluations on $E = \mathrm{C}[a, b]$ are *not bounded* for the norms $\|\cdot\|_p$ with $p < \infty$.

We sketch a proof for the special case that $p = 1$. Consider functions f_n as in Figure 3. Then $|f_n(x_0)| = n \to \infty$. On the other hand, $\|f_n\|_1 \le \frac{1}{2} \cdot n \cdot \frac{2}{n} = 1$ for all $n \in \mathbb{N}$. Hence there is no $c \ge 0$ such that $|\delta_{x_0}(f_n)| \le c \|f_n\|_1$ for all n, and the point evaluation at x_0 is not bounded with respect to $\|\cdot\|_1$. Ex.2.11

Example 2.23 (Inner Products). If $(H, \langle \cdot, \cdot \rangle)$ is an inner product space, and $g \in H$, then

$$\psi_g : H \longrightarrow \mathbb{K}, \qquad \psi_g(f) := \langle f, g \rangle$$

is a linear functional. By Cauchy–Schwarz,

$$|\psi_g(f)| = |\langle f, g \rangle| \le \|f\| \, \|g\|$$

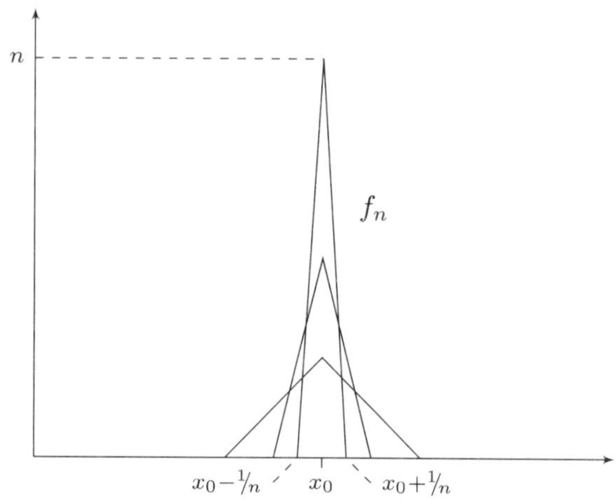

Figure 3. Functions that are $\|\cdot\|_1$-small but with a large value at x_0.

for all $f \in H$. This yields $\|\psi_g\| \le \|g\|$. We claim that actually $\|\psi_g\| = \|g\|$, i.e.,

$$\sup_{\|f\|\le 1} |\langle f, g\rangle| = \|g\| . \tag{2.6}$$

This is clear if $g = 0$. In case $g \neq 0$ we can put $f = (1/\|g\|)g$ and find $\|f\| = 1$ and $\psi_g(f) = (1/\|g\|)\langle g, g\rangle = \|g\|$.

Example 2.24 (Shifts). On $E = \ell^2$ the **left shift** L and the **right shift** R are defined by

$$\begin{aligned} L &: (x_1, x_2, x_3 \dots) \longmapsto (x_2, x_3, \dots), \\ R &: (x_1, x_2, x_3 \dots) \longmapsto (0, x_1, x_2, x_3, \dots). \end{aligned}$$

Thus, in function notation,

$$(Lf)(n) := f(n+1), \quad (Rf)(n) = \begin{cases} 0 & \text{if} \quad n = 1, \\ f(n-1) & \text{if} \quad n \ge 2. \end{cases}$$

It is easy to see that

$$\|Rf\|_2 = \|f\|_2 \qquad (f \in \ell^2)$$

so R is an *isometry* and hence $\|R\| = 1$. Turning to L, we obtain for $f \in \ell^2$,

$$\begin{aligned} \|Lf\|_2^2 &= \sum\nolimits_{n=1}^{\infty} |(Lf)(n)|^2 = \sum\nolimits_{n=1}^{\infty} |f(n+1)|^2 \\ &= \sum\nolimits_{n=2}^{\infty} |f(n)|^2 \le \sum\nolimits_{n=1}^{\infty} |f(n)|^2 = \|f\|_2^2 , \end{aligned}$$

which implies that $\|L\| \le 1$. Inserting e_2, the second standard unit vector, we have $Le_2 = e_1$ and so $\|Le_2\|_2 = \|e_1\|_2 = 1 = \|e_2\|_2$, which implies that $\|L\| = 1$.

Shift operators can be defined also on the sequence spaces ℓ^1, ℓ^∞.

Note that $LR = \mathrm{I}$, R is injective and L is surjective. But R is not surjective and L is not injective. Such a situation cannot occur in finite dimensions!

Example 2.25 (Multiplication Operators). A bounded sequence $\lambda = (\lambda_n)_{n\in\mathbb{N}}$ induces a **multiplication operator** $A_\lambda : \ell^2 \to \ell^2$ by

$$(A_\lambda f)(n) := \lambda_n f(n) \qquad (n \in \mathbb{N}, f \in \ell^2).$$

The sequence $\lambda \in \ell^\infty$ is called the **multiplier**. If you consider elements of ℓ^2 as infinite column vectors, then A_λ amounts to multiplying with an infinite *diagonal matrix* with the main diagonal formed by the λ_j.

Since λ is a bounded sequence, we obtain

$$(2.7) \qquad \|A_\lambda f\|_2^2 = \sum\nolimits_{n=1}^{\infty} |\lambda_n f(n)|^2 \le \sum\nolimits_{n=1}^{\infty} \|\lambda\|_\infty^2 |f(n)|^2 = \|\lambda\|_\infty^2 \|f\|_2^2$$

for every $f \in \ell^2$. Therefore, A_λ is bounded and $\|A_\lambda\| \le \|\lambda\|_\infty$. On the other hand,

$$\|A_\lambda e_n\|_2 = \|\lambda_n e_n\|_2 = |\lambda_n| \, \|e_n\|_2 = |\lambda_n|$$

where e_n is the n-th standard unit vector. Hence $\|A_\lambda\| \ge |\lambda_n|$ for every $n \in \mathbb{N}$, and thus $\|A_\lambda\| \ge \sup_n |\lambda_n| = \|\lambda\|_\infty$. Combining both estimates yields $\|A_m\| = \|m\|_\infty$.

Similarly, one can define multiplication operators on ℓ^1 and ℓ^∞ and on the space $(\mathrm{C}[a,b], \|\cdot\|_p)$, with $p \in \{1, 2, \infty\}$. Ex.2.12 Ex.2.13

Recall the occurence of a *supremum* in the definition of the operator norm:

$$\|A\| = \sup\{\|Af\|_F \mid f \in E, \ \|f\|_E \le 1\}.$$

In almost all the examples above where we computed a norm, this **norm is attained**, i.e., the supremum was actually a maximum.

However, the last example shows that this need not be the case. Indeed, in order that the norm of the multiplication operator A_λ from above is attained at f, say, one must have equality in the estimate (2.7). But this requires $|\lambda_n| = \|\lambda\|_\infty$ whenever $f(n) \neq 0$. *So the norm is attained if and* *only if $|\lambda|$ attains its supremum.* In particular, if $\lambda_k = \frac{k}{k+1}$ for all $k \in \mathbb{N}$, then the norm of the multiplication operator A_λ is not attained!

Ex.2.14
Ex.2.15

Example 2.26 (Orthogonal Projections). Let $(E, \langle\cdot,\cdot\rangle)$ be an inner product space, and let $e_1, \dots, e_n$ be a finite orthonormal system in E. Consider the orthogonal projection

$$Pf := \sum\nolimits_{j=1}^{n} \langle f, e_j\rangle\, e_j \qquad (f \in E)$$

as in Lemma 1.10. Then $\|P\| \le 1$ by Lemma 1.10.b). Since $Pf = f$ for $f \in \mathrm{span}\{e_1, \dots, e_n\} \neq \{0\}$, we conclude that $\|P\| = 1$.

Example 2.27. Consider $\mathrm{C}[a, b]$ with the supremum norm. Fix a function $m \in \mathrm{C}[a, b]$ and consider the linear operator

$$T_m : \mathrm{C}[a, b] \longrightarrow \mathbb{K}, \qquad T_m f := \int_a^b m(s) f(s)\,\mathrm{d}s.$$

Then T_m is bounded, and

$$\|T_m\|_{\mathcal{L}(\mathrm{C}[a,b];\mathbb{K})} = \|m\|_1 = \int_a^b |m(s)|\,\mathrm{d}s.$$

Ex.2.16 The proof is left as Exercise 2.16.

2.5. *The ℓ^p-Spaces $(1 \le p < \infty)$

The spaces ℓ^1 and ℓ^2 introduced above are special cases of a more general construction. We fix a real number $p \in [1, \infty)$. The ***p*-norm** of a scalar sequence $x = (x_n)_{n \in \mathbb{N}}$ is

$$\|x\|_p := \Big(\sum\nolimits_{n=1}^{\infty} |x_n|^p\Big)^{1/p} \in [0, \infty],$$

and $x = (x_n)_{n\in\mathbb{N}}$ is called ***p*-summable** if $\|x\|_p < \infty$. We let

$$\ell^p = \ell^p(\mathbb{N}) := \Big\{(x_n)_{n\in\mathbb{N}} \mid x_n \in \mathbb{K} \text{ for all } n \in \mathbb{N} \text{ and } \sum\nolimits_{n=1}^{\infty} |x_n|^p < \infty\Big\}$$

be the set of all p-summable scalar sequences. The aim of this section is to prove the following result.

Theorem 2.28. *For each $1 \le p < \infty$ the set ℓ^p is a vector space and $\|\cdot\|_p$ is a norm on it.*

The main task in the proof of Theorem 2.28 is to establish the triangle inequality for the p-norm

$$(2.8) \qquad \Big(\sum_{n=1}^{\infty} |x_n + y_n|^p\Big)^{1/p} \le \Big(\sum_{n=1}^{\infty} |x_n|^p\Big)^{1/p} + \Big(\sum_{n=1}^{\infty} |y_n|^p\Big)^{1/p}$$

also called **Minkowski's inequality**.[5] The key is the following auxiliary result from calculus.

Lemma 2.29. *For $1 \le p < \infty$ and $a, b \ge 0$,*

$$(a+b)^p = \inf_{0<t<1} t^{1-p}a^p + (1-t)^{1-p}b^p.$$

Proof. The assertion is clearly true if $p = 1$ or $ab = 0$. Hence we may suppose that $1 < p < \infty$ and $a, b > 0$. The smooth function $f(t) := t^{1-p}a^p + (1-t)^{1-p}b^p$ for $t \in (0,1)$ satisfies $\lim_{t\searrow 0} f(t) = +\infty = \lim_{t\nearrow 1} f(t)$. Hence f has a minimum somewhere in $(0,1)$. To find it we let $f'(t) = 0$ and find (after some computation) $t = a/(a+b)$. Now we insert this back into f and we are done. □

Proof of Theorem 2.28. It is straightforward to show that for a scalar sequence $x = (x_n)_{n\in\mathbb{N}}$ we have $\|x\|_p = 0$ if and only if $x = 0$ and $\|\lambda x\|_p = |\lambda| \, \|x\|_p$ for any $\lambda \in \mathbb{K}$. In particular, $0 \in \ell^p$ and if $x \in \ell^p$, then also $\lambda x \in \ell^p$. For the remaining part, let $x, y \in \ell^p$. Then for each $n \in \mathbb{N}$ and $t \in (0,1)$,

$$|x_n + y_n|^p \le (|x_n| + |y_n|)^p \le t^{1-p} |x_n|^p + (1-t)^{1-p} |y_n|^p,$$

by Lemma 2.29. Summing over $n \in \mathbb{N}$ we obtain

$$(2.9) \qquad \|x+y\|_p^p \le t^{1-p} \|x\|_p^p + (1-t)^{1-p} \|y\|_p^p \quad \text{for all } t \in (0,1).$$

In particular, since $x, y \in \ell^p$, $\|x+y\|_p^p < \infty$, i.e., $x + y \in \ell^p$ as well. If we take the infimum with respect to $t \in (0,1)$ in (2.9) and apply Lemma 2.29 again, we obtain

$$\|x+y\|_p^p \le \big(\|x\|_p + \|y\|_p\big)^p,$$

which is Minkowski's inequality. □

One can show that the p-norm on ℓ^p comes from an inner product if and only if $p = 2$ (Exercise 2.29). As a replacement for the inner product one has a "duality" between certain pairs of ℓ^p-spaces. To make this more precise, fix again $p \in [1, \infty)$ and let $q \in (1, \infty]$ be the **dual** or **conjugate exponent**, defined by $1/p + 1/q = 1$ (where we let $1/\infty = 0$).

[5]Hermann Minkowski (1864–1909), German mathematician.

Theorem 2.30 (Hölder's Inequality[6]). *Let $1 \le p < \infty$ with dual exponent $q \in (1,\infty]$, and let $x = (x_n)_{n\in\mathbb{N}}$ and $y = (y_n)_{n\in\mathbb{N}}$ be scalar sequences. Then*

$$\sum\nolimits_{n=1}^{\infty} |x_n y_n| \le \|x\|_p \, \|y\|_q . \tag{2.10}$$

In particular, if $x \in \ell^p$ and $y \in \ell^q$, then $x \cdot y \in \ell^1$ and

$$\Big|\sum\nolimits_{n=1}^{\infty} x_n y_n\Big| \le \|x \cdot y\|_1 \le \|x\|_p \, \|y\|_q .$$

The proof of Hölder's inequality is similar to the proof of Minkowski's inequality. It rests on the following auxiliary result from calculus.

Lemma 2.31. *Let $p, q \in (1,\infty)$ such that ${}^1\!/_p + {}^1\!/_q = 1$, and let $a, b \ge 0$. Then*

$$ab = \inf_{t>0} \left[\frac{t^p}{p} a^p + \frac{t^{-q}}{q} b^q \right] .$$

Proof. The proof is left as Exercise 2.30. □

Proof of Theorem 2.30. If $\|x\|_p = \infty$ or $\|y\|_q = \infty$ the assertion is trivial: either the right-hand side is infinite or the left-hand side is zero. So we may suppose that $x \in \ell^p$ and $y \in \ell^q$. In the case $p = 1$ we have $q = \infty$ and (2.10) is straightforward. So let us suppose that $1 < p, q < \infty$.
Then, by Lemma 2.31, for each $n \in \mathbb{N}$ and $t > 0$ we have

$$|x_n y_n| = |x_n| \, |y_n| \le \frac{t^p}{p} |x_n|^p + \frac{t^{-q}}{q} |y_n|^q .$$

Summing over $n \in \mathbb{N}$ we obtain

$$\sum\nolimits_{n=1}^{\infty} |x_n y_n| \le \frac{t^p}{p} \|x\|_p^p + \frac{t^{-q}}{q} \|y\|_q^q .$$

Now we take the infimum over $t > 0$ and apply Lemma 2.31 again, and we are done. □

The idea behind Lemmas 2.29 and 2.31 is that a convex (concave) function is the pointwise supremum (infimum) of affine functions. Or, in more geometric terms, that a closed convex set in $\mathbb{R}^d$ is the intersection of closed hyperspaces. Confer Theorem 16.6 below and also [**Haa07**].

The notion of "duality" is the topic of Chapter 16.

[6] Otto Hölder (1859–1937), German mathematician.

Exercises 2A

Exercise 2.1. Write down the instances of the triangle inequality in the standard inner product spaces $\mathbb{K}^d$ and $\mathrm{C}[a, b]$.

Exercise 2.2. Complete the proof of the claim in the Example 2.4.

Exercise 2.3. Mimic the proof in Example 2.4 to show that the set

$$E := \{f \in \mathrm{C}[0, \infty) \mid \int_0^\infty |f(t)|^2 \, \mathrm{d}t < \infty\}$$

is a vector space and that

$$\langle f, g \rangle_{\mathrm{L}^2} := \int_0^\infty f(t)\overline{g(t)} \, \mathrm{d}t$$

defines an inner product on it.

Exercise 2.4. Let $(E, \|\cdot\|)$ be a normed space. Show (by induction on $n \in \mathbb{N}$) that

$$\|f_1 + \cdots + f_n\| \le \|f_1\| + \cdots + \|f_n\|$$

holds for each collection of elements $f_1, \dots, f_n \in E$, $n \in \mathbb{N}$.

Exercise 2.5. Show that 1-norm and maximum norm on $\mathbb{K}^d$ are indeed norms.

Exercise 2.6. Complete the proof of the norm properties of the 1-norm and the supremum norm on $\mathrm{C}[a, b]$. Where is it important that we deal with *continuous* functions here?

Exercise 2.7. Show that $(\mathcal{B}(\Omega), \|\cdot\|_\infty)$ is indeed a normed vector space. (Mimic the proof in the $\mathrm{C}[a, b]$-case.)

Exercise 2.8. Show that ℓ^1 is a vector space and that $\|\cdot\|_1$ is a norm on it. (Mimic the ℓ^2-case treated in Example 2.4)

Exercise 2.9. Prove Lemma 2.16.

Exercise 2.10. Let E, F be inner product spaces, let $T : E \to F$ be a linear isometry. Show that

$$\langle Tf, Tg \rangle_F = \langle f, g \rangle_E \qquad \text{for all } f, g \in E.$$

[Hint: Polarization identity in case $\mathbb{K} = \mathbb{R}$ and Exercise 1.5 in case $\mathbb{K} = \mathbb{C}$.]

Exercise 2.11. Let $E := \mathrm{C}[a, b]$ with the supreum norm and let $t_0 \in [a, b]$. Show that point evaluation $\delta_{t_0}(f) := f(t_0)$ is a bounded linear mapping from $(\mathrm{C}[a, b], \|\cdot\|_\infty) \to (\mathbb{K}, |\cdot|)$ with norm $\|\delta_{t_0}\| = 1$.

Show that the point evaluation $\delta_{t_0} : \mathrm{C}[a, b] \to \mathbb{K}$ is unbounded, if one considers $\|\cdot\|_2$ on $\mathrm{C}[a, b]$.

Exercise 2.12. Consider on ℓ^1 the multiplication operator A_λ induced by the sequence $\lambda = (1 - 1/n)_{n \in \mathbb{N}}$. What is its norm? Is it attained?

Exercise 2.13. Let $m \in \mathrm{C}[a, b]$. Consider on $(\mathrm{C}[a, b], \|\cdot\|_\infty)$ the multiplication operator

$$A : \mathrm{C}[a, b] \longrightarrow \mathrm{C}[a, b] \qquad Af = mf.$$

Prove that $\|A\| = \|m\|_\infty$.

Exercise 2.14. Let $E := \{f \in \mathrm{C}[0,1] \mid f(1) = 0\}$, with supremum norm. Consider the functional φ on E defined by

$$\varphi(f) := \int_0^1 f(x)\,\mathrm{d}x \qquad (f \in E).$$

Show that φ is bounded with norm $\|\varphi\| = 1$. Then show that for every $0 \neq f \in E$ one has $|\varphi(f)| < \|f\|_\infty$.

Exercise 2.15. For a continuous function $f \in \mathrm{C}[a,b]$ let

$$(Jf)(t) := \int_a^t f(s)\,\mathrm{d}s \qquad (t \in [a,b]).$$

Show that

$$J : (\mathrm{C}[a,b], \|\cdot\|_1) \longrightarrow (\mathrm{C}[a,b], \|\cdot\|_\infty)$$

is a bounded linear mapping. What is its kernel, what is its range, what is its norm?

Exercise 2.16 (Example 2.27). Fix $m \in \mathrm{C}[a,b]$ and consider the linear functional

$$T : \mathrm{C}[a,b] \to \mathbb{K} \qquad Tf := \int_0^1 m(s)f(s)\,\mathrm{d}s$$

on $E = \mathrm{C}[a,b]$ with the supremum norm.

a) Prove that $\|T\| = \|m\|_1$. (For the nontrivial estimate you may wish to consider the functions $f_\epsilon = \frac{\overline{m}}{|m|+\epsilon}$ for $\epsilon > 0$.)

b) Show that the norm $\|T\|$ *is attained* if m has no zeros.

c) (tricky) Give an example of m such that $\|T\|$ is *not* attained. Prove your claims!

Exercises 2B

Exercise 2.17. Make a sketch of the unit balls of $\|\cdot\|_1$ and $\|\cdot\|_\infty$ on $\mathbb{R}^2$.

Exercise 2.18. In the examples, $\mathbb{K}^d$ and $\mathrm{C}[a,b]$, find pairs of vectors violating the parallelogram law for the 1-norm, and the supremum norm, respectively.

Exercise 2.19. Mimic the proof of Example 2.4 to show that the set

$$E := \{f \in \mathrm{C}(\mathbb{R}) \mid \int_{-\infty}^{\infty} |f(t)|^2\,\mathrm{d}t < \infty\}$$

is a vector space and that

$$\langle f, g\rangle_{\mathrm{L}^2} := \int_{-\infty}^{\infty} f(t)\overline{g(t)}\,\mathrm{d}t$$

defines an inner product on it.

Exercise 2.20. Mimic the proof of Example 2.4 to show that the set

$$E := \{f \in \mathrm{C}[0,\infty) \mid \int_0^\infty |f(t)|^2 e^{-t}\,\mathrm{d}t < \infty\}$$

is a vector space and that

$$\langle f, g\rangle := \int_0^\infty f(t)\overline{g(t)}e^{-t}\,\mathrm{d}t$$

defines an inner product on it.

Exercise 2.21. The set of scalar **null sequences** is

$$\mathrm{c}_0 := \{(x_j)_{j\in\mathbb{N}} \mid x_j \in \mathbb{K} \text{ for all } j \in \mathbb{N} \text{ and } \lim_{j\to\infty} x_j = 0\}.$$

Show that c_0 is a linear subspace of ℓ^∞, and it contains ℓ^2. (More on this space in Exercise 3.10 below.)

Exercise 2.22. Show that $\ell^1 \subseteq \ell^2$ and that this inclusion is proper.

Exercise 2.23. Let $w : (0,\infty) \to \mathbb{R}$ be continuous, $w(t) > 0$ for all $t > 0$. Show that the set

$$E := \left\{f \in \mathrm{C}(0,\infty) \mid \int_0^\infty |f(t)|\, w(t)\,\mathrm{d}t < \infty\right\}$$

is a vector space and

$$\|f\|_1 := \int_0^\infty |f(t)|\, w(t)\,\mathrm{d}t$$

is a norm on it.

Exercise 2.24. Let $w : (0,\infty) \to \mathbb{R}$ be continuous, $w(t) > 0$ for all $t > 0$. Show that the set

$$E := \left\{f \in \mathrm{C}(0,\infty) \mid \sup_{t>0} |f(t)|\, w(t) < \infty\right\}$$

is a vector space and

$$\|f\|_\infty := \sup_{t>0} |f(t)|\, w(t)$$

defines a norm on it.

Exercise 2.25. Let $(F, \|\cdot\|_F)$ be a normed space, let E be any vector space and $T : E \to F$ an *injective* linear mapping. Show that

$$\|f\|_E := \|Tf\|_F \qquad (f \in E)$$

defines a norm on E, and T becomes an isometry with respect to this norm.

Exercise 2.26. Let E consist of all functions $f : \mathbb{R}_+ \to \mathbb{K}$ constant on each interval $[n-1, n)$, $n \in \mathbb{N}$, and such that

$$\|f\|_1 := \int_0^\infty |f(t)|\,\mathrm{d}t < \infty.$$

Show that E is a vector space, $\|\cdot\|_1$ is a norm on it and describe an isometric isomorphism $T : \ell^1 \to E$.

Exercise 2.27. Let, as in Exercise 2.14, $E := \{f \in \mathrm{C}[0,1] \mid f(1) = 0\}$ with supremum norm. Consider the multiplication operator A defined by $(Af)x = xf(x)$, $x \in [0,1]$. Show that $\|Af\|_\infty < 1$ for every $f \in E$ such that $\|f\|_\infty \leq 1$, but nevertheless $\|A\| = 1$.

Exercise 2.28. Consider the space $\mathrm{C}[a,b]$, endowed with the $\|\cdot\|_2$. Let $m \in \mathrm{C}[a,b]$ be fixed and consider the multiplication operator

$$T_m : \mathrm{C}[a,b] \longrightarrow \mathrm{C}[a,b], \qquad T_m f := mf \quad \text{(pointwise multiplication)}.$$

a) Show that $\|T_m\| = \|m\|_\infty$.

b) Give a concrete example where the norm $\|T_m\|$ of T_m is not attained (justification!).

c) Characterize those functions $m \in \mathrm{C}[a,b]$ for which the norm $\|T_m\|$ of T_m is (not) attained.

Exercises 2C

Exercise 2.29. Show that for $1 \le p \le \infty$ the p-norm on ℓ^p does not satisfy the parallelogram law unless $p = 2$.

Exercise 2.30. Let $p, q \in (1, \infty)$ such that $1/p + 1/q = 1$, and let $a, b \ge 0$. Show that

$$ab = \inf_{t>0} \left[\frac{t^p}{p} a^p + \frac{t^{-q}}{q} b^q \right].$$

Exercise 2.31. Let $1 \le p < \infty$. Show that by

$$\|f\|_p := \left(\int_a^b |f(t)|^p \, \mathrm{d}t \right)^{1/p}$$

a norm is defined on $\mathrm{C}[a,b]$. [Hint: Use Lemma 2.29 as in the proof of Theorem 2.28.]

Exercise 2.32. Let $1 \le p \le \infty$ and let q be defined by $1/p + 1/q = 1$. Show that

$$\left| \int_a^b f(t) g(t) \, \mathrm{d}t \right| \le \|f\|_p \|g\|_q$$

for all $f, g \in \mathrm{C}[a,b]$. [Hint: Use Lemma 2.31 as in the proof of Theorem 2.30.]

Exercise 2.33. Let $(E, \|\cdot\|_E)$ be a normed space and Ω a nonempty set. A function $f : \Omega \to E$ is bounded if its image $f(\Omega)$ is a bounded subset of Ω. Equivalently, f is bounded if

$$\|f\|_\infty := \sup\{\|f(t)\|_E \mid t \in \Omega\} < \infty.$$

Show that the set

$$\mathcal{B}(\Omega; E) := \{f : \Omega \longrightarrow E \mid \|f\|_\infty < \infty\}$$

of bounded E-valued functions on Ω is a linear space and $\|\cdot\|_\infty$ is a norm on it.

Exercise 2.34. Let $(E, \|\cdot\|_E)$ be a normed space. A function $f : [a,b] \to E$ is called **of bounded variation** if

$$\|f\|_v := \sup \sum_{j=1}^n \|f(t_j) - f(t_{j-1})\|_E < \infty$$

where the sup is taken over all decompositions $a = t_0 < t_1 < \cdots < t_n = b$ with $n \in \mathbb{N}$ arbitrary. Denote by

$$\mathrm{BV}([a,b]; E) := \{f : [a,b] \longrightarrow E \mid \|f\|_v < \infty\}.$$

Show that $\mathrm{BV}([a,b];E)$ is a linear space and $\|\cdot\|_v$ satisfies the triangle inequality and is homogeneous. Is it a norm? How about

$$\|f\| := \|f(a)\|_E + \|f\|_v\,?$$

Exercise 2.35. Let $(E,\|\cdot\|_E)$ be a normed space and $\alpha \in (0,1]$. A function $f : [a,b] \to E$ is called **Hölder continuous** of order α if there is a finite number $c = c(f) \geq 0$ such that

$$\|f(t) - f(s)\|_E \leq c\,|t-s|^\alpha$$

for all $s,t \in [a,b]$. Hölder continuous functions of order $\alpha = 1$ are also called **Lipschitz[7] continuous**. Let

$$\mathrm{C}^\alpha([a,b];E) := \{f : [a,b] \longrightarrow E \mid f \text{ is Hölder continuous of order } \alpha\}.$$

Show that $\mathrm{C}^\alpha([a,b];E)$ is a linear space and that

$$\|f\|_{(\alpha)} := \sup\left\{\, \|f(t) - f(s)\|_E \,/\, |s-t|^\alpha \;\mid\; s,t \in [a,b], s \neq t\right\}$$

satisfies the triangle inequality and is homogeneous. Is it a norm? How about

$$\|f\| := \|f(a)\|_E + \|f\|_{(\alpha)}\,?$$

Exercise 2.36. Let $(E,\|\cdot\|_E)$ be a normed space. A function $f : [a,b] \to E$ is a E-valued **step function** if there is a partition $a = t_0 < t_1 < \cdots < t_n = b$ of the interval $[a,b]$ such that f is constant on each open interval (t_{j-1}, t_j), $j = 1,\dots,n$. Show that the set $\mathrm{St}([a,b];E)$ of E-valued step functions is a linear space.

Let $f \in \mathrm{St}([a,b];E)$ and choose a partition $a = t_0 < t_1 < \cdots < t_n = b$ of $[a,b]$ such that there are vectors $x_j \in E$ with $f \equiv x_j$ on (t_{j-1}, t_j) for $j = 1,\dots,n$. Define

$$\int_a^b f(s)\,\mathrm{d}s := \sum\nolimits_{j=1}^n (t_j - t_{j-1})x_j \in E.$$

Show that $\int_a^b f(s)\,\mathrm{d}s$ does not depend on the chosen partition, and that

$$\left\|\int_a^b f(s)\,\mathrm{d}s\right\|_E \leq \int_a^b \|f(s)\|_E\,\mathrm{d}s \leq (b-a)\,\|f\|_\infty\,.$$

[7]Rudolf Lipschitz (1832–1903), German mathematician.

Chapter 3

Distance and Approximation

Approximation is at the heart of analysis. In this chapter we shall see how a norm naturally induces a notion of distance, and how this leads to the notion of convergent sequences.

3.1. Metric Spaces

Originally, i.e., in three-dimensional geometry, a "vector" is a translation of (affine) three-space, a model for physical motion or displacement. The length of a vector is just the length of this displacement. By introducing a coordinate system in three-space, points can be identified with vectors: with each point P one associates the vector which "moves" the origin O to P. Given this identification, vectors "are" points and the *distance* between the two points x and y is the *length* of the vector $x - y$.

These remarks lead us to introduce the **distance** between two elements $f, g \in E$ of a normed space $(E, \|\cdot\|)$ as

$$d_E(f, g) := \|f - g\| .$$

The function

$$d_E : E \times E \longrightarrow \mathbb{R}_+, \qquad (f, g) \longmapsto \|f - g\|$$

is called the **associated metric** and has the following properties:

$$\begin{aligned} &d_E(f, g) = 0 \iff f = g, \\ &d_E(f, g) = d_E(g, f), \\ &d_E(f, g) \leq d_E(f, h) + d_E(h, g) \end{aligned}$$

Ex.3.1 with f, g, h being arbitrary elements of E.

We should better write $d_{\|\cdot\|}(f, g)$ instead of $d_E(f, g)$, since the distance depends evidently on the norm. However, our notation is more convenient, and we shall take care that no confusion arises.

Definition 3.1. A **metric** on a set Ω is a mapping $d : \Omega \times \Omega \to [0, \infty)$ satisfying the following three conditions:

1) $d(x, y) = 0$ if and only if $x = y$ (**definiteness**),
2) $d(x, y) = d(y, x)$ (**symmetry**),
3) $d(x, y) \le d(x, z) + d(z, y)$ (**triangle inequality**),

for all $x, y, z \in \Omega$. A **metric space** is a pair (Ω, d) with Ω being a set and d a metric on it. For $x \in \Omega$ and $r > 0$ the set

$$\mathrm{B}(x, r) := \{y \in \Omega \mid d(x, y) < r\}$$

is called the **open ball** of radius r around x.

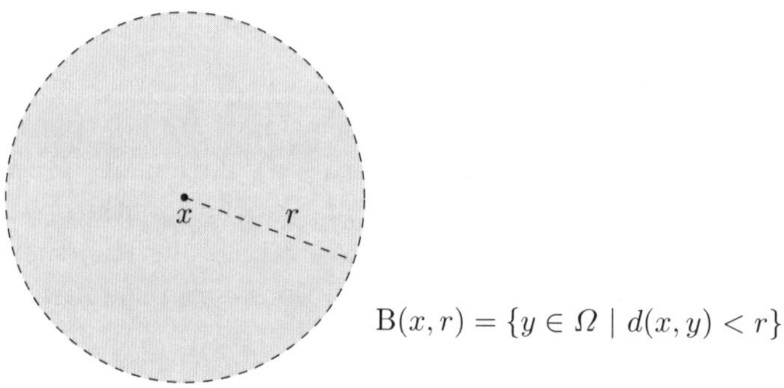

Figure 4. The open ball of radius r.

We immediately note important examples.

Example 3.2 (Normed spaces). Every normed space $(E, \|\cdot\|)$ is a metric space with respect to the associated metric $d_E(x, y) = \|x - y\|$, $x, y \in E$.

Example 3.3 (Discrete metric). Every set Ω becomes a metric space with respect to the **discrete metric**, defined by

$$d(x, y) := \begin{cases} 0 & \text{if } x = y, \\ 1 & \text{if } x \neq y. \end{cases}$$

(Check that the axioms for a metric are indeed satisfied!)

Example 3.4. The interval $[0, \infty]$ becomes a metric space under

$$d(x, y) := \left| \frac{1}{1+x} - \frac{1}{1+y} \right| \qquad (x, y \in [0, \infty]),$$

where we use the convention that $1/\infty = 0$. Ex.3.2

Example 3.5 (Induced metric). If (Ω, d) is a metric space and $A \subseteq \Omega$ is an arbitrary subset, then A becomes a metric space by just restricting the metric d to $A \times A$. This metric on A is called the **induced metric**. For example, the interval $(0, 1]$ is a metric space in its own right by setting $d(x, y) := |x - y|$, $x, y \in (0, 1]$. This metric is induced by the usual metric on $\mathbb{R}$.

A metric space need not carry any algebraic structure. (Indeed, the discrete metric can be defined on any set.) So one should not be misled by the terminology "metric *space*". The word "space" appears in several contexts in mathematics but needs further determination to be meaningful, like in "vector (linear) space", "metric space", "topological space", ...

3.2. Convergence

Recall from your undergraduate analysis course that a sequence of real numbers $(x_n)_{n \in \mathbb{N}}$ converges to a real number $x \in \mathbb{R}$ if

$$\forall \epsilon > 0 \ \exists N \in \mathbb{N} \text{ such that } \forall n \geq N : \ |x - x_n| < \epsilon,$$

and this is abbreviated by writing

$$\lim_{n \to \infty} x_n = x.$$

We now generalize this concept of convergence to general metric spaces. Note that a scalar sequence $(x_n)_{n \in \mathbb{N}}$ converges to $x \in \mathbb{R}$ if and only if $\lim_{n \to \infty} |x - x_n| = 0$, and $|x - x_n| = d_{\mathbb{R}}(x, x_n)$ is the natural distance (metric) on $\mathbb{R}$.

Definition 3.6. Let (Ω, d) be a metric space. A sequence $(x_n)_{n \in \mathbb{N}}$ in Ω **converges** to an element $x \in \Omega$ if $\ d(x_n, x) \to 0$ as $n \to \infty$, i.e.,

$$\forall \epsilon > 0 \ \exists N \in \mathbb{N} \text{ such that } \forall n \geq N : \ d(x, x_n) < \epsilon.$$

This is abbreviated by

$$x_n \to x \quad \text{or} \quad \lim_{n\to\infty} x_n = x,$$

and x is called a **limit** of the sequence $(x_n)_{n\in\mathbb{N}}$.

We say that a sequence $(x_n)_{n\in\mathbb{N}}$ in Ω is **convergent** (in Ω) if it has a limit, i.e., if there is $x \in \Omega$ such that $x_n \to x$.

The convergence $x_n \to x$ can be rephrased in the following way: for every $\epsilon > 0$ one has $x_n \in \mathrm{B}(x, \epsilon)$ *eventually*, (i.e., for all sufficiently large $n \in \mathbb{N}$)

Example 3.7. In a discrete metric space a sequence is convergent if and only if it is *eventually constant.*

Example 3.8. If $(E, \|\cdot\|)$ is a normed space, then $x_n \to x$ in E is equivalent to $\|x_n - x\| \to 0$. For instance, if $E = \mathbb{K}^d$ with the Euclidean metric and writing $x_n = (x_{n1}, \dots, x_{nd})$ for all $n \in \mathbb{N}$, then $(x_n)_{n\in\mathbb{N}}$ is convergent in $\mathbb{K}^d$ if and only if each coordinate sequence $(x_{nj})_{n\in\mathbb{N}}$, $j = 1, \dots, d$, is convergent in $\mathbb{K}$ (see also the beginning of the next section).

We shall see more examples of convergent and nonconvergent sequences shortly. Let us return to the theory.

Lemma 3.9. *Limits are unique. More precisely, let $(x_n)_{n\in\mathbb{N}}$ be a sequence in Ω such that $x_n \to x \in \Omega$ and $x_n \to x' \in \Omega$. Then $x = x'$.*

Proof. By the triangle inequality we have

$$d(x, x') \le d(x, x_n) + d(x_n, x')$$

for every $n \in \mathbb{N}$. Since both $d(x, x_n) \to 0$ and $d(x_n, x') \to 0$, it follows that $d(x, x') = 0$. Hence, by definiteness of d, $x = x'$. □

By the lemma, we shall henceforth speak of "the" limit of a convergent sequence.

Note that the concept of convergence is always *relative* to a given metric space. The assertion "The sequence $(x_n)_{n\in\mathbb{N}}$ is convergent" to be meaningful requires a metric *and* a space Ω where this metric is defined. When the reference space is in doubt, we may say for clarification that $(x_n)_{n\in\mathbb{N}}$ is convergent *in* Ω. This remark is not as trivial as it sounds, see the following example.

Example 3.10. Consider the sequence $x_n = {}^1\!/_n$, $n \in \mathbb{N}$. Does it converge? Well, that depends:

1) yes, in $\Omega = \mathbb{R}$ with the standard metric $d(x, y) = |x - y|$, and one has $\lim_{n\to\infty} x_n = 0$;
2) no, in $\Omega = (0, 1]$, $d(x, y) = |x - y|$;
3) no, in $\Omega = [0, 1]$, d the discrete metric;
4) yes, in $\Omega = (0, 1]$ with the metric

$$d(x, y) := \left|e^{2\pi i x} - e^{2\pi i y}\right| \qquad (x, y \in (0, 1]).$$

Here one has $\lim_{n\to\infty} x_n = 1$. Ex.3.3

3.3. Uniform, Pointwise and (Square) Mean Convergence

We now turn to the metric spaces we are really interested in, namely subsets of normed spaces. In particular, we shall look at the examples we encountered in Chapter 2: the Euclidean space $(\mathbb{K}^d, \|\cdot\|_2)$, the sequence spaces $(\ell^p, \|\cdot\|_p)$, and the spaces $(\mathrm{C}[a, b], \|\cdot\|_p)$, for $p = 1, 2, \infty$.

Finite-dimensional case. Let $E := (\mathbb{K}^d, \|\cdot\|_2)$ be the d-dimensional Euclidean space. Each element $f \in E$ is a d-vector with scalar components, hence a sequence $(f_n)_{n\in\mathbb{N}}$ in $\mathbb{K}^d$ can be written as

$$f_n = (x_{n1}, x_{n2}, \dots, x_{nd}) \qquad (n \in \mathbb{N}).$$

Now, as $n \to \infty$ we have

$$\begin{aligned} f_n \to f \quad &\iff \quad \|f - f_n\|_2 \to 0 \quad \iff \quad \sum_{j=1}^{d} |x_j - x_{nj}|^2 \to 0 \\ &\iff \quad x_{nj} \to x_j \qquad \text{for each } j = 1, \dots, d, \end{aligned}$$

by elementary analysis. The latter assertion is called **convergence in each component**. Consequently, the study of convergence in finite dimensions can be reduced to convergence of finitely many scalar sequences.

One could ask what happens if one replaces the Euclidean norm by another one. The answer — nothing! — gives Theorem 4.29 below: Convergence with respect to *any* norm on $\mathbb{K}^d$ is the same as componentwise convergence.

Sequence spaces. Let us now consider the infinite-dimensional analogue of the Euclidean space, the sequence space ℓ^2. Each element $f \in \ell^2$ is a vector with infinitely many components:

$$f = (x_1, x_2, x_3, \dots).$$

Hence a sequence $(f_n)_{n\in\mathbb{N}}$ in ℓ^2 can be written as

$$\begin{aligned} f_1 &= (x_{11},\ x_{12},\ x_{13}\dots),\\ f_2 &= (x_{21},\ x_{22},\ x_{23}\dots),\\ &\ \vdots \\ f_n &= (x_{n1},\ x_{n2},\ x_{n3}\dots).\\ &\ \vdots \end{aligned} \tag{3.4}$$

Note that we are dealing with a sequence of (scalar) sequences here, and this may be a little confusing in the beginning.

The notation in (3.4) may be suggestive, but the letter "x" appearing there seems rather unmotivated. Note that x_{nj} is just the j-component of the vector f_n, and there are different generic notations for that, e.g., f_{nj} or $[f_n]_j$ or $f_n(j)$. I prefer the latter one, since it draws on the actual definition of a sequence/infinite vector as a *function on* $\mathbb{N}$, and hence $f_n(j)$ is the value of this function at point j, i.e., the j-th component of the sequence.

In the scheme (3.4), the "vertical sequence" $(x_{nj})_{n\in\mathbb{N}}$ is just the sequence of j-th components of the vectors f_n. *Componentwise convergence* of $(f_n)_{n\in\mathbb{N}}$ to f as $n \to \infty$ hence means that *each* of these vertical sequences converges:

$$x_{nj} \to x_j \qquad \text{as } n \to \infty, \text{ for each } j \in \mathbb{N},$$

and convergence in norm means that

$$\|f_n - f\|_2 \to 0, \quad \text{i.e.,} \quad \sum_{j=1}^{\infty} |x_j - x_{nj}|^2 \to 0 \qquad \text{as } n \to \infty.$$

Note that for each fixed component $j \in \mathbb{N}$ we have the trivial inequality

$$|x_j - x_{nj}| \le \|f - f_n\|_2 \,.$$

The sandwich theorem from elementary analysis hence tells us that if $f_n \to f$ in $\|\cdot\|_2$, then also $x_{nj} \to x_j$ for each $j \in \mathbb{N}$. In short: *Convergence in* $\|\cdot\|_2$ *implies componentwise convergence.*

We have seen above that in finite dimensions the converse also holds: componentwise convergence implies convergence in norm. However, this is false in infinite dimensions, as the next example shows.

Example 3.11. Consider the sequence of standard unit vectors $(e_n)_{n\in\mathbb{N}}$, defined by

$$e_n = (0, 0, \ldots, 0, 1, 0, \ldots)$$

where the 1 is located at the nth place. Note that one has $e_n(j) = 0$ if $n > j$, and hence

$$\lim_{n\to\infty} e_n(j) = 0$$

for each component $j \in \mathbb{N}$. On the other hand, $\|e_n\|_2 = 1$ for all $n \in \mathbb{N}$, and hence $e_n \not\to 0$ in $\|\cdot\|_2$.

Could $(e_n)_{n\in\mathbb{N}}$ converge in $\|\cdot\|_2$ to something other than 0? Well, no, since we have seen that $\|\cdot\|_2$-convergence implies convergence in every component. Hence we conclude that the sequence $(e_n)_{n\in\mathbb{N}}$ of standard unit vectors *converges componentwise, but not in* $\|\cdot\|_2$.

Now let us look at the other ℓ^p-spaces.

Example 3.12 (Scale of ℓ^p-Spaces)**.** Let $f : \mathbb{N} \to \mathbb{K}$ be any scalar sequence. We claim that

$$\|f\|_\infty \le \|f\|_2 \le \|f\|_1 \qquad \text{in } [0, \infty]. \tag{3.5}$$

For the first inequality we note that we have $|f(j)| \le \|f\|_2$ for every $j \in \mathbb{N}$, and hence we can take the supremum over j to obtain $\|f\|_\infty \le \|f\|_2$. For the second inequality we estimate

$$\|f\|_2^2 = \sum_{j=1}^{\infty} |f(j)|^2 = \sum_{j=1}^{\infty} |f(j)| \cdot |f(j)| \le \sum_{j=1}^{\infty} |f(j)| \cdot \|f\|_1 = \|f\|_1^2$$

if we let $0 \cdot \infty = 0$. From (3.5) we obtain

$$\ell^1 \subseteq \ell^2 \subseteq \ell^\infty.$$

Furthermore, by replacing f by $f - f_n$ in (3.5) and using the sandwich theorem we see that *convergence in* $\|\cdot\|_1$ *implies convergence in* $\|\cdot\|_2$*, which in turn implies convergence in* $\|\cdot\|_\infty$.

As in the ℓ^2 case one sees that convergence in ℓ^p-norm implies componentwise convergence, but not vice versa. Ex.3.4

Uniform vs. Pointwise Convergence. We now turn to the supremum norm $\|\cdot\|_\infty$ on the space $\mathcal{B}(\Omega)$ of bounded functions on some nonempty set Ω (introduced in Example 2.8). Recall the definition

$$\|f\|_\infty = \sup\{|f(t)| \mid t \in \Omega\}$$

of the supremum norm of a bounded function $f : \Omega \to \mathbb{K}$. Then

$$\|f - g\|_\infty \leq \epsilon \quad \iff \quad \forall\, t \in \Omega : |f(t) - g(t)| \leq \epsilon.$$

So $f_n \to f$ in the norm $\|\cdot\|_\infty$ may be written as

$$\forall \epsilon > 0 \ \exists N \in \mathbb{N} \text{ s.t. } \forall n \geq N \ \forall t \in \Omega : |f(t) - f_n(t)| \leq \epsilon.$$

Note that the chosen $N = N(\epsilon)$ may depend on ϵ but it is the same (= "uniform") for every $t \in \Omega$. Therefore we say that $(f_n)_{n\in\mathbb{N}}$ converges to f **uniformly** (on Ω); and the supremum norm $\|\cdot\|_\infty$ is also called the **uniform norm**.

A weaker notion of convergence (and the analogue of componentwise convergence above) is the notion of **pointwise convergence**. We say that a sequence $(f_n)_{n\in\mathbb{N}}$ of functions on Ω **converges pointwise** to a function $f : \Omega \to \mathbb{K}$, if

$$f_n(t) \to f(t) \qquad \text{in } \mathbb{K} \quad \text{as } n \to \infty$$

for every $t \in \Omega$. In formal notation

$$\forall t \in \Omega \ \forall \epsilon > 0 \ \exists N \in \mathbb{N} \ \text{ s.t. } \forall n \geq N : |f(t) - f_n(t)| \leq \epsilon.$$

Here the $N = N(\epsilon, t)$ may depend on ϵ *and* the point $t \in \Omega$.

A function $f : \Omega \to \mathbb{K}$ can be seen as a "vector" with components indexed by points in Ω, i.e., $f(t)$ is the t-component of the vector f. This terminology is coherent with the case $\Omega = \mathbb{N}$, in which case we have $\mathcal{B}(\Omega) = \ell^\infty$, and pointwise convergence as introduced in this paragraph is the same as componentwise convergence as considered above.

Clearly, uniform convergence implies pointwise convergence, as follows also from the (trivial) estimate

$$|f(t) - f_n(t)| \leq \sup_{s\in\Omega} |(f - f_n)(s)| = \|f - f_n\|_\infty$$

for each $t \in \Omega$. The converse is not true.

Example 3.13. Consider $\Omega = (0, \infty)$ and the functions $f_n(x) = \mathrm{e}^{-nx}$. Then clearly $f_n(x) \to 0$ as $n \to \infty$ for each $x \in \Omega$. However,

$$\|f_n\|_\infty = \sup_{x>0} \left|\mathrm{e}^{-nx}\right| = 1$$

for each $n \in \mathbb{N}$, and thus $f_n \not\to 0$ uniformly.

If we consider $g_n := nf_n$, then $\|g_n\|_\infty = n \to \infty$, but still $g_n(x) \to 0$ for each $x > 0$. This shows that pointwise convergence of a sequence does not even imply its boundedness, let alone convergence, in the supremum norm. Ex.3.5 Ex.3.6

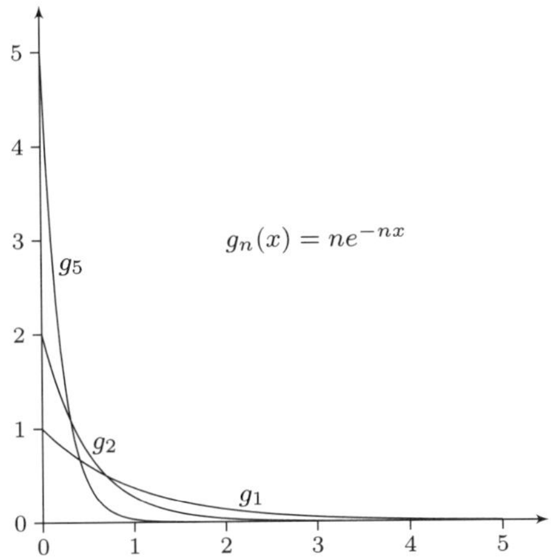

Figure 5. A plot of the graphs of g_1, g_2 and g_5.

That uniform convergence implies pointwise convergence is a special instance of the general logical fact

$$\exists x \; \forall y \, P(x,y) \quad \Longrightarrow \quad \forall y \; \exists x \, P(x,y)$$

where $P(x,y)$ is an arbitrary assertion containing x, y as free variables. But obviously, in general,

$$\forall y \; \exists x \, P(x,y) \quad \text{does not imply} \quad \exists x \; \forall y \, P(x,y).$$

Now let us look at the other p-norms on $\mathrm{C}[a,b]$.

Example 3.14 (p-Norms on $\mathrm{C}[a,b]$). *For each interval* $[a,b] \subseteq \mathbb{R}$ *we have*

$$\|f\|_1 \le \sqrt{b-a}\,\|f\|_2 \quad \textit{and} \quad \|f\|_2 \le \sqrt{b-a}\,\|f\|_\infty \tag{3.6}$$

for all $f \in \mathrm{C}[a,b]$.

Proof. The first inequality follows from Cauchy–Schwarz and

$$\|f\|_1 = \int_a^b |f| = \langle |f|, \mathbf{1} \rangle \le \|f\|_2 \, \|\mathbf{1}\|_2 = \sqrt{b-a}\,\|f\|_2,$$

where we have written **1** for the function which is constantly equal to 1. The second inequality follows from

$$\|f\|_2^2 = \int_a^b |f|^2 \leq \int_a^b \|f\|_\infty^2 \mathbf{1} = (b-a)\|f\|_\infty^2 . \qquad \square$$

Convergence in the norms $\|\cdot\|_1$ and $\|\cdot\|_2$ are often called **convergence in mean** and **convergence in square mean**, respectively. Hence, by (3.6), uniform convergence implies convergence in square mean, which in

Ex.3.7 turn implies convergence in mean.

How about the converse implications? Consider as an example $[a,b] = [0,1]$ and in $\mathrm{C}[0,1]$ the sequence of functions $f_n(t) := t^n$, for $t \in [0,1]$ and $n \in \mathbb{N}$. Then $f_n(t) \to 0$ as $n \to \infty$ for each $t < 1$, but $f_n(1) = 1$ for all $n \in \mathbb{N}$. This means that $(f_n)_{n\in\mathbb{N}}$ converges pointwise to the function f given by

$$f(t) := \begin{cases} 0 & \text{for } 0 \leq t < 1, \\ 1 & \text{for } t = 1, \end{cases}$$

which is, however, not contained in $\mathrm{C}[0,1]$. A fortiori, $(f_n)_{n\in\mathbb{N}}$ has also no uniform (i.e., $\|\cdot\|_\infty$-) limit in $\mathrm{C}[0,1]$.

Now let us consider the norms $\|\cdot\|_1$ and $\|\cdot\|_2$ instead. For $n \in \mathbb{N}$ we compute

$$\|f_n\|_1 = \int_0^1 t^n \, \mathrm{d}t = \frac{1}{n+1} \to 0 \qquad \text{and}$$

$$\|f_n\|_2^2 = \int_0^1 |f_n(t)|^2 \, \mathrm{d}t = \int_0^1 t^{2n} \, \mathrm{d}t = \frac{1}{2n+1} \to 0.$$

This shows that $f_n \to 0$ in $\mathrm{C}[a,b]$ with the standard inner product *and* in $\mathrm{C}[a,b]$ with respect to $\|\cdot\|_1$.

To complete the picture, let $g_n := \sqrt{2n+1}\, f_n$. Then

$$\|g_n\|_2 = \frac{\sqrt{2n+1}}{\sqrt{2n+1}} = 1, \quad \text{but} \quad \|g_n\|_1 = \frac{\sqrt{2n+1}}{n+1} \to 0.$$

Ex.3.8 Hence $\|\cdot\|_1$-convergence does not imply $\|\cdot\|_2$-convergence.

The previous example shows also that convergence in (square) mean on an interval $[a,b]$ does not imply pointwise convergence.

3.4. The Closure of a Subset

One of the major goals of analysis is to describe whether and how a 'complicated' object may be *approximated* by 'simple' ones. The protoypical example here is the approximation of real numbers (complicated) by certain rational numbers (simple). In general what is to be considered a good approximation can vary considerably, and we restrict our discussion to the metrical aspects. To stay in the example, suppose that $A \subseteq \mathbb{Q}$ is a set of rational numbers, e.g., A is the set

$$A = \{ n/10^k \mid n, k \in \mathbb{N}_0 \}$$

of all positive numbers with a finite decimal expansion. Asking how "well" one can approximate a given number $x \in \mathbb{R}$ by elements of A then means to ask how small the distance $|x - a|$ can be made when we vary $a \in A$. So we are actually interested in the number

$$\inf\{|x - a| \mid a \in A\}.$$

Note that in our example this number is zero (by decimal expansion), but if $x = 1/3$, say, then the infimum is not a minimum.

Definition 3.15. Let (Ω, d) be a metric space and $A \subseteq \Omega$. For an element $x \in \Omega$ the number

$$d(x, A) := \inf\{d(x, a) \mid a \in A\}$$

is called the **distance of x to A.**

As seen above, the distance $d(x, A)$ need not be attained, i.e., the infimum need not be a minimum. We shall have to say more about this in Chapter 8. For the moment we are interested in the case that $d(x, A)$ is zero.

Lemma 3.16. *Let (Ω, d) be a metric space, let $A \subseteq \Omega$ and $x \in \Omega$. The following assertions are equivalent:*

(i) $d(x, A) = 0$.

(ii) *For each $\epsilon > 0$ there is $a \in A$ such that $d(x, a) < \epsilon$.*

(iii) *There exists a sequence $(a_n)_{n\in\mathbb{N}}$ in A such that $a_n \to x$.*

Proof. Suppose that (i) holds and $\epsilon > 0$. Then $d(x, A) = 0 < \epsilon$, and by the definition of an infimum, there exists $a \in A$ such that $d(x, a) < \epsilon$.

Now suppose that (ii) holds. For $n \in \mathbb{N}$ choose $a_n \in A$ such that $d(x, a_n) < 1/n$. Then $a_n \to x$, and $(a_n)_{n\in\mathbb{N}}$ is a sequence in A.

Finally, suppose that (iii) holds, and let $a_n \in A$ and $a_n \to x$. Then, $0 \leq d(x, A) \leq d(x, a_n)$ for each $n \in \mathbb{N}$, and hence $d(x, A) = 0$. □

Each of the statements (i)–(iii) of Lemma 3.16 specifies the intuitive idea that x can be *arbitrarily well approximated* by elements of A.

Definition 3.17. Let (Ω, d) be a metric space, and let $A \subseteq \Omega$. The **closure** of A (in Ω) is the set

$$\overline{A} := \{x \in \Omega \mid \text{there is a sequence } (a_n)_{n\in\mathbb{N}} \text{ in } A \text{ with } a_n \to x\}.$$

The set A is called **dense** (in Ω) if $\overline{A} = \Omega$.

So $\overline{A} = \{x \in \Omega \mid d(x, A) = 0\}$ consists of all the points in Ω that can
Ex.3.9 be arbitrarily well approximated by elements from A, and A is dense in Ω if every point of Ω has this property.

The classical example of a dense subset of a metric space is $A = \mathbb{Q}$ in $\Omega = \mathbb{R}$ with the standard metric. Here is a similar example in infinite dimensions.

Example 3.18. The space of **finite sequences** is defined as

$$c_{00} := \{(x_j)_{j\in\mathbb{N}} \mid x_j = 0 \text{ eventually}\} = \operatorname{span}\{e_j \mid j \in \mathbb{N}\},$$

where $\{e_j \mid j \in \mathbb{N}\}$ are the standard unit vectors introduced in Example 3.11. Clearly, it is a subspace of ℓ^2.

Claim: *The space c_{00} is dense in ℓ^2; i.e., every element $f \in \ell^2$ is the $\|\cdot\|_2$-limit of a sequence $(f_n)_{n\in\mathbb{N}}$ in c_{00}.*

Proof. Fix

$$f = (x_j)_{j\in\mathbb{N}} = (x_1, x_2, x_3, \dots) \in \ell^2.$$

Then it is natural to try as approximants the finite sequences created from f by "cutting off the tail"; i.e., we define

$$f_n := (x_1, x_2, \dots, x_n, 0, 0, \dots) \in c_{00}$$

for $n \in \mathbb{N}$. Now

$$\|f - f_n\|_2^2 = \sum_{j=1}^{\infty} |f_n(j) - f_n(j)|^2 = \sum_{j=n+1}^{\infty} |x_j|^2 \to 0 \qquad (n \to \infty)$$

since $f \in \ell^2$. This yields $\|f - f_n\|_2 \to 0$ as claimed. □

Example 3.19. The space c_{00} of finite scalar sequences is *not* dense in ℓ^∞. Indeed, let $\mathbf{1} = (1, 1, 1, \dots)$ be the vector having all its entries equal to one. Then $\|\mathbf{1} - f\|_\infty \geq 1$ for each $f \in c_{00}$, and hence $d(\mathbf{1}, c_{00}) = 1$. What is the $\|\cdot\|_\infty$-closure of c_{00} in ℓ^∞?

Claim: *The $\|\cdot\|_\infty$-closure of c_{00} in ℓ^∞ is the space of scalar* **null sequences**

$$c_0 = \{(x_n)_{n\in\mathbb{N}} \mid \lim_{n\to\infty} x_n = 0\}.$$

The proof is left as an exercise (see also Exercise 2.21). Ex.3.10

The closure of a subset A of a metric space Ω is somewhat analogous to the linear span of a subset B of a vector space E: In span(B) one collects all the vectors that can be produced by performing the operation "form a finite linear combination" on elements of B, and in $\overline{A}$ one collects all elements of Ω that can be produced by performing the operation "take the limit of a convergent sequence" on members of A.

The following properties of the closure operation are sometimes helpful.

Lemma 3.20. *Let (Ω, d) be a metric space, and let $A, B \subseteq \Omega$. Then the following assertions hold:*

a) $\overline{\emptyset} = \emptyset$, $\overline{\Omega} = \Omega$,

b) $A \subseteq \overline{A}$,

c) $A \subseteq B \quad \Longrightarrow \quad \overline{A} \subseteq \overline{B}$,

d) $\overline{\overline{A}} = \overline{A}$,

e) $\overline{A \cup B} = \overline{A} \cup \overline{B}$.

Proof. Assertions a)–c) are pretty obvious, so we leave them as an exercise.

For the proof of d), note that $\overline{A} \subseteq \overline{\overline{A}}$ follows from b). For the converse, suppose that $x \in \overline{\overline{A}}$, and let $\epsilon > 0$. By Lemma 3.16 there is $y \in \overline{A}$ such that $d(x, y) < \epsilon/2$; again by Lemma 3.16 there is $z \in A$ such that $d(y, z) < \epsilon/2$. Then $d(x, z) \le d(x, y) + d(y, z) < \epsilon$, and by Lemma 3.16 we have $x \in \overline{A}$.

For the proof of e), note that since $A \subseteq A \cup B$ it follows from c) that $\overline{A} \subseteq \overline{A \cup B}$, and likewise for B. This yields the inclusion "$\supseteq$". To prove the converse inclusion, take $x \in \overline{A \cup B}$; then by definition there is a sequence $(x_n)_{n\in\mathbb{N}}$ in $A \cup B$ such that $x_n \to x$. One of the sets $\{n \in \mathbb{N} \mid x_n \in A\}$ and $\{n \in \mathbb{N} \mid x_n \in B\}$ is infinite (as they partition $\mathbb{N}$), and without loss of generality we may suppose it is the first one. Then this defines a subsequence $(x_{n_k})_{k\in\mathbb{N}}$ in A. Since also $x = \lim_{k\to\infty} x_{n_k}$, we conclude that $x \in \overline{A}$. □ Ex.3.11
Ex.3.12

As with convergence, the closure $\overline{A}$ of a set is taken with respect to a surrounding metric space. This makes using the notation "$\overline{A}$" (where the dependence on d and Ω is not visible) a source of potential mistakes. Closures of the same set in different metric spaces usually differ.

Example 3.21. Consider $A = \{q \in \mathbb{Q} \mid 0 < q < 1\}$. Then

1) $\overline{A} = [0, 1]$ in $\Omega = \mathbb{R}$ with the standard metric;
2) $\overline{A} = (0, 1)$ in $\Omega = (0, 1)$ with the standard metric;
3) $\overline{A} = A$ in $\Omega = \mathbb{R}$ with the discrete metric. Ex.3.13

In the course of this book we shall need a famous and important density result with respect to the supremum norm due to Weierstrass.[1] We postpone the proof to Chapter 9.

Theorem 3.22 (Weierstrass). *Let $[a, b]$ be a compact interval in $\mathbb{R}$. Then the space of polynomials $\mathrm{P}[a, b]$ is dense in $\mathrm{C}[a, b]$ with respect to the supremum norm.*

We let, for $k \in \mathbb{N}$ or $k = \infty$,

$$\mathrm{C}^k[a, b] := \{f : [a, b] \longrightarrow \mathbb{K} \mid f \text{ is } k\text{-times continuously differentiable}\}.$$

Since polynomials are infinitely differentiable, Weierstrass' theorem implies that $\mathrm{C}^\infty[a, b]$ is dense in $\mathrm{C}[a, b]$.

Exercises 3A

Exercise 3.1. Let $(E, \|\cdot\|)$ be a normed space, with associated distance function d_E defined by $d_E(f, g) = \|f - g\|$. Show that d_E has the three properties claimed for it on page 37.

Exercise 3.2. Show that by

$$d(x, y) := \left| \frac{1}{1+x} - \frac{1}{1+y} \right| \qquad (x, y \in [0, \infty])$$

is a metric on $[0, \infty]$, where $1/\infty := 0$. Then prove the following assertions for a sequence $(x_n)_{n\in\mathbb{N}} \subseteq [0, \infty]$ and $x \in [0, \infty)$.

a) $x_n \to x$ if and only if eventually $x_n \in \mathbb{R}$ and $|x_n - x| \to 0$.
b) $x_n \to \infty$ if and only if

$$\forall K \geq 0 \ \exists N \in \mathbb{N} \ \forall n \geq N \text{ such that } x_n \geq K.$$

(This shows that the usual convergence in $\mathbb{R}$ and the divergent sequences with limit ∞ — one of the very awkward terminological features of undergraduate analysis — can be subsumed under one notion of metric convergence.)

Exercise 3.3. Prove the convergence assertions 1)–4) of Example 3.10.

[1] Karl Weierstrass (1815–1897), German mathematician.

Exercise 3.4. Show that the inclusions

$$\ell^1 \subseteq \ell^2 \subseteq \ell^\infty,$$

are all strict. Give an example of sequences $(f_n)_{n\in\mathbb{N}}, (g_n)_{n\in\mathbb{N}}$ in ℓ^1 with

$$\|f_n\|_\infty \to 0, \ \|f_n\|_2 \to \infty \quad \text{and} \quad \|g_n\|_2 \to 0, \ \|g_n\|_1 \to \infty.$$

Exercise 3.5. Let $f_n(t) := (1 + nt)^{-1}$ for $t \in (0, \infty)$ and $n \in \mathbb{N}$.

a) Show that for each $\epsilon > 0$, $(f_n)_{n\in\mathbb{N}}$ is uniformly convergent on $[\epsilon, \infty)$ (to which function?).

b) Show that $(f_n)_{n\in\mathbb{N}}$ is not uniformly convergent on $(0, \infty)$.

Exercise 3.6. Let $h \in \mathcal{B}(0, \infty)$ such that $\lim_{t\searrow 0} h(t) = 0$. Define $g_n(t) := e^{-nt}h(t)$ for $t > 0$. Show that $g_n \to 0$ uniformly on $(0, \infty)$.

Exercise 3.7. Show with the help of the inequalities (3.6) that on an interval $[a, b]$ uniform convergence implies convergence in $\|\cdot\|_2$, and $\|\cdot\|_2$-convergence implies convergence in $\|\cdot\|_1$.

Exercise 3.8. Consider on $[0, 1]$ the function

$$f_n(t) := \begin{cases} 2n^3 t & \text{if } \ 0 \le t \le \frac{1}{2n}, \\ 2n^2 - 2n^3 t & \text{if } \ \frac{1}{2n} \le t \le \frac{1}{n}, \\ 0 & \text{if } \ \frac{1}{n} \le t \le 1. \end{cases}$$

Make a sketch of the graph of f_n. Does $(f_n)_{n\in\mathbb{N}}$ converge pointwise (uniformly, in mean, in square mean) as $n \to \infty$?

Exercise 3.9. Let $A \subseteq \mathbb{R}$ be a nonempty set of real numbers, bounded from above. Show that $\sup A \in \overline{A}$.

Exercise 3.10. Show that

$$\overline{\mathrm{c}_{00}} = \mathrm{c}_0 = \{(x_j)_{j\in\mathbb{N}} \mid \lim_{j\to\infty} x_j = 0\}$$

(closure in ℓ^∞ with respect to the supremum norm).

Exercise 3.11. Find an example of a metric space (Ω, d) and subsets $A, B \subseteq \Omega$ such that

$$\overline{A} \cap \overline{B} \neq \overline{A \cap B}.$$

Exercise 3.12 ("dense in dense is dense"). Let (Ω, d) be a metric space, and let $A, B \subseteq \Omega$ such that $A \subseteq \overline{B}$ and A is dense in Ω. Show that also B is dense in Ω.

Exercise 3.13. Show that

$$\mathrm{C}_0[a, b] := \{f \in \mathrm{C}[a, b] \mid f(a) = f(b) = 0\}$$

is $\|\cdot\|_2$-dense in $\mathrm{C}[a, b]$. Is it also $\|\cdot\|_\infty$-dense?

Exercises 3B

Exercise 3.14. Let (Ω, d) be a metric space, $(x_n)_{n\in\mathbb{N}} \subseteq \Omega$, $x \in \Omega$. Show that the following assertions are equivalent:

(i) $x_n \not\to x$.

(ii) There is $\epsilon > 0$ and a *subsequence* $(x_{n_k})_{k\in\mathbb{N}}$ with $d(x, x_{n_k}) \geq \epsilon$ for all $k \in \mathbb{N}$.

Conclude that the following assertions are equivalent:

(i) $x_n \to x$.

(ii) Each subsequence of $(x_n)_{n\in\mathbb{N}}$ has a subsequence that converges to x.

Exercise 3.15. Let (Ω, d) be a metric space. A subset $A \subseteq \Omega$ is **bounded** if

$$\operatorname{diam}(A) := \sup\{d(x,y) \mid x, y \in A\} < \infty.$$

Show that a subset A of a normed space $(E, \|\cdot\|)$ is bounded in this sense if and only if it is bounded in the sense of Definition 2.15, i.e., if $\sup\{\|x\| \mid x \in A\} < \infty$.

Exercise 3.16. Let $(x_n)_{n\in\mathbb{N}}$ be a sequence in the metric space (Ω, d) and let $x \in \Omega$. Show that there is a subsequence of $(x_n)_{n\in\mathbb{N}}$ converging to x if and only if

$$x \in \bigcap\nolimits_{n\in\mathbb{N}} \overline{\{x_k \mid k \geq n\}}.$$

Exercise 3.17. A **double sequence** in a metric space (Ω, d) is a function $\mathbb{N}\times\mathbb{N} \to \Omega$, written $(x_{n,m})_{n,m\in\mathbb{N}}$. We say that $x \in \Omega$ is the **limit** of this double sequence, written

$$\lim_{n,m\to\infty} x_{n,m} = x \quad \text{or} \quad x_{n,m} \to x \quad (n, m \to \infty)$$

if for each $\epsilon > 0$ there is $N \in \mathbb{N}$ such that $d(x_{n,m}, x) \leq \epsilon$ whenever $n, m \geq N$.

a) Show that a double sequences can have at most one limit.

b) Show that if $x_{n,m} \to x$ as $n, m \to \infty$, and $n_k, m_k \in \mathbb{N}$ are such that $n_k \to \infty$ and $m_k \to \infty$ as $k \to \infty$, then $\lim_{k\to\infty} x_{n_k, m_k} = x$, too.

c) Give an example of a double sequence $(x_{n,m})_{n,m\in\mathbb{N}}$ such that $\lim_{n\to\infty} x_{n,n}$ exists, but $\lim_{n,m\to\infty} x_{n,m}$ does not.

Exercise 3.18. A scalar sequence $x = (x_n)_{n\in\mathbb{N}}$ is called of **bounded variation** if

$$\|x\|_{\mathrm{bv}} := |x_1| + \sum_{k=1}^{\infty} |x_{k+1} - x_k| < \infty.$$

Define $\mathrm{bv} := \{x \mid x \text{ is a scalar sequence with } \|x\|_{\mathrm{bv}} < \infty\}$.

a) Show that $\|x\|_\infty \leq \|x\|_{\mathrm{bv}}$ for every sequence $x = (x_n)_{n\in\mathbb{N}}$.

b) Show that bv is a linear space and $\|\cdot\|_{\mathrm{bv}}$ is a norm on it.

c) Show that every sequence of bounded variation is convergent, but not every convergent sequence is of bounded variation.

d) Consider the (linear!) operator

$$T : (x_1, x_2, \ldots) \longmapsto (x_1, x_2 - x_1, x_3 - x_2, \ldots)$$

Show that $T : \mathrm{bv} \to \ell^1$ is an isometric isomorphism, and give a formula for its inverse T^{-1}.

Exercise 3.19. The **quadratic variation** of a scalar sequence $x = (x_n)_{n\in\mathbb{N}}$ is

$$\|x\|_{\mathrm{qv}} := \left(|x_1|^2 + \sum_{k=1}^{\infty} |x_{k+1} - x_k|^2 \right)^{1/2} \in [0, \infty].$$

Let $\mathrm{qv} := \{x \mid \|x\|_{\mathrm{qv}} < \infty\}$ be the space of all sequences of **finite quadratic variation**.

a) Show that qv is a linear space and $\|\cdot\|_{\mathrm{qv}}$ is a norm on it. (Does this norm come from an inner product?)

b) Show that $\ell^2 \subseteq \mathrm{qv}$ and compute the norm of the inclusion operator

$$T : \ell^2 \longrightarrow \mathrm{qv}, \qquad Tx := x.$$

c) Show that $\mathrm{qv} \neq \ell^2$, in fact, even $\mathrm{qv} \not\subseteq \ell^\infty$.

Exercises 3C

Exercise 3.20. Let $1 \le p < q \le \infty$. Show that $\|f\|_q \le \|f\|_p$ for each scalar sequence $f = (f(j))_{j\in\mathbb{N}}$. Show that $\ell^q \subseteq \ell^p$ with a proper inclusion.

Exercise 3.21. Show that c_{00}, the space of finite sequences, is $\|\cdot\|_p$-dense in ℓ^p for each $1 \le p < \infty$.

Exercise 3.22. Let $f \in \mathrm{C}[0,1]$ such that $f(0) = 0$. Define for $n \in \mathbb{N}$

$$(T_n f)(t) := \int_0^t \mathrm{e}^{-n(t-s)} f(s)\, ds \qquad (t \in [0,1]).$$

a) Show that $\|T_n f\|_\infty \le \frac{(1-\mathrm{e}^{-n})}{n} \|f\|_\infty$.

b) Show that

$$nT_n f(t) - f(t) = \int_0^t n\mathrm{e}^{-ns}(f(t-s) - f(t))\, \mathrm{d}s + \mathrm{e}^{-nt} f(t)$$

for all $t \in [0,1]$ and $n \in \mathbb{N}$.

c) Show that $\lim_{n\to\infty} nT_n f = f$ uniformly on $[0,1]$. [Hint: b) and Exercise 3.6.]

d) Conclude that $\mathrm{C}^1[0,1]$ is dense in $\mathrm{C}[0,1]$.

Exercise 3.23. Define $p_0 := 0$ and $p_{n+1}(t) := p_n(t) + (1/2)(t - p_n(t)^2)$. Show by induction that

$$0 \le \sqrt{t} - p_n(t) \le \frac{2\sqrt{t}}{2 + n\sqrt{t}} \qquad (t \in [0,1]).$$

Conclude that $p_n(t) \to \sqrt{t}$ uniformly in $t \in [0,1]$. Use this to show that $p_n(t^2) \to |t|$ uniformly in $t \in [-1,1]$.

Exercise 3.24. Let $f_n(x) := \sum_{k=0}^{n} {}^{(-x)^k}\!/_{k!}$ for $x \geq 0$. Show that $e^{-x} f_n(x) \to e^{-2x}$ uniformly on $\mathbb{R}_+$. [Hint: Taylor's theorem.[2]]

Exercise 3.25. Let $\mathrm{s} := \mathcal{F}(\mathbb{N}) = \{x = (x_j)_{j\in\mathbb{N}} \mid x_j \in \mathbb{K}\ \forall j \in \mathbb{N}\}$ be the space of all scalar sequences. Show that

$$d(x,y) := \sum_{j=1}^{\infty} 2^{-j} \frac{|x_j - y_j|}{1 + |x_j - y_j|}$$

is a metric on s. Show that $x_n = (x_{nj})_{j\in\mathbb{N}}$ converges to $y = (y_j)_{j\in\mathbb{N}}$ with respect to this metric if and only if $x_n \to y$ componentwise, i.e., $x_{nj} \to y_j$ for all $j \in \mathbb{N}$.

Remark: Pointwise convergence on $\mathcal{F}[a,b]$ is not a metric convergence. A proof of this fact can be found in books on topology.

[2]Brook Taylor (1685–1731), English mathematician.

Chapter 4

Continuity and Compactness

We continue our introduction to metric topology with the fundamental concepts of open and closed sets, continuity of mappings and compactness of metric spaces. Finally, we discuss the equivalence of metrics.

4.1. Open and Closed Sets

You may know the concepts of open and closed subsets of $\mathbb{R}^d$ from undergraduate analysis. Here is a definition for general metric spaces.

Definition 4.1. Let (Ω, d) be a metric space. A subset $A \subseteq \Omega$ is **closed** if $\overline{A} \subseteq A$, i.e., if for *every* sequence $(x_n)_{n\in\mathbb{N}}$ in Ω the implication

$$x_n \to x \in \Omega \text{ and } x_n \in A \text{ for all } n \in \mathbb{N} \quad \Longrightarrow \quad x \in A.$$

holds. A subset $O \subseteq \Omega$ is called **open** if

$$\forall x \in O \; \exists \epsilon > 0 \quad \text{such that} \quad \mathrm{B}(x, \epsilon) \subseteq O.$$

Closedness means that every point that can be approximated with elements of A is already contained in A, so forming the closure cannot make the set larger. Note that, since always $A \subseteq \overline{A}$, a set A is closed if and only if $A = \overline{A}$.

Openness means that for each point x there is some positive distance $\epsilon = \epsilon(x)$ with the property that if you deviate from x not more than ϵ, you will remain inside the set.

Lemma 4.2. *Let (Ω, d) be a metric space. Then the following assertions hold.*

a) *$\emptyset$ and Ω are both open and closed.*

b) *Each singleton $\{x\}$, $x \in \Omega$, is closed.*

c) *$\overline{A}$ is closed for each $A \subseteq \Omega$.*

d) *The open ball with center $x \in \Omega$ and radius $r > 0$,*

$$\mathrm{B}(x, r) = \{y \in \Omega \mid d(x, y) < r\},$$

is open.

e) *The* **closed ball** *with center $x \in \Omega$ and radius $r > 0$,*

$$\mathrm{B}[x, r] := \{y \in \Omega \mid d(x, y) \le r\},$$

is closed.

Proof. a) and b) are straightforward. For c), let $B := \overline{A}$. Then $\overline{B} = \overline{\overline{A}} = \overline{A} = B$, by Lemma 3.20. To prove d), fix $x \in \Omega$ and $r > 0$. If $y \in \mathrm{B}(x, r)$ define $0 < \epsilon := d(x, y) < r$. By the triangle inequality, for each $z \in \mathrm{B}(y, r - \epsilon)$,

$$d(x, z) \le d(x, y) + d(y, z) = \epsilon + d(y, z) < \epsilon + (r - \epsilon) = r.$$

Hence $z \in \mathrm{B}(x, r)$, which shows that $\mathrm{B}(y, r - \epsilon) \subseteq \mathrm{B}(x, \epsilon)$ (cf. Figure 6).

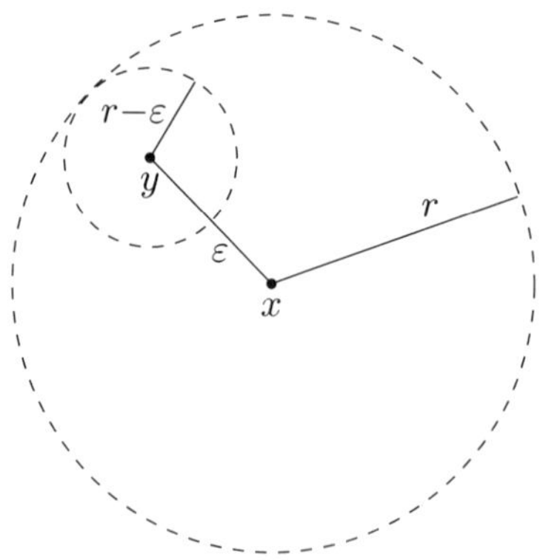

Figure 6. An "open ball" is indeed open.

Finally, to prove e) let $(y_n)_{n \in \mathbb{N}}$ be a sequence in $\mathrm{B}[x, r]$ such that $y_n \to y \in \Omega$. Then for all $n \in \mathbb{N}$,

$$d(x, y) \le d(x, y_n) + d(y_n, y) \le r + d(y_n, y).$$

Since $d(y_n, y) \to 0$, it follows that $d(x, y) \le r$, that is $y \in \mathrm{B}[x, r]$. $\square$

A small subtlety: in general, for example in a discrete metric space, it may happen that $\mathrm{B}[x,r] \neq \overline{\mathrm{B}(x,r)}$, i.e., the closed ball of radius r is in general *not* the closure of the open ball with same radius. In normed spaces, however, one always has $\mathrm{B}[x,r] = \overline{\mathrm{B}(x,r)}$; see Exercise 15.1.

Examples 4.3. 1) In a discrete metric space, *every* subset is both open and closed.

2) Consider $\Omega = \ell^\infty$ with the metric induced by the norm $\|\cdot\|_\infty$. The open ball with center $f = 0$ and radius $r = 1$ is

$$\begin{aligned}\mathrm{B}(0,1) &= \{x \in \ell^\infty \mid \|x\|_\infty < 1\} \\ &= \{x = (x_j)_{j\in\mathbb{N}} \mid \exists \delta \in (0,1)\ \forall j \in \mathbb{N} : |x_j| \le \delta\}.\end{aligned}$$ Ex.4.1

As 'convergence' and 'closure', the notions 'open' and 'closed' are meaningful only relative to a metric space. To avoid ambiguities one uses the more precise phrase "A is open/closed *in* (Ω, d)". If the metric is understood, one can equivalently say "A is open/closed in Ω" or "A is an open/closed subset of Ω".

Example 4.4. Consider the set $A := (0,1]$. Is it open or closed? As with convergence, the answer is: it depends.

1) A is neither open nor closed in $\Omega = \mathbb{R}$ with the standard metric.
2) A is open, but not closed in $\Omega = [0,1]$ with the standard metric.
3) A is open and closed in $\Omega = (0,1]$ with any metric.
4) A is open and closed in $\Omega = [-1,1] \setminus \{0\}$ with the standard metric.
5) A is closed, but not open in $\Omega = (0,\infty)$ with the standard metric.

Ex.4.2
Ex.4.3

The following result states the basic properties of open and closed subsets.

Theorem 4.5. *Let (Ω, d) be a metric space. A subset $A \subseteq \Omega$ is closed if and only if its complement $A^c = \Omega \setminus A$ is open.*

Moreover, the collection of closed subsets of Ω has the following properties:

a) *$\emptyset$ and Ω are closed.*

b) *If* $(A_\iota)_\iota$ *is any nonempty collection of closed sets, then* $\bigcap_\iota A_\iota$ *is closed.*

c) *If* A *and* B *are closed, then* $A \cup B$ *is closed.*

The collection of open subsets has the following properties.

d) $\emptyset$ *and* Ω *are open.*

e) *If* $(O_\iota)_\iota$ *is any nonempty collection of open sets, then* $\bigcup_\iota O_\iota$ *is open.*

f) *If* O *and* W *are open, then* $O \cap W$ *is open.*

Ex.4.4 **Proof.** The proof of the first assertion is left as Exercise 4.4.

The assertions about open sets follow from the ones about closed sets by De Morgan's laws from set theory. So we prove only the latter ones. Assertion a) and b) follow directly from Lemma 3.20, a) and d) and the definition of a closed set. For c), let $A = \bigcap_\iota A_\iota$ with closed sets $A_\iota \subseteq \Omega$. To show that A is closed, let $(x_n)_{n\in\mathbb{N}} \subseteq A$ and suppose that $x_n \to x \in \Omega$. For every ι, $(x_n)_{n\in\mathbb{N}} \subseteq A_\iota$, and as A_ι is closed, $x \in A_\iota$. Hence $x \in A$, and this had to be
Ex.4.5 shown. $\square$

Attention: In general an infinite union $A = \bigcup_{n\in\mathbb{N}} A_n$ of closed sets A_n need *not* be closed: $(0,1] = \bigcup_{n\in\mathbb{N}}[1/n, 1]$.

4.2. Continuity

Another central notion of metric topology is continuity, which we introduce next.

Definition 4.6. Let $(\Omega', d'), (\Omega, d)$ be two metric spaces. A mapping $f : \Omega \to \Omega'$ is called **continuous at** $x \in \Omega$, if for *every* sequence $(x_n)_{n\in\mathbb{N}}$ in Ω the implication

$$x_n \to x \quad \Longrightarrow \quad f(x_n) \to f(x) \tag{4.1}$$

holds. The mapping f is simply called **continuous** if it is continuous at every point $x \in \Omega$.

Note that "$f(x_n) \to f(x)$" in (4.1) could be equivalently written as

$$\lim_{n\to\infty} f(x_n) = f(\lim_{n\to\infty} x_n).$$

So by its very definition, continuity allows us to interchange the application of a mapping with a limit process.

Our definition of continuity is different from the $\epsilon-\delta$-definition common in undergraduate analysis courses, but the following lemma shows that they are equivalent. Ex.4.6

Lemma 4.7. *Let $f : \Omega \to \Omega'$ be a mapping between two metric spaces $(\Omega', d'), (\Omega, d)$. Then for $x \in \Omega$ the following assertions are equivalent:*

(i) *f is continuous at x.*

(ii) $\forall \epsilon > 0\ \exists \delta > 0\ \forall y \in \Omega : d(x,y) < \delta \implies d(f(x), f(y)) < \epsilon.$

Furthermore, also the following assertions are equivalent:

(iii) *f is continuous.*

(iv) *$f^{-1}(U)$ is open in Ω for each open set $U \subseteq \Omega'$.*

(v) *$f^{-1}(A)$ is closed in Ω for each closed set $A \subseteq \Omega'$.*

Proof. We only prove that (iii) implies (v) and leave the remaining implications as exercise. Suppose that f is continuous and that $A \subseteq \Omega'$ is closed. Take a sequence $(x_n)_{n\in\mathbb{N}}$ in $f^{-1}(A)$ with $x_n \to x \in \Omega$. Then $f(x_n) \in A$ for all $n \in \mathbb{N}$, and $f(x_n) \to f(x)$ by continuity. As A is closed, $f(x) \in A$, i.e., $x \in f^{-1}(A)$. □ Ex.4.7

Parts (iv) and (v) of Lemma 4.7 are used frequently to recognize open or closed sets. For instance, if $f : \Omega \to \mathbb{R}$ is continuous, then for $a \in \mathbb{R}$ the sets $\{x \in \Omega \mid f(x) < a\}$ and $\{x \in \Omega \mid f(x) > a\}$ are open, and the sets $\{x \in \Omega \mid f(x) \le a\}$, $\{x \in \Omega \mid f(x) \ge a\}$, and $\{x \in \Omega \mid f(x) = a\}$ are closed.

Continuity is preserved under composition.

Lemma 4.8. *Let (Ω_j, d_j) be a metric space for $j = 1, 2, 3,$ and let*

$$f : \Omega_1 \longrightarrow \Omega_2 \quad \text{and} \quad g : \Omega_2 \longrightarrow \Omega_3$$

be mappings. If f is continuous at $x \in \Omega_1$ and g is continuous at $f(x)$, then $g \circ f$ is continuous at x.

Hence, if f and g are continuous, then so is $g \circ f$.

Proof. If $x_n \to x$ in Ω_1, then $f(x_n) \to f(x)$ in Ω_2 by continuity of f at x. Hence $g(f(x_n)) \to g(f(x))$ in Ω_3, by continuity of g at $f(x)$. □

Let us turn to some important examples of continuous mappings.

Continuity of Metric and Norm. Many continuity proofs are built on suitable estimates, and this is the reason why inequalities play such a central role in analysis. To exemplify this, we note the following lemma.

Lemma 4.9 (Second triangle inequality). *Let (Ω, d) be a metric space. Then*

$$|d(x,z) - d(y,w)| \le d(x,y) + d(z,w) \qquad \textit{for all } x,y,z,w \in \Omega.$$

Proof. The triangle inequality $d(x,z) \le d(x,y) + d(y,w) + d(w,z)$ yields

$$d(x,z) - d(y,w) \le d(x,y) + d(w,z).$$

Interchanging the roles of x, y and z, w leads to

$$-(d(x,z) - d(y,w)) = d(y,w) - d(x,z) \le d(y,x) + d(z,w).$$

Combining both inequalities concludes the proof. □

As announced, the estimate translates into a continuity statement.

Corollary 4.10. *Let $(x_n)_{n\in\mathbb{N}}$ and $(y_n)_{n\in\mathbb{N}}$ be sequences in a metric space (Ω, d) such that $x_n \to x \in \Omega$ and $y_n \to y \in \Omega$. Then $d(x_n, y_n) \to d(x,y)$. In particular, for each fixed $y \in \Omega$ the function*

$$\Omega \longrightarrow \mathbb{R}_+, \qquad x \longmapsto d(x,y)$$

is continuous.

Proof. By Lemma 4.9, $|d(x,y) - d(x_n,y_n)| \le d(x,x_n) + d(y,y_n) \to 0$. By the sandwich theorem, $|d(x,y) - d(x_n,y_n)| \to 0$, i.e., $d(x_n,y_n) \to d(x,y)$. The second assertion follows by taking the constant sequence $y_n = y$ for all $n \in \mathbb{N}$. □

The previous proof is typical for convergence proofs in analysis. In order to prove that $f_n \to f$ with respect to some metric d, one tries to estimate $d(f_n, f) \le a_n$ with some sequence of real numbers $(a_n)_{n\in\mathbb{N}}$ converging to 0. The *sandwich theorem* from undergraduate real analysis then finishes the job. In this way tedious $\epsilon - N$ arguments are avoided.

Remark 4.11. Corollary 4.10 combined with (iv) and (v) of Lemma 4.7 yields another proof that an open ball $\mathrm{B}(y,r) = \{x \mid d(x,y) < r\}$ is open and a closed ball $\mathrm{B}[y,r] = \{x \mid d(x,y) \le r\}$ is closed; cf. the grey box after the proof of Lemma 4.7.

Let us specialize these general observations to the case of normed spaces.

Example 4.12 (Normed Spaces). Let $(E, \|\cdot\|)$ be a normed space. Specializing $\|f\| = d_E(f, 0)$ in Lemma 4.9 yields the "second triangle inequality"

$$\big|\, \|f\| - \|g\| \,\big| \le \|f - g\| \qquad (f, g \in E).$$

By Corollary 4.10 the norm mapping $E \to \mathbb{R}_+,\quad f \mapsto \|f\|$ is continuous, i.e.,

$$f_n \to f \quad \text{in E} \quad \Longrightarrow \quad \|f_n\| \to \|f\| \quad \text{in } \mathbb{R},$$

and by Lemma 4.2.e) the unit ball $B_E = \{f \in E \mid \|f\| \le 1\}$ is closed.

Continuity of Algebraic Operations. A normed space is not just a metric space, but one has in addition the algebraic operations of sum, scalar multiple and — in case the norm comes from an inner product — the inner product. These are continuous, as the next theorem shows.

Theorem 4.13. *Let $(E, \|\cdot\|)$ be a normed space. Then the addition mapping, the scalar multiplication and — if present — the inner product are continuous in the following sense:*
If $f_n \to f$ and $g_n \to g$ in E and $\lambda_n \to \lambda$ in $\mathbb{K}$, then

$$f_n + g_n \to f + g \quad \textit{and} \quad \lambda_n f_n \to \lambda f.$$

If in addition the norm is induced by an inner product $\langle \cdot, \cdot \rangle$, then

$$\langle f_n, g_n \rangle \to \langle f, g \rangle \qquad \textit{in } \mathbb{K}.$$

Proof. Continuity of addition follows from

$$\begin{aligned} d_E(f_n + g_n, f + g) &= \|(f_n + g_n) - (f + g)\| = \|(f_n - f) + (g_n - g)\| \\ &\le \|f_n - f\| + \|g_n - g\| = d_E(f_n, f) + d_E(g_n, g) \to 0. \end{aligned}$$

For the scalar multiplication note that

$$\lambda_n f_n - \lambda f = (\lambda_n - \lambda)(f_n - f) + \lambda(f_n - f) + (\lambda_n - \lambda)f;$$

taking norms and using the triangle inequality yields

$$d_E(\lambda_n f_n, \lambda f) \le |\lambda_n - \lambda|\, \|f_n - f\| + |\lambda|\, \|f_n - f\| + |\lambda_n - \lambda|\, \|f\| \to 0.$$

Suppose that the norm is induced by an inner product. Then

$$\langle f_n, g_n \rangle - \langle f, g \rangle = \langle f_n - f, g_n - g \rangle + \langle f, g_n - g \rangle + \langle f_n - f, g \rangle .$$

Taking absolute values and estimating with the triangle and the Cauchy–Schwarz inequality yields

$$\begin{aligned} |\langle f_n, g_n \rangle - \langle f, g \rangle| &= |\langle f_n - f, g_n - g \rangle| + |\langle f, g_n - g \rangle| + |\langle f_n - f, g \rangle| \\ &\le \|f_n - f\|\, \|g_n - g\| + \|f\|\, \|g_n - g\| + \|f_n - f\|\, \|g\| . \end{aligned}$$

As each of these summands tends to 0 as $n \to \infty$, so does the left-hand
Ex.4.8 side. □

If E, F are normed spaces, then the Cartesian product $E \times F$ is again a normed space in a natural way; see Exercise 4.20. In view of that exercise we see that Theorem 4.13 simply says that the mappings

$$\begin{aligned} E \times E &\longrightarrow E, & (f, g) &\longmapsto f + g, \\ \mathbb{K} \times E &\longrightarrow E, & (\lambda, f) &\longmapsto \lambda f, \\ E \times E &\longrightarrow \mathbb{K}, & (f, g) &\longmapsto \langle f, g \rangle, \end{aligned}$$

are continuous in the sense of Definition 4.6.

Let us note the following important consequences of Theorem 4.13.

Corollary 4.14. *The following assertions hold:*

a) *Let $(E, \|\cdot\|)$ be a normed space, and let $F \subseteq E$ be a linear subspace of E. Then the closure $\overline{F}$ of F in E is also a linear subspace of E.*

b) *Let $(E, \langle \cdot, \cdot \rangle)$ be an inner product space and $S \subseteq E$. Then $S^\perp$ is a closed linear subspace of E and $S^\perp = \overline{S}^\perp$.*

Ex.4.9 **Proof.** The proof is an exercise. □
Ex.4.10

If E is a normed space and $A \subseteq E$ is a sub*set* of E, then $\operatorname{span}(A)$ is a linear subspace of E. Hence by Corollary 4.14, its closure is a closed linear subspace of E. It is common to write

$$\overline{\operatorname{span}}(A) := \overline{\operatorname{span}(A)}$$

for this space, and we shall do so from now on.

Here is a useful application of Theorem 4.13.

Example 4.15. *Let* $\mathrm{C}_0[a, b] := \{f \in \mathrm{C}[a, b] \mid f(a) = f(b) = 0\}$ *and*

$$\mathrm{C}_0^1[a, b] := \{f \in \mathrm{C}^1[a, b] \mid f(a) = f(b) = 0\}.$$

Then $\mathrm{C}_0^1[a, b]$ *is dense in* $\mathrm{C}_0[a, b]$ *with respect to the supremum norm.*

Proof. Let $f \in \mathrm{C}[a, b]$ with $f(a) = f(b) = 0$. By the polynomial Weierstrass theorem or Exercise 3.22 we find a sequence $(p_n)_{n\in\mathbb{N}}$ in $\mathrm{C}^1[a, b]$ such that $p_n \to f$ uniformly on $[a, b]$. Since uniform convergence implies pointwise convergence, $a_n := p_n(a) \to f(a) = 0$ and $b_n := p_n(b) \to f(b) = 0$. We subtract from p_n a linear polynomial to make it zero at the boundary:

$$q_n(t) := p_n(t) - a_n \frac{b-t}{b-a} - b_n \frac{t-a}{b-a}.$$

Then $q_n(a) = 0 = q_n(b)$ and hence $q_n \in \mathrm{C}_0^1[a, b]$. But since $a_n, b_n \to 0$, $\lim_{n\to\infty} q_n = \lim_{n\to\infty} p_n = f$ uniformly. (Note that if p_n is a polynomial, then also q_n is, and hence — supposing Weierstrass' theorem — we have proved the stronger result: $\mathrm{P}[a, b] \cap \mathrm{C}_0[a, b]$ *is dense in* $\mathrm{C}_0[a, b]$.) □

The results of Theorem 4.13 are frequently used without explicit reference.

Bounded Linear Mappings. The natural mappings between vector spaces are the linear mappings. Considering *normed* spaces it is then natural to ask which linear mappings are continuous. The following theorem shows that we already have encountered them!

Theorem 4.16. *A linear mapping $T : E \to F$ between two normed spaces $(E, \|\cdot\|_E)$ and $(F, \|\cdot\|_F)$ is continuous if and only if it is bounded.*

Proof. Suppose that T is bounded. Then, if $f_n \to f$ in E is an arbitrary convergent sequence in E,

$$\|Tf_n - Tf\|_F = \|T(f_n - f)\|_F \le \|T\|_{F\leftarrow E} \|f_n - f\|_E \to 0$$

as $n \to \infty$. So $Tf_n \to Tf$, hence T is continuous.

For the converse, suppose that T is *not* bounded. Then $\|T\|$ as defined in (2.2) is not finite. Hence there is a sequence of vectors $(g_n)_{n\in\mathbb{N}}$ in E such that

$$\|g_n\| \le 1 \quad \text{and} \quad \|Tg_n\| \ge n \qquad (n \in \mathbb{N}).$$

Define $f_n := (1/n)g_n$. Then $\|f_n\| = (1/n)\|g_n\| \le 1/n \to 0$, but

$$\|Tf_n\| = \|T((1/n)g_n)\| = (1/n)\|Tg_n\| \ge 1$$

for all $n \in \mathbb{N}$. Hence $Tf_n \not\to 0$ and therefore T is not continuous. □

Although boundedness and continuity are the same for linear mappings between normed spaces, in functional analysis the term "bounded linear mapping" is preferred to "continuous linear mapping".

The **kernel** of a bounded linear mapping $T : E \to F$,

$$\ker(T) = \{f \in E \mid Tf = 0\}$$

is a *closed* linear subspace of E, since $\ker(T) = T^{-1}\{0\}$ is the inverse image of the (closed!) singleton set $\{0\}$.

Example 4.17. *The space* $\mathrm{C}_0[a, b]$ *is a closed subspace of* $\mathrm{C}[a, b]$ *with respect to the supremum norm.*

Proof. $\mathrm{C}_0[a, b]$ is the kernel of the bounded linear mapping

$$T : \mathrm{C}[a, b] \longrightarrow \mathbb{K}^2, \qquad Tf := (f(a), f(b)). \qquad \square$$

Ex.4.11 Ex.4.12 On the other hand, the **range** of a bounded linear mapping $T : E \to F$,

$$\operatorname{ran}(T) = \{Tf \mid f \in E\} = \{g \in F \mid \exists f \in E : Tf = g\}$$

need not be closed in F. Its closure, usually denoted by $\overline{\operatorname{ran}}(T)$ is a closed linear subspace of F.

Example 4.18. Let $E = c_{00}$ and $F = \ell^2$, both endowed with $\|\cdot\|_2$, and let $T : E \to F$, $Tx := x$ for $x \in E$. Being an isometry, T is bounded. Its range is not closed, since $\overline{\operatorname{ran}}(T) = \overline{c_{00}} = \ell^2$ is the whole of ℓ^2 (see Example 3.18).

4.3. Sequential Compactness

Already from undergraduate courses you know that compactness is an important feature of certain sets in finite dimensions. We extend the concept to general metric spaces.

Definition 4.19. A metric space (Ω, d) is called **(sequentially) compact** if every sequence in Ω has a convergent subsequence.

Ex.4.13 Note that compactness, in contrast to closedness or openness, is not a relative notion. If one speaks of a compact subset A of a metric space Ω, it is meant that A is compact with respect to the induced metric. That is to say, $A \subseteq \Omega$ is (sequentially) compact if every sequence in A has a subsequence that converges to a point in A. (However, cf. also Exercise 4.22.)

Lemma 4.20. *Let* $A \subseteq \Omega$ *be subset of a metric space* (Ω, d). *If* A *is compact, then* A *is closed in* Ω*; and if* Ω *is compact and* A *is closed in* Ω*, then* A *is compact.*

Proof. For the first assertion, suppose that A is compact and let $(x_n)_{n\in\mathbb{N}}$ be a sequence in A with $x_n \to x \in \Omega$. By compactness, there is a subsequence $(x_{n_k})_{k\in\mathbb{N}}$ converging to some element $y \in A$. But also $x_{n_k} \to x$, and as limits are unique, $x = y \in A$. The second statement is left as an (easy) exercise. □

It can be shown that for metric spaces sequential compactness is the same as compactness, defined by a property involving open covers. We do not need this notion for the main part of the book but include it in the optional Section 4.5 below. Because of the equivalence (Theorem 4.41) we often use the word "compact" instead of "sequentially compact".

From elementary analysis courses the reader knows already a wealth of examples of sequentially compact metric spaces.

Theorem 4.21 (Bolzano[1]–Weierstrass). *With respect to the Euclidean metric on $\mathbb{K}^d$ a subset $A \subseteq \mathbb{K}^d$ is (sequentially) compact if and only if it is closed and bounded.*

This theorem is very close to the *axioms* of the real numbers. For the reader's convenience we have included a discussion of these axioms and a proof of the Bolzano–Weierstrass theorem in Appendix A.5. The following example shows that the finite-dimensionality in Theorem 4.21 is *essential*.

Example 4.22. The closed unit ball of ℓ^2 is not compact. Indeed, the canonical unit vectors $(e_n)_{n\in\mathbb{N}}$ satisfy $\|e_n - e_m\| = \sqrt{2}$ if $n \neq m$. Hence no subsequence of this sequence can be convergent.

This example extends to *each* infinite-dimensional inner product space H. Indeed, by the Gram–Schmidt procedure, H must contain an infinite orthonormal system $(e_n)_{n\in\mathbb{N}}$; cf. also Corollary 4.34.

Let us return to the general theory.

Theorem 4.23. *Let (Ω, d) be a compact metric space, and let $f : (\Omega, d) \to (\Omega', d')$ be a continuous mapping. Then the following assertions hold.*

a) *The metric space $(f(\Omega), d')$ is compact, and $f(\Omega)$ is closed in Ω'.* Ex.4.14

b) *If $f : \Omega \to \Omega'$ bijective, then $f^{-1} : \Omega' \to \Omega$ is continuous.*

c) *The mapping f is even* **uniformly continuous**, *i.e., it satisfies*

$$\forall \epsilon > 0\ \exists \delta > 0\ \forall x, y \in \Omega : d(x, y) < \delta \implies d'(f(x), f(y)) < \epsilon.$$

d) *Let $\emptyset \neq A_n \subseteq \Omega$ be a closed subset of Ω for each $n \in \mathbb{N}$ such that $A_1 \supseteq A_2 \supseteq A_3 \ldots$. Then $\bigcap_{n\in\mathbb{N}} A_n \neq \emptyset$.*

[1] Bernard Bolzano (1781–1848), Bohemian mathematician, philosopher and priest.

Proof. a) Let $(y_n)_{n\in\mathbb{N}}$ be a sequence in $f(\Omega)$. By definition, for each $n \in \mathbb{N}$ there is $x_n \in \Omega$ such that $f(x_n) = y_n$. By compactness of Ω there is a subsequence $(x_{n_k})_{k\in\mathbb{N}}$ converging to some $x \in \Omega$. Then, by continuity of f, $y_{n_k} = f(x_{n_k}) \to f(x) \in f(\Omega)$. Lemma 4.20 concludes the proof of a).
The remaining statements b)–d) are left as Exercises 4.23, 4.24, and 4.25. □

Here is a well-known corollary.

Corollary 4.24. *Let (Ω, d) be compact and $f : \Omega \to \mathbb{R}$ a continuous mapping. Then f is bounded and attains its supremum $\sup_{x\in\Omega} f(x)$ and infimum $\inf_{x\in\Omega} f(x)$.*

Proof. By Theorem 4.23, $f(\Omega) \subseteq \mathbb{R}$ is compact, hence bounded and closed. In particular, it must contain its supremum and its infimum. □

4.4. Equivalence of Norms

We have already seen examples of different norms defined on the *same* underlying vector space, for instance, the norms $\|\cdot\|_p$ on $\mathrm{C}[a, b]$ for $p=1, 2, \infty$. Each norm carries with it its own notion of convergence, but often there are relations between them. E.g., we have seen that for a continuous function $f \in \mathrm{C}[a, b]$ one has the inequalities

$$\|f\|_1 \le \sqrt{b-a}\,\|f\|_2 \le (b-a)\,\|f\|_\infty .$$

As a consequence we have the implications

$$\|f - f_n\|_\infty \to 0 \quad\Longrightarrow\quad \|f - f_n\|_2 \to 0 \quad\Longrightarrow\quad \|f - f_n\|_1 \to 0.$$

This motivates the following definition.

Definition 4.25. Given two norms $\|\cdot\|_w$ and $\|\cdot\|_s$ on a vector space E, then $\|\cdot\|_w$ is called **weaker** than $\|\cdot\|_s$ if there is $c > 0$ such that

$$\|f\|_w \le c\,\|f\|_s \qquad \text{for all } f \in E. \tag{4.2}$$

In this case, $\|\cdot\|_s$ is called **stronger** than $\|\cdot\|_w$.

So on $\mathrm{C}[a, b]$, the supremum norm $\|\cdot\|_\infty$ is stronger than both $\|\cdot\|_2$ and $\|\cdot\|_1$, and $\|\cdot\|_2$ is stronger than $\|\cdot\|_1$.

We shall now give a characterization of this weaker/stronger relation.

Theorem 4.26. *Let two norms $\|\cdot\|_w$ and $\|\cdot\|_s$ on a vector space $E \neq \{0\}$ be given. We denote by*

$$B_w := \{f \in E \mid \|f\|_w \le 1\} \quad \textit{and} \quad B_s := \{f \in E \mid \|f\|_s \le 1\}$$

the corresponding unit balls. The following statements are equivalent:

(i) $\|\cdot\|_w$ *is weaker than* $\|\cdot\|_s$.

(ii) *For each sequence* $(f_n)_{n\in\mathbb{N}}$ *in* E *and* $f \in E$ *one has*

$$\|f - f_n\|_s \to 0 \quad \Longrightarrow \quad \|f - f_n\|_w \to 0.$$

(iii) *The identity mapping*

$$\mathrm{I} : (E, \|\cdot\|_s) \longrightarrow (E, \|\cdot\|_w), \qquad f \longmapsto f$$

is continuous.

(iv) *There is* $c > 0$ *such that* $B_s \subseteq c\, B_w$.

(v) $\inf\{\|f\|_s \mid f \in E,\ \|f\|_w = 1\} > 0$.

Proof. (i) $\Rightarrow$ (ii): This follows by replacing f by $f - f_n$ in (4.2) and by the sandwich theorem.

(ii) $\Rightarrow$ (iii): This is just the definition of continuity via sequences.

(iii) $\Rightarrow$ (i): The identity mapping is linear. By (iii) it is continuous, hence bounded by Theorem 4.16.

(i) $\Rightarrow$ (iv): Take $f \in B_s$. Then, by (iii), $\|f\|_w \le c\,\|f\|_s \le c \cdot 1 = c$. Hence $f = c(1/c)f \in cB_s$.

(iv) $\Rightarrow$ (v): Take $f \in E$ with $\|f\|_w = 1$. Then $f \neq 0$ and hence $\|f\|_s \neq 0$. Therefore, $(1/\|f\|_s)f \in B_s$. By (iv) there is $g \in B_w$ such that $(1/\|f\|_s)f = cg$. This yields

$$1 = \|f\|_w = \big\|c\,\|f\|_s\, g\big\|_w = c\,\|f\|_s\,\|g\|_w \le c\,\|f\|_s\,,$$

from which we conclude that $\|f\|_s \ge 1/c$.

(v) $\Rightarrow$ (i): Let $\delta := \inf\{\|f\|_s \mid \|f\|_w = 1\}$ and $c := 1/\delta > 0$. If $f = 0$, then trivially

$$\|f\|_w = 0 \le c \cdot 0 = c\,\|f\|_s\,.$$

If $f \neq 0$, then $g := (1/\|f\|_w)f$ satisfies $\|g\|_w = 1$ and by (v) it follows that

$$\delta \le \|g\|_s = \big\|(1/\|f\|_w)f\big\| = \frac{\|f\|_s}{\|f\|_w}.$$

Consequently, $\|f\|_w \le (1/\delta)\,\|f\|_s = c\,\|f\|_s$. □

Theorem 4.26 is remarkable for several reasons. First, it tells us that *qualitative* relations (ii), (iii) concerning convergence of sequences can be expressed in *quantitative* terms (i), (iv), (v). [Of course, nothing is said about the precise value of c there, and so the statement itself is qualitative in nature.] Second, a *topological* relation (ii), (iii) is characterized by a *geometric* one (iv).

Definition 4.27. Two norms $\|\cdot\|_1, \|\cdot\|_2$ on a vector space E are called **equivalent** if each one is weaker/stronger than the other, i.e., there are constants $m_1, m_2 \geq 0$ such that

$$\|f\|_1 \leq m_1 \|f\|_2 \quad \text{and} \quad \|f\|_2 \leq m_2 \|f\|_1 \quad \text{for all } f \in E.$$

Example 4.28. Consider the norms $\|\cdot\|_1$ and $\|\cdot\|_\infty$ on $E = \mathbb{K}^d$. We have $|x_j| \leq \|x\|_1$ for all $j = 1, \dots, d$, and hence $\|x\|_\infty \leq \|x\|_1$. On the other hand, we have

$$\|x\|_1 = \sum_{j=1}^{d} |x_j| \leq \sum_{j=1}^{d} \|x\|_\infty = d \cdot \|x\|_\infty .$$

Hence the two norms are equivalent. The inequalities translate into the inclusions

$$\mathrm{B}_1 \subseteq \mathrm{B}_\infty \subseteq d \cdot \mathrm{B}_1,$$

where B_1 and B_∞ are the unit balls for $\|\cdot\|_1$ and $\|\cdot\|_\infty$, respectively. See Figure 7 for the case $d = 2$.

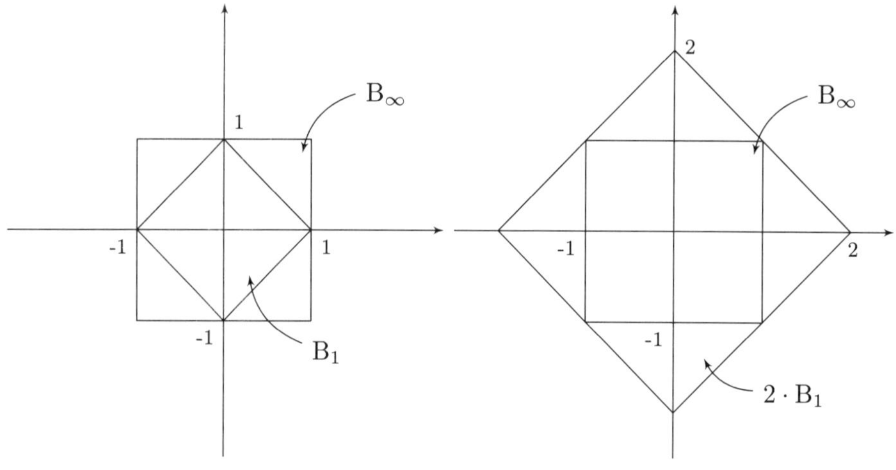

Figure 7. Equivalence of $\|\cdot\|_1$ and $\|\cdot\|_\infty$ on $\mathbb{R}^2$.

The previous example is not surprising in view of the following theorem.

Theorem 4.29. *Let E be a finite dimensional linear space. Then all norms on E are equivalent.*

Proof. By choosing an algebraic basis $e_1, \dots, e_d$ in E we may suppose that $E = \mathbb{K}^d$ and $\{e_1, \dots, e_d\}$ is the canonical basis.

Let $\|\cdot\|$ be any norm on $\mathbb{K}^d$. We shall prove that $\|\cdot\|$ and $\|\cdot\|_2$ are equivalent, the latter being the Euclidean norm. Define

$$m_1 := \left(\sum\nolimits_{j=1}^{d} \|e_j\|^2\right)^{1/2}.$$

Then $\|x\| \le m_1 \|x\|_2$ for all $x \in E = \mathbb{K}^d$; see Example 2.19. By the second triangle inequality

$$\Big| \|x\| - \|y\| \Big| \le \|x - y\| \le m_1 \|x - y\|_2$$

we obtain that the norm mapping

$$(\mathbb{K}^d, \|\cdot\|_2) \longrightarrow \mathbb{R}_+, \qquad x \longmapsto \|x\|$$

is continuous. By Bolzano–Weierstrass (Theorem 4.21), the Euclidean unit sphere

$$\mathbb{S}^{d-1} = \{x \in \mathbb{K}^d \mid \|x\|_2 = 1\}$$

is compact. Hence by Corollary 4.24 there is $x' \in \mathbb{S}^{d-1}$ such that

$$\|x'\| = \inf\{\|y\| \mid y \in \mathbb{S}^{d-1}\}.$$

Now, because $\|x'\|_2 = 1$ we must have $x' \neq 0$ and since $\|\cdot\|$ is a norm, $\|x'\| > 0$. By Theorem 4.26, implication (v)$\Rightarrow$(i), we conclude that there is $m_2 \ge 0$ such that

$$\|x\|_2 \le m_2 \|x\| \quad \text{for all } x \in \mathbb{K}^d.$$ □

Corollary 4.30. a) *Let F be a finite-dimensional normed space, and let E be an arbitrary normed space. Then every linear mapping $T : F \to E$ is bounded.*

b) *Each finite-dimensional subspace F of a normed space E is closed.*

Proof. a) We identify E with $\mathbb{K}^d$ by choosing a basis. By Example 2.19 and Theorem 4.29 there are constants m_1, m_2 such that

$$\|Tx\|_F \le m_1 \|x\|_2 \le m_1 m_2 \|x\|$$

for all $x \in \mathbb{K}^d$. Hence T is bounded.

b) This is left as Exercise 4.15. □ Ex.4.15

Finally, we note that the restriction to finite dimensions in Theorem 4.29 is necessary.

Example 4.31. On $\mathrm{C}[a, b]$ no two of the norms $\|\cdot\|_1$, $\|\cdot\|_2$ and $\|\cdot\|_\infty$ are equivalent. This follows from Example 3.14. Ex.4.16

***Compactness in Infinite-Dimensional Spaces.** By the Bolzano-Weierstrass theorem and the equivalence of all norms, "compact = closed + bounded" holds for subsets of finite-dimensional normed spaces. We shall see in the following that it *fails* whenever the space is infinite-dimensional!

Lemma 4.32 (Riesz[2]). *Let F be closed linear subspace of a normed space E with $F \neq E$. Then*

$$\sup_{f\in E, \|f\|=1} d(f, F) = 1.$$

In other words, for each $\lambda \in (0,1)$ there is $f \in E$ with $\|f\| = 1$ and $\|f - g\| \geq \lambda$ for all $g \in F$.

Proof. For $f \in E \setminus F$ and $u, v \in F$ we have

$$\frac{f-u}{\|f-u\|} - v = \frac{f - (u + \|f-u\|\, v)}{\|f-u\|}.$$

If v passes through all of F, so does $u + \|f - u\|\, v$, and hence taking norms and then the infimum with respect to $v \in F$ yields

$$d\Big(\frac{f-u}{\|f-u\|}, F\Big) = \frac{d(f,F)}{\|f-u\|}.$$

By varying $u \in F$ here, the right-hand side can be made arbitrarily close to 1, so the proof is complete. □

Using compactness of the unit ball in finite dimensions, we obtain the following sharpening of Riesz's lemma.

Corollary 4.33. *Let F be a finite-dimensional subspace of a normed space E with $F \neq E$. Then there is a vector $f \in E$ such that $\|f\| = 1$ and $\|f - g\| \geq 1$ for all $g \in F$.*

Proof. By Corollary 4.30, F is a closed. Without loss of generality we may suppose that E is finite-dimensional. Riesz's lemma then yields, for each $n \in \mathbb{N}$, a vector $f_n \in E$, $\|f_n\| = 1$ and $\|f_n - g\| \geq 1 - \frac{1}{n}$ for all $g \in F$. Since E has finite dimension, there is a subsequence $(n_k)_k$ such that $\lim_{k\to\infty} f_{n_k} =: f$ exists in E. It then follows from the continuity of the norm mapping that f has the desired properties. □

By applying the previous result inductively, we obtain the following.

Corollary 4.34. *In each infinite-dimensional normed space E there is a sequence of unit vectors $(f_n)_{n\in\mathbb{N}}$ such that $\|f_n - f_m\| \geq 1$ for all $n, m \in \mathbb{N}$ with $n \neq m$.*

---

[2]Frigyes Riesz (1880–1956), Hungarian mathematician.

As in Example 4.22 we see that a sequence as in Corollary 4.34 does not have any convergent subsequence. Hence the closed unit ball of any infinite-dimensional normed space is not compact.

4.5. *Separability and General Compactness

This (optional) section is devoted to two topological properties encoding that a metric space is "small" in some sense (if not finite).

Definition 4.35. A metric space is called **separable** if it contains a countable dense set.

The first simple lemma states that subsets inherit separability from the surrounding space.

Lemma 4.36. *Let (Ω, d) be a separable metric space, and let $A \subseteq \Omega$ be any subset. Then A with the induced metric is also separable.*

Proof. Let $D \subseteq \Omega$ be countable and dense in Ω. For each $x \in D$ and $n \in \mathbb{N}$ pick a point $a_{x,n} \in A \cap \mathrm{B}(x, 1/n)$ if there is such a point at all. Collect all these points $a_{x,n}$ in a set E. Then E is countable and dense in A. Indeed, if $a \in A$ is arbitrary and $\epsilon > 0$ we can find $x \in D$ and $n \in \mathbb{N}$ with $d(a, x) < 1/n < \epsilon/2$. But then $A \cap \mathrm{B}(x, 1/n) \neq \emptyset$ and hence $d(a, a_{x,n}) < 2/n < \epsilon$. □

For a normed space, separability is often easy to test by virtue of the following result.

Theorem 4.37. *A normed space E is separable if and only if there is a countable set $M \subseteq E$ such that* $\operatorname{span}(M)$ *is dense in E.*

Proof. One implication is trivial: if E is separable it contains a countable dense set D, and hence $\operatorname{span}(D)$ is dense a fortiori. For the converse suppose that M is countable such that $\operatorname{span}(M)$ is dense in E. We let Q be a countable dense set in $\mathbb{K}$, e.g., $Q = \mathbb{Q}$ if $\mathbb{K} = \mathbb{R}$ or $Q = \mathbb{Q} + \mathrm{i}\mathbb{Q}$ if $\mathbb{K} = \mathbb{C}$. Then

$$D := \Big\{ \sum\nolimits_{j=1}^{n} \lambda_j x_j \mid n \in \mathbb{N},\ \lambda_1, \dots, \lambda_n \in Q,\ x_1, \dots, x_n \in M \Big\}$$

is countable. We claim that $\operatorname{span}(M) \subseteq \overline{D}$. Indeed, if $x \in \operatorname{span}(M)$ we find $n \in \mathbb{N}$, $\lambda_1, \dots, \lambda_n \in \mathbb{K}$ and $x_1, \dots, x_n \in M$ with $x = \sum_{j=1}^{n} \lambda_j x_j$. For each $1 \le j \le n$ we can pick a sequence $\lambda_{j,m} \in Q$ with $\lambda_{j,m} \to \lambda_j$ as $m \to \infty$. Then

$$\sum\nolimits_{j=1}^{n} \lambda_{j,m} x_j \to \sum\nolimits_{j=1}^{n} \lambda_j x_j \qquad (m \to \infty)$$

by Theorem 4.13, and the claim is proved. It follows that $E \subseteq \overline{\operatorname{span}(M)} \subseteq \overline{\overline{D}} = \overline{D}$, hence E is separable. □

Examples 4.38. Each space ℓ^p, $1 \le p < \infty$, is separable by Exercise 3.21, and c_0 is separable by Exercise 3.10. The space $(C[a,b], \|\cdot\|_\infty)$ is separable (Exercise 4.35). More generally, each space $C(K)$, K a compact metric space, is separable (Exercise 4.39). The space ℓ^∞ is not separable (Exercise 4.36). (See page 282 for an extended list of examples.)

An **ϵ-mesh** in a metric space (Ω, d) is a finite subset $\{x_1, \dots, x_n\} \subseteq \Omega$ such that each point of Ω is ϵ-close to at least one x_j, i.e.,

$$\Omega \subseteq \bigcup\nolimits_{j=1}^{n} B(x_j, \epsilon).$$

A metric space (Ω, d) is called **precompact** if it has an ϵ-mesh for each $\epsilon > 0$.

Lemma 4.39. *Each compact metric space is precompact and each precompact space is separable.*

Proof. We leave the proof of the first statement as Exercise 4.37. Suppose that (Ω, d) is precompact. For each $n \in \mathbb{N}$ let $F_n \subseteq \Omega$ be a finite $1/n$-mesh of Ω, i.e., $\Omega = \bigcup_{x \in F_n} B(x, 1/n)$. Then we let $D := \bigcup_{n \in \mathbb{N}} F_n$, which is a countable set. If $y \in \Omega$ and $\epsilon > 0$ are arbitrary, we can find $n \in \mathbb{N}$ and a point $x \in F_n$ with $d(y, x) < 1/n < \epsilon$. Hence D is dense in Ω. □

In the following we shall show the equivalence of sequential compactness and usual compactness, a notion yet to be defined.

An **open cover** of a metric space is a family $(U_\alpha)_{\alpha \in I}$ of open subsets of Ω such that $\Omega \subseteq \bigcup_{\alpha \in I} U_\alpha$. A **subcover** of an open cover $(U_\alpha)_{\alpha \in I}$ is each subfamily $(U_\alpha)_{\alpha \in F}$, $F \subseteq I$, which is still a cover of Ω. The subcover is called **finite subcover** if F is a finite set. A metric space (Ω, d) is called **compact** if every open cover of Ω has a finite subcover.

Lemma 4.40. *A metric space (Ω, d) is compact if and only if it has the following property. If $(F_\alpha)_{\alpha \in I}$ is a family of closed subsets of Ω such that $\bigcap_{\alpha \in F} F_\alpha \neq \emptyset$ for every finite subset $F \subseteq I$, then $\bigcap_{\alpha \in I} F_\alpha \neq \emptyset$.*

Proof. This follows from the definition of compactness by passing to complements. (Note that complements of open sets are closed and complements of unions are intersections.) □

We can now prove the main characterization result.

Theorem 4.41. *A metric space is compact if and only if it is sequentially compact.*

Proof. Suppose that (Ω, d) is sequentially compact. Then it is precompact and hence separable by Lemma 4.39. Let $D \subseteq \Omega$ be a countable dense subset, and let $(U_\alpha)_{\alpha \in I}$ be an open cover of Ω, with an arbitrary index set I.

In the first step of the proof we show that *there is a countable subcover.* To establish this we consider the countable set

$$M := \{(x, n) \in D \times \mathbb{N} \mid \exists \alpha \in I : \mathrm{B}(x, 1/n) \subseteq U_\alpha\} \subseteq D \times \mathbb{N}.$$

Then for each $(x, n) \in M$ we can pick $\alpha = \alpha(x, n)$ with $\mathrm{B}(x, 1/n) \subseteq U_\alpha$. We claim that

$$\Omega \subseteq \bigcup_{(x,n)\in M} U_{\alpha(x,n)}.$$

Indeed, for $y \in \Omega$ we find $\alpha \in I$ with $y \in U_\alpha$; then $\epsilon > 0$ such that $\mathrm{B}(y, \epsilon) \subseteq U_\alpha$; finally $x \in D$ and $n \in \mathbb{N}$ with $d(x, y) < 1/n < \epsilon/2$. By the triangle inequality

$$y \in \mathrm{B}(x, 1/n) \subseteq \mathrm{B}(y, \epsilon) \subseteq U_\alpha,$$

whence $(x, n) \in M$. But then $y \in \mathrm{B}(x, 1/n) \subseteq U_{\alpha(x,n)}$, and the claim is proved.

For the remaining step we may hence assume that I is countable, so $I = \mathbb{N}$ without loss of generality. Suppose that no finite subcover exists. Then for each $n \in \mathbb{N}$ we find

$$x_n \in \Omega \setminus \bigcup\nolimits_{j=1}^{n} U_j.$$

By sequential compactness we can pass to a subsequence and may hence suppose that $x_n \to x$ for some $x \in \Omega$. Then there is $m \in \mathbb{N}$ such that $x \in U_m$, and hence $x_n \in U_m$ for eventually all $n \in \mathbb{N}$. This is a contradiction.

To prove the converse assertion, suppose that (Ω, d) is compact and let $(x_n)_{n\in\mathbb{N}}$ be a sequence in Ω. We consider the descending family $\overline{\{x_k \mid k \geq n\}}$, $n \in \mathbb{N}$, of nonempty closed sets. Because the family is descending, the intersection of any finite number of these sets is not empty. Hence by Lemma 4.40 we conclude that there is $x \in \Omega$ with

$$x \in \bigcap\nolimits_{n\in\mathbb{N}} \overline{\{x_k \mid k \geq n\}}.$$

But this means that $(x_n)_{n\in\mathbb{N}}$ has a subsequence converging to x; see Exercise 3.16. □

In Exercise 5.28 below we shall encounter another characterization of compactness: a metric space is compact if and only if it is precompact and complete.

Exercises 4A

Exercise 4.1. Prove the assertions about the spaces in Examples 4.3.

Exercise 4.2. Prove the assertions in Example 4.4.

Exercise 4.3. Show that each of the sets

$$A := \{f \in \ell^\infty \mid \exists k \in \mathbb{N} \mid |f(k)| = \|f\|_\infty\},$$
$$B := \{f \in \ell^\infty \mid \forall\, k \in \mathbb{N} : |f(k)| < 1\},$$

is neither open nor closed in ℓ^∞ (with the supremum norm).

Exercise 4.4. Give a proof of the first assertion from Theorem 4.5, i.e., show that a subset A of a metric space (Ω, d) is closed if and only if its complement $A^c = \Omega \setminus A$ is open.

Exercise 4.5. Prove assertions d)–f) of Theorem 4.5 directly from the definition of an open set.

Exercise 4.6. Let (Ω, d) and (Ω', d') be two metric spaces, and let $f, g : \Omega \to \Omega'$ be continuous mappings. Show that if $A \subseteq \Omega$ and $f(x) = g(x)$ for all $x \in A$, then $f(x) = g(x)$ even for all $x \in \overline{A}$.

Exercise 4.7. Prove the remaining implications of Lemma 4.7.

Exercise 4.8. Let Ω be any set. Suppose that $f_n \to f$ and $g_n \to g$ uniformly, with all functions being contained in the space $\mathcal{B}(\Omega)$. Show that $f_n g_n \to fg$ uniformly as well.

Exercise 4.9 (Corollary 4.14.a)**.** Let F be a subset of a normed space $(E, \|\cdot\|)$. Suppose that F is a *linear* sub*space*. Show that $\overline{F}$ is a linear subspace as well.

Exercise 4.10 (Corollary 4.14.b)**.** Let $(E, \langle\cdot,\cdot\rangle)$ be an inner product space and $S \subseteq E$. Then $S^\perp$ is a closed linear subspace of E and $S^\perp = \overline{S}^\perp$.

Exercise 4.11. Consider $E := \mathrm{C}[a, b]$ with the supremum norm, and let

$$\mathrm{C}_{\mathrm{per}}[a, b] := \{f \in \mathrm{C}[a, b] \mid f(a) = f(b)\}.$$

Show that $\mathrm{C}_{\mathrm{per}}[a, b]$ is a closed linear subspace of $\mathrm{C}[a, b]$ and that the space $\mathrm{P}[a, b] \cap \mathrm{C}_{\mathrm{per}}[a, b]$ is dense in $\mathrm{C}_{\mathrm{per}}[a, b]$.

Exercise 4.12. Consider $E = \mathrm{C}[a, b]$ with the supremum norm, and let $F := \{f \in \mathrm{C}[a, b] \mid \int_a^b f(t)\, \mathrm{d}t = 0\}$. Show that F is a closed linear subspace of $\mathrm{C}[a, b]$ and prove that the space $\mathrm{P}[a, b] \cap F$ is dense in F.

Exercise 4.13. Characterize the sets Ω such that Ω is compact with respect to the discrete metric.

Exercise 4.14. Let (Ω, d) and (Ω', d') be two metric spaces, let $f : \Omega \to \Omega'$ be continuous, and let $A \subseteq \Omega$. Show that

$$f(\overline{A}) \subseteq \overline{f(A)}$$

with equality if (but not necessarily only if) $\overline{A}$ is compact.

Exercise 4.15. Let F be a finite-dimensional subspace of a normed space E. Show that F is closed in E. [Hint: Bolzano–Weierstrass.]

Exercise 4.16. Show that no two of the norms $\|\cdot\|_1$, $\|\cdot\|_2$ and $\|\cdot\|_\infty$ are equivalent on c_{00}. [Hint: Consider the vectors $f_n := e_1 + \cdots + e_n$, $n \in \mathbb{N}$, where e_n denotes the n-th canonical unit vector.] See also Exercise 3.4.

Exercises 4B

Exercise 4.17. Let F be a closed linear subspace of a normed space E and $h \in E \setminus F$. Show that the subspace $F \oplus \mathbb{K} \cdot h$ is also closed in E.

Exercise 4.18. Let (Ω_1, d_1) and (Ω_2, d_2) be a metric spaces and let $\Omega := \Omega_1 \times \Omega_2$. Show that

$$d : \Omega \longrightarrow \mathbb{R}_+, \qquad d((x_1, x_2), (y_1, y_2)) := d_1(x_1, x_2) + d_2(x_2, y_2)$$

is a metric on Ω. Let $x_n := (x_{1n}, x_{2n}) \in \Omega$ for $n \in \mathbb{N}$ and $x = (x_1, x_2) \in \Omega$. Show that

$$x_n \to x \quad \text{in } (\Omega, d) \qquad \Leftrightarrow \qquad x_{jn} \to x_j \quad \text{in } (\Omega_j, d_j) \text{ for } j = 1 \text{ and } j = 2.$$

The new metric space (Ω, d) is called the **product** of the original metric spaces (Ω_j, d_j), $j = 1, 2$.

Exercise 4.19. Let (Ω, d) be a metric space. Show that the first assertion in Corollary 4.10 is equivalent to the continuity of the mapping

$$d : \Omega \times \Omega \longrightarrow \mathbb{R}_+,$$

where on the Cartesian product $\Omega \times \Omega$ we use the metric defined in Exercise 4.18.

Exercise 4.20. In the situation of Exercise 4.18 suppose $\Omega_j = E_j$ is a vector space and its metric d_j is associated with a norm $\|\cdot\|_j$, for $j = 1, 2$. Show that $\Omega = E_1 \times E_2$ is a linear space with respect to the componentwise operations; then show that

$$\|(f_1, f_2)\|_E := \|f_1\|_1 + \|f_2\|_2 \qquad (f_1 \in E_1,\ f_2 \in E_2)$$

defines a norm on E which has d (as in Exercise 4.18) as its associated metric.

Exercise 4.21. Let $(H, \langle \cdot, \cdot \rangle)$ be an inner product space. Show that the following assertions are equivalent for $f, f_n \in H$ $(n \in \mathbb{N})$:

(i) $\|f_n\| \to \|f\|$ and $\langle f_n, h \rangle \to \langle f, h \rangle$ for all $h \in H$;

(ii) $f_n \to f$.

(The second property in (i) is called the **weak convergence** of the sequence $(f_n)_{n \in \mathbb{N}}$ to f.)

Exercise 4.22. A subset $A \subseteq \Omega$ of a metric space (Ω, d) is called **relatively compact** if $\overline{A}$ is compact. Show that A is relatively compact if and only if every sequence in A has a subsequence that converges to a point in Ω.

Exercise 4.23 (Theorem 4.23.b). Let $f : (\Omega, d) \to (\Omega', d')$ be a *bijective* and continuous mapping between metric spaces. Show that if (Ω, d) is compact, then f^{-1} is continuous. Then show that without the compactness of Ω the previous conclusion may fail.

Exercise 4.24 (Theorem 4.23.c). Let $f : (\Omega, d) \to (\Omega', d')$ be a continuous mapping between metric spaces. Show that if (Ω, d) is compact, then f is *uniformly* continuous. (See Theorem 4.23 for the definition.)

Exercise 4.25 (Theorem 4.23.d). Show that a metric space (Ω, d) is (sequentially) compact if and only if it has the following property: If $\emptyset \neq A_n \subseteq \Omega$ is a closed subset of Ω for each $n \in \mathbb{N}$ such that $A_1 \supseteq A_2 \supseteq A_3 \dots$, then $\bigcap_{n\in\mathbb{N}} A_n \neq \emptyset$. [Hint for the converse: Exercise 3.16.]

Exercise 4.26. Two metrics d, d' on a set Ω are called **equivalent** if

$$d(x_n, x) \to 0 \quad \Longleftrightarrow \quad d'(x_n, x) \to 0$$

for *every* sequence $(x_n)_{n\in\mathbb{N}}$ in Ω and every point $x \in \Omega$. In short: two metrics are equivalent if they produce the same convergent sequences. Then clearly two norms on a vector space are equivalent if and only if the associated metrics are.

a) Suppose that d_1, d_2 are equivalent metrics on a set Ω. Show that a set $A \subseteq \Omega$ is closed/open/compact with respect to d_1 if and only if it is closed/open/compact with respect to d_2. Then show that a mapping is continuous with respect to both metrics d_1, d_2, or none of them.

b) Show that the discrete metric and the Euclidean metric on $\mathbb{R}^d$ are not equivalent.

Exercise 4.27. Let d be a metric on Ω. Show that

$$d'(x,y) := \min\{d(x,y), 1\} \qquad (x, y \in \Omega)$$

defines a metric equivalent to d. (This shows that when passing to an equivalent metric the boundedness of a subset may be lost; cf. Exercise 3.15. However, boundedness of a subset of a *normed* space is preserved when passing to an equivalent norm (why?).)

Exercise 4.28. Show that "equivalence of norms" is indeed an equivalence relation on the set of all norms on a given vector space E.

Exercise 4.29. On $\mathbb{K}^d$ consider the mapping

$$\alpha(x) := \int_0^1 \left| \sum\nolimits_{j=1}^d x_j t^j \right| \mathrm{d}t \qquad (x = (x_1, \dots, x_d) \in \mathbb{K}^d).$$

Show that α is a norm on $\mathbb{K}^d$. (You can use Exercise 2.25. Do you see, how?) Then prove that

$$\inf\{\alpha(x) \mid x_1 + \dots + x_d = 1\} > 0.$$

Exercise 4.30. Let $E := \{f \in \mathrm{C}^1[0,1] \mid f(0) = 0\}$ and note that E is a linear subspace of $\mathrm{C}^1[0,1]$.

a) Find a constant $c \geq 0$ such that $\|f\|_\infty \leq c \|f'\|_1$ for all $f \in E$.

b) Are the two norms(!) on E

$$f \longmapsto \|f\|_\infty \quad \text{and} \quad f \longmapsto \|f'\|_1$$

equivalent? [Hint: Consider a sequence of functions of the form $f_n(x) := g(nx)$, $x \in [0,1]$, where g is a suitable function on $\mathbb{R}_+$.]

Exercise 4.31. Let (Ω, d) be a metric space. Show that if $A \subseteq \Omega$ is closed and $(x_{n,m})_{n,m\in\mathbb{N}}$ is a double sequence (cf. Exercise 3.17) in A converging to $x \in \Omega$, then $x \in A$.

Exercises 4C

Exercise 4.32. Show that the only subsets of $\mathbb{R}$ that are both open and closed are $\emptyset$ and $\mathbb{R}$.

Exercise 4.33 (Dini's Theorem[3]). Let (K, d) be a (sequentially) compact metric space, and let $(f_n)_{n\in\mathbb{N}}$ be a sequence in $\mathrm{C}(K)$ such that for every $x \in K$ one has

$$0 \le f_{n+1}(x) \le f_n(x) \quad (n \in \mathbb{N}) \qquad \text{and} \qquad \lim_{n\to\infty} f_n(x) = 0.$$

Show that $\|f_n\|_\infty \to 0$. (Bonus: Give two separate proofs: one employing sequential compactness, the other employing open covers.)

Exercise 4.34. Let $(E, \|\cdot\|)$ be a normed space, and let $\varphi : E \to \mathbb{K}$ be a linear functional. Show that the following assertions are equivalent:

(i) φ is bounded.

(ii) φ is continuous.

(iii) $\ker(\varphi)$ is closed.

Exercise 4.35. Show that the space $(\mathrm{C}[a,b], \|\cdot\|_\infty)$ is separable.

Exercise 4.36. Show that the space ℓ^∞ is *not* separable. [Hint: Consider the subset of all $\{0,1\}$-sequences.]

Exercise 4.37. Show that a compact metric space is precompact.

Exercise 4.38. Let $[a,b] \subseteq \bigcup_{j=1}^m U_j$ be a finite open cover of the compact interval $[a,b]$. Show that there is a partition $a = t_0 < t_1 < \cdots < t_n = b$ such that for each $1 \le i \le n$ there is $1 \le j \le m$ with $[t_{i-1}, t_i] \subseteq U_j$.

Exercise 4.39. Let K be a compact metric space. Show that the Banach space $(\mathrm{C}(K), \|\cdot\|_\infty)$ is separable. (You will need the abstract Stone–Weierstrass Theorem[4]; see [**Con90**, V.8] or [**Lax02**, 13.3].)

[3]Ulisse Dini (1845–1918), Italian mathematician.

[4]Marshall H. Stone (1903–1989), American mathematician.

Exercise 4.40. Let E, F, G be normed spaces, and let $B : E \times F \to G$ be a *bilinear* mapping. Show that B is continuous (cf. Exercise 4.20) if and only if there is a constant $c \geq 0$ such that

$$\|B(x,y)\|_G \leq c \|x\|_E \, \|y\|_F \, .$$

[Hint: For the "if" part mimic the proof of continuity of scalar multiplication in Theorem 4.13.]

Chapter 5

Banach Spaces

In this chapter we shall discuss the important concepts of a Cauchy sequence and the completeness of a metric space. Then we introduce one of the most important notions of functional analysis, that of a Banach space.

5.1. Cauchy Sequences and Completeness

Consider the interval $(0,1]$ and forget for the moment that you know about the existence of the surrounding space $\mathbb{R}$. The sequence $(1/n)_{n\in\mathbb{N}}$ does not converge in $(0,1]$ neither with respect to the standard metric nor the discrete metric, but — in a sense — for different reasons. In the first case, by looking at the distances $d(x_n, x_m)$ for large $n, m \in \mathbb{N}$ one has the feeling that the sequence "should" converge, however, the space $(0,1]$ lacks a possible limit point. In the second case (discrete metric) one has $d(x_n, x_m) = 1$ for all $n \neq m$, and so one feels that there is no chance to make this sequence convergent by enlarging the space. This leads to the following definition.

Definition 5.1. A sequence $(x_n)_{n\in\mathbb{N}}$ in a metric space (Ω, d) is called a **Cauchy sequence** if $d(x_n, x_m) \to 0$ as $n, m \to \infty$, i.e., if

$$\forall \epsilon > 0 \ \exists N \in \mathbb{N} \text{ such that } \forall n, m \geq N : \ d(x_n, x_m) < \epsilon.$$

Here are some properties.

Lemma 5.2. *Let (Ω, d) be a metric space. Then the following assertions hold.*

a) *Each convergent sequence in Ω is a Cauchy sequence.*

b) *Each Cauchy sequence is bounded.*

c) *If a Cauchy sequence has a convergent subsequence, then it converges.*

Proof. a) Let $(x_n)_{n\in\mathbb{N}}$ be convergent, with limit $x \in \Omega$. Then by the triangle inequality,

$$d(x_n, x_m) \le d(x_n, x) + d(x, x_m) \qquad (n, m \in \mathbb{N}).$$

If $\epsilon > 0$ is fixed, by hypothesis one has $d(x_n, x) < \epsilon/2$ for eventually all $n \in \mathbb{N}$, and so $d(x_n, x_m) < \epsilon$ for eventually all $n, m \in \mathbb{N}$.

b) By definition there is $N \in \mathbb{N}$ such that $d(x_N, x_n) \le 1$ for all $n \ge N$. Define

$$M := \max\{1, d(x_1, x_N), \dots, d(x_{N-1}, x_N)\} < \infty.$$

If $n, m \in \mathbb{N}$ are arbitrary, then

$$d(x_n, x_m) \le d(x_n, x_N) + d(x_N, x_m) \le M + M = 2M.$$

This proves the claim.

c) Let $(x_n)_{n\in\mathbb{N}}$ be a Cauchy sequence and suppose that the subsequence $(x_{n_k})_{k\in\mathbb{N}}$ converges to $x \in \Omega$. Fix $\epsilon > 0$ and choose $N \in \mathbb{N}$ such that $d(x_n, x_m) < \epsilon/2$ for $n, m \ge N$. Choose k so large that $n_k \ge N$ and $d(x, x_{n_k}) < \epsilon/2$. Then, if $n \ge N$,

$$d(x_n, x) \le d(x_n, x_{n_k}) + d(x_{n_k}, x) < \epsilon/2 + \epsilon/2 = \epsilon. \qquad \square$$

Ex.5.1
Ex.5.2

By our introductory example, there are metric spaces with Cauchy sequences that do not converge. The following definition singles out those spaces where such an 'unpleasant' situation cannot occur.

Definition 5.3. A metric d on a set Ω is called **complete** if every d-Cauchy sequence converges. A metric space (Ω, d) is called **complete** if d is a complete metric on Ω.

Coming back to the introductory example, we may say that $(0, 1]$ with the standard metric is not complete. The space $\mathbb{R}$ with the standard metric is complete. This is almost an axiom about real numbers; see Appendix A.5. Using this fact, we go to higher (but finite) dimensions.

Theorem 5.4. *The Euclidean metric on* $\mathbb{K}^d$ *is complete.*

Proof. This follows from Corollary A.4 since $\mathbb{K}^d = \mathbb{R}^{2d}$ as metric spaces, when considered with the Euclidean metrics. $\square$

Ex.5.3 The following is a very useful fact when one wants to prove the completeness of a subspace of a given metric space.

Lemma 5.5. *Let (Ω, d) be a metric space, and let $A \subseteq \Omega$.*

a) *If (Ω, d) is complete and A is closed in Ω, then A with respect to the induced metric is complete.*

b) *If A is complete with respect to the induced metric, then it is closed in Ω.*

c) *If A is (sequentially) compact, then it is complete with respect to the induced metric.*

Proof. We prove a) and leave the proof of b) and c) as Exercise 5.4. Suppose that $(x_n)_{n\in\mathbb{N}} \subseteq A$ is a Cauchy sequence with respect to the induced metric. Then it is (trivially) a Cauchy sequence in Ω. By assumption, it has a limit $x \in \Omega$. Since $A \subseteq \Omega$ is closed, it follows that $x \in A$, whence $x_n \to x$ in A (again trivially). □ Ex.5.4

From the previous lemma we conclude immediately that every closed subset of the Euclidean space $\mathbb{K}^d$ is complete with respect to the induced (=Euclidean) metric.

Incompleteness of a metric space is not a substantial problem, because every metric space can be viewed as a dense subset of a complete metric space. Such a "surrounding" space is called a **completion**, and it can be constructed by the same methods that Cantor[1] and Heine[2] used to construct the real numbers from the rationals; see Appendix B. For normed spaces there is a more explicit construction (Corollary 16.11).

5.2. Hilbert Spaces

Let $(H, \langle\cdot,\cdot\rangle)$ be an inner product space. Recall that the inner product induces a natural norm $\|\cdot\|$ by

$$\|f\| = \sqrt{\langle f, f\rangle} \qquad (f \in H)$$

and with this norm a natural metric d is associated via

$$d(f,g) := \|f-g\| = \sqrt{\langle f-g, f-g\rangle} \qquad (f,g \in H).$$

The discussion and the example of the previous section motivate the following definition.

Definition 5.6. An inner product space $(H, \langle\cdot,\cdot\rangle)$ is called a **Hilbert space** if H is complete with respect to the metric associated with the inner product.

[1] Georg Cantor (1845–1918), German mathematician.

[2] Eduard Heine (1821–1881), German mathematician.

From Theorem 5.4 from above we see that $\mathbb{K}^d$ with its standard inner product is a Hilbert space. Here is the infinite-dimensional version of it.

Theorem 5.7. *The space ℓ^2 with its standard inner product is a Hilbert space.*

Proof. For convenience we use function notation, i.e., we write elements from ℓ^2 as functions on $\mathbb{N}$.

Take a Cauchy sequence $f_1, f_2, f_3 \ldots$ in ℓ^2. Note that each f_n is now a function on $\mathbb{N}$. The proof follows a standard procedure: First find the limit function by looking at what the sequence does in each component. Then prove that the alleged limit function is indeed a limit in the given metric.

Fix $j \in \mathbb{N}$. Then obviously

$$|f_n(j) - f_m(j)| \le \|f_n - f_m\|_2 \qquad (n, m \in \mathbb{N}).$$

Hence the sequence $(f_n(j))_{n\in\mathbb{N}}$ is a Cauchy sequence in $\mathbb{K}$. By the completeness of $\mathbb{K}$, it has a limit, say

$$f(j) := \lim_{n\to\infty} f_n(j).$$

This yields a candidate $f : \mathbb{N} \to \mathbb{K}$ for the limit of the sequence f_n. But we still have to prove that $f \in \ell^2$ and $\|f - f_n\|_2 \to 0$.

Fix $\epsilon > 0$ and $M = M(\epsilon) \in \mathbb{N}$ such that $\|f_n - f_m\|_2 < \epsilon$ if $n, m > M$. For fixed $N \in \mathbb{N}$ we obtain

$$\sum_{j=1}^{N} |f_n(j) - f_m(j)|^2 \le \sum_{j=1}^{\infty} |f_n(j) - f_m(j)|^2 = \|f_n - f_m\|_2^2 \le \epsilon^2$$

for all $n, m \ge M$. Letting $m \to \infty$ yields

$$\sum_{j=1}^{N} |f_n(j) - f(j)|^2 \le \epsilon^2$$

for all $n \ge M$ and all $N \in \mathbb{N}$. Letting $N \to \infty$ gives

$$\|f_n - f\|_2^2 = \sum_{j=1}^{\infty} |f_n(j) - f(j)|^2 \le \epsilon^2,$$

i.e., $\|f_n - f\|_2 \le \epsilon$ for $n \ge M$. In particular, by the triangle inequality,

$$\|f\|_2 \le \|f_M - f\|_2 + \|f_M\|_2 \le \epsilon + \|f_M\|_2 < \infty,$$

whence $f \in \ell^2$. Moreover, since $\epsilon > 0$ was arbitrary, $\|f_n - f\|_2 \to 0$ as $n \to \infty$, and this was left to prove. $\square$

After this positive result, here is a negative one.

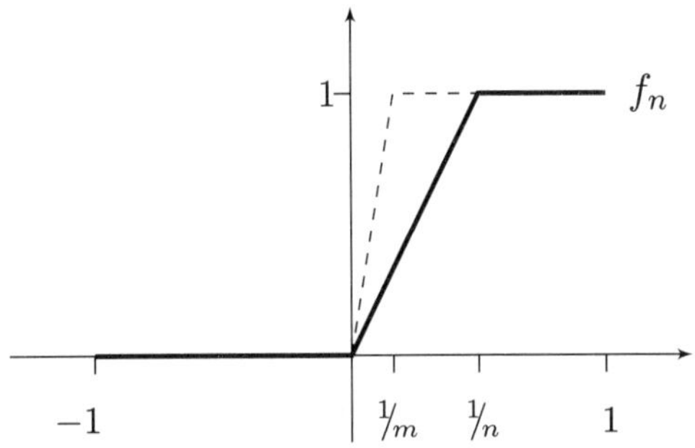

Figure 8. The sequence $(f_n)_{n\in\mathbb{N}}$ is $\|\cdot\|_2$-Cauchy.

Theorem 5.8. *The space* $\mathrm{C}[a,b]$*, endowed with the standard inner product, is not complete, i.e., not a Hilbert space.*

Proof. We show this for $[a,b]=[-1,1]$, the general case being similar. One has to construct a $\|\cdot\|_2$-Cauchy sequence that is not convergent. To this aim consider the functions

$$f_n : [-1,1] \longrightarrow \mathbb{R}, \quad f_n(t) := \begin{cases} 0 & t \in [-1,0], \\ nt & t \in [0, 1/n], \\ 1 & t \in [1/n, 1]; \end{cases}$$

see Figure 8. Then for $m \ge n$ we have $f_n = f_m$ on $[1/n, 1]$ and on $[-1,0]$, hence

$$\|f_n - f_m\|_2^2 = \int_{-1}^{1} |f_n(t) - f_m(t)|^2 \,\mathrm{d}t = \int_0^{1/n} |f_n(t) - f_m(t)|^2 \,\mathrm{d}t \le 4/n$$

since $|f_n|, |f_m| \le 1$. It follows that $(f_n)_{n\in\mathbb{N}}$ is a $\|\cdot\|_2$-Cauchy sequence. We show by contradiction that it does not converge: Suppose that the limit is $f \in \mathrm{C}[-1,1]$. Then

$$\begin{aligned} \int_{-1}^{0} |f(t)|^2 \,\mathrm{d}t &= \int_{-1}^{0} |f(t) - f_n(t)|^2 \,\mathrm{d}t \le \int_{-1}^{1} |f(t) - f_n(t)|^2 \,\mathrm{d}t \\ &= \|f - f_n\|_2^2 \to 0 \end{aligned}$$

as $n \to \infty$. Hence $\int_{-1}^{0} |f|^2 = 0$, and by Lemma 1.3, $f = 0$ on $[-1,0]$. On the other hand, for $0 < a < 1$ and $n > 1/a$ we have

$$\begin{aligned} \int_a^1 |f(t) - 1|^2 \,\mathrm{d}t &= \int_a^1 |f(t) - f_n(t)|^2 \,\mathrm{d}t \\ &\le \int_{-1}^{1} |f(t) - f_n(t)|^2 \,\mathrm{d}t = \|f - f_n\|_2^2 \to 0 \end{aligned}$$

as $n \to \infty$. Hence $\int_a^1 |f - \mathbf{1}|^2 = 0$, and again by Lemma 1.3, $f = \mathbf{1}$ on $[a, 1]$.
Ex.5.5 Since $a \in (0, 1)$ was arbitrary, f is discontinuous at 0, a contradiction. □

Our proof that the sequence $(f_n)_{n \in \mathbb{N}}$ from above does not converge in C$[-1, 1]$ may look a little awkward. Isn't it clear from sketching the graph that $(f_n)_{n \in \mathbb{N}}$ converges to the noncontinuous function f that is 0 on $[-1, 0]$ and 1 on $(0, 1]$? Think about it, and then explain why our "awkward" proof is in fact necessary.

5.3. Banach Spaces

The notion of completeness of an inner product space is actually a property of the norm, not of the inner product. So it makes sense to coin an analogous notion for normed spaces.

Definition 5.9. A normed space $(E, \|\cdot\|)$ is called a **Banach space**[3] if it is complete with respect to its associated metric.

So Hilbert spaces are special cases of Banach spaces. However, we again want to stress that there are many more Banach spaces which are not Hilbert, due to the failing of the parallelogram law; cf. Remark 2.10.

Example 5.10. *Every finite-dimensional normed space is a Banach space.*

Proof. All norms on a finite-dimensional space are equivalent. It is easy to see (Exercise 5.2) that equivalent norms have the same Cauchy sequences. As we know completeness for the Euclidean norm, we are done. □

Example 5.11. *Let Ω be a nonempty set. Then $(\mathcal{B}(\Omega), \|\cdot\|_\infty)$ is a Banach space.*

Proof. Let $(f_n)_{n \in \mathbb{N}} \subseteq \mathcal{B}(\Omega)$ be a $\|\cdot\|_\infty$-Cauchy sequence. We need to find $f \in \mathcal{B}(\Omega)$ such that $\|f_n - f\|_\infty \to 0$. Since we know that uniform convergence implies pointwise convergence, we should find f by defining

$$f(x) := \lim_n f_n(x) \qquad (x \in \Omega).$$

This is possible for the following reason. For fixed $x \in \Omega$ we have

$$|f_n(x) - f_m(x)| = |(f_n - f_m)(x)| \le \|f_n - f_m\|_\infty \to 0 \quad \text{as} \quad n, m \to \infty$$

[3]Stefan Banach (1892-1945), Polish mathematician.

by hypothesis. So $(f_n(x))_{n\in\mathbb{N}}$ is a Cauchy sequence in $\mathbb{K}$, and since $\mathbb{K}$ is complete, the limit — which we call $f(x)$ — exists.

Having defined our tentative limit function f we have to show two things: first that $f \in \mathcal{B}(\Omega)$, i.e., f is indeed a bounded function; second that indeed $\|f_n - f\|_\infty \to 0$. To this end we can use no information other than the Cauchy property of the sequence $(f_n)_{n\in\mathbb{N}}$. So fix $\epsilon > 0$. Then there is $N \in \mathbb{N}$ such that

$$|f_n(x) - f_m(x)| \le \|f_n - f_m\|_\infty \le \epsilon$$

for all $x \in \Omega$ and all $n, m \ge N$. Now fix $x \in \Omega$ and $n \ge N$; as $m \to \infty$ we obtain

$$|f_n(x) - f(x)| = \lim_m |f_n(x) - f_m(x)| \le \epsilon$$

since the function $t \longmapsto |f_n(x) - t|$ is continuous on $\mathbb{K}$. The inequality above holds for all $x \in \Omega$ and all $n \ge N$. Taking the supremum over x we therefore obtain

$$\|f_n - f\|_\infty \le \epsilon \quad \text{for all } n \ge N.$$

In particular, $\|f_n - f\|_\infty < \infty$, and so $f = f_n - (f_n - f) \in \mathcal{B}(\Omega)$. Summarizing the considerations above, we have shown that to each $\epsilon > 0$ there is $N \in \mathbb{N}$ such that $\|f_n - f\|_\infty \le \epsilon$ for all $n \ge N$; but this is just a reformulation of $\|f_n - f\|_\infty \to 0$, as desired. □

Example 5.12. The space ℓ^∞ of bounded scalar sequences is a Banach space with respect to the supremum norm $\|\cdot\|_\infty$. Ex.5.6

We have seen above that $\mathrm{C}[a, b]$ is not complete with respect to $\|\cdot\|_2$. (The same is true for the norm $\|\cdot\|_1$.) Things are different for the supremum norm. Ex.5.7

Example 5.13. *The space* $\mathrm{C}[a, b]$ *is a Banach space with respect to the supremum norm* $\|\cdot\|_\infty$.

Proof. We know already that the space $\mathcal{B}[a, b]$ of bounded functions is complete with respect to $\|\cdot\|_\infty$. Hence, by Lemma 5.5, it suffices to show that $\mathrm{C}[a, b]$ is *closed* in $(\mathcal{B}[a, b], \|\cdot\|_\infty)$.

To this end, take $(f_n)_{n\in\mathbb{N}} \subseteq \mathrm{C}[a, b]$ and $f_n \to f$ uniformly on $[a, b]$, for some bounded function $f \in \mathcal{B}[a, b]$. We fix an arbitrary $x \in [a, b]$ and have to show that f is continuous at x. Using our definition of continuity, we take a sequence $(x_m)_{m\in\mathbb{N}} \subseteq [a, b]$ with $x_m \to x$ and have to show that $f(x_m) \to f(x)$ as $m \to \infty$. By the scalar triangle inequality we may write

$$\begin{aligned} |f(x) - f(x_m)| &\le |f(x) - f_n(x)| + |f_n(x) - f_n(x_m)| + |f_n(x_m) - f(x_m)| \\ &\le 2\|f_n - f\|_\infty + |f_n(x) - f_n(x_m)| \end{aligned}$$

for all $n, m \in \mathbb{N}$. Given $\epsilon > 0$ choose n so large that $\|f_n - f\|_\infty < \epsilon$. For this n, since f_n is continuous at x, we find N such that

$$|f_n(x) - f_n(x_m)| < \epsilon \quad \text{whenever} \quad m \geq N.$$

Then, for all $m \geq N$,

$$|f(x) - f(x_m)| \leq 2\|f_n - f\|_\infty + |f_n(x) - f_n(x_m)| < 3\epsilon. \qquad \square$$

The heart of the proof above is called a "3ϵ"-argument. One can shorten it a little in the following way. As above, derive the inequality

$$|f(x) - f(x_m)| \leq 2\|f_n - f\|_\infty + |f_n(x) - f_n(x_m)|$$

for all $n, m \in \mathbb{N}$. Then take the lim sup with respect to m and obtain

$$\begin{aligned}\limsup_{m\to\infty} |f(x) - f(x_m)| &\leq 2\|f_n - f\|_\infty + \limsup_{m\to\infty} |f_n(x) - f_n(x_m)| \\ &= 2\|f - f_n\|_\infty .\end{aligned}$$

The left-hand side does not depend on n, so letting $n \to \infty$ shows that $\limsup_{m\to\infty} |f(x) - f(x_m)| = 0$, i.e., $f(x_m) \to f(x)$ as $m \to \infty$.

The following example generalizes Example 5.13.

Example 5.14. Let (Ω, d) be a metric space. Denote by

$$\mathrm{C_b}(\Omega) = \mathrm{C_b}(\Omega; \mathbb{K}) := \{f \in \mathcal{B}(\Omega) \mid f \text{ is continuous}\}$$

the space of functions on Ω that are bounded *and* continuous. *Then* $\mathrm{C_b}(\Omega)$ *is a closed subspace of* $(\mathcal{B}(\Omega), \|\cdot\|_\infty)$, *hence is a Banach space with respect*
Ex.5.8 *to the supremum norm.*

Ex.5.9
Ex.5.10

As mentioned at the end of the previous section, every noncomplete metric space has a "completion". In case the original space is normed, its completion is a Banach space in a natural way; see Theorem B.5. But such a completion is a fairly abstract object and it is natural to ask whether the completion of $(\mathrm{C}[a, b], \|\cdot\|_2)$, e.g., can be viewed as a space of functions on $[a, b]$. The answer is basically yes, but we need Lebesgue integration theory for it. This will be the topic of Chapter 7.

5.4. Series in Banach Spaces

Let $(E, \|\cdot\|)$ be a normed vector space and let $(f_n)_{n\in\mathbb{N}} \subseteq E$ be a sequence of elements of E. As in the scalar case, known from undergraduate courses, the formal series

$$\sum_{n=1}^{\infty} f_n \tag{5.1}$$

denotes the *sequence of partial sums* $(s_n)_{n\in\mathbb{N}}$ defined by

$$s_n := \sum\nolimits_{j=1}^{n} f_j \qquad (n \in \mathbb{N}).$$

If $\lim_{n\to\infty} s_n$ exists in E we call the series (5.1) (simply) **convergent** and use the symbol $\sum_{n=1}^{\infty} f_n$ also to denote its limit.

Definition 5.15. Let $(E, \|\cdot\|)$ be a normed vector space. A sequence $(f_n)_{n\in\mathbb{N}}$ of elements of E is called **absolutely summable** if the series $\sum_{n=1}^{\infty} f_n$ converges **absolutely**, i.e., if

$$\sum\nolimits_{n=1}^{\infty} \|f_n\| < \infty.$$

It is known from undergraduate analysis that if $E = \mathbb{K}$ is the scalar field, then absolute convergence implies (simple) convergence. This is a general fact, due to completeness.

Theorem 5.16. *Let $(E, \|\cdot\|)$ be a Banach space and let $(f_n)_{n\in\mathbb{N}}$ be a sequence in E such that $\sum_{n=1}^{\infty} \|f_n\| < \infty$. Then $\sum_{n=1}^{\infty} f_n$ converges in E and*

$$\left\| \sum\nolimits_{n=1}^{\infty} f_n \right\| \le \sum\nolimits_{n=1}^{\infty} \|f_n\|.$$

Proof. The claim is that the sequence $(s_n)_{n\in\mathbb{N}}$ of partial sums converges in E. Since E is a Banach space, i.e., complete, it suffices to show that $(s_n)_{n\in\mathbb{N}}$ is a Cauchy sequence. To this end, take $m > n$ and observe that

$$\|s_m - s_n\| = \left\| \sum\nolimits_{j=n+1}^{m} f_j \right\| \le \sum\nolimits_{j=n+1}^{m} \|f_j\| \le \sum\nolimits_{j=n+1}^{\infty} \|f_j\| \to 0$$

as $n \to \infty$. The continuity of the norm then yields

$$\begin{aligned}\left\| \sum\nolimits_{n=1}^{\infty} f_n \right\| &= \left\| \lim_{n\to\infty} \sum\nolimits_{j=1}^{n} f_j \right\| = \lim_{n\to\infty} \left\| \sum\nolimits_{j=1}^{n} f_j \right\| \le \lim_{n\to\infty} \sum\nolimits_{j=1}^{n} \|f_j\| \\ &= \sum\nolimits_{j=1}^{\infty} \|f_j\|.\end{aligned}$$

□

Example 5.17. The so-called **Weierstrass M-test** says the following: *Let $(f_n)_{n\in\mathbb{N}}$ be a sequence of continuous functions on $[a, b]$ such that there exists a sequence of real numbers $M_n \ge 0$ with the property that*

$$|f_n(x)| \le M_n \text{ for all } x \in [a, b],\ n \in \mathbb{N} \quad \text{and} \quad \sum_{n=1}^{\infty} M_n < \infty.$$

Then $\sum_{n=1}^{\infty} f_n(x)$ converges absolutely and uniformly on $[a, b]$ to a continuous function on $[a, b]$.

In effect, this is just a special case of Theorem 5.16: the hypotheses imply that $\|f_n\|_\infty \le M_n$ for each $n \in \mathbb{N}$ and hence $\sum_{n=1}^\infty \|f_n\|_\infty \le \sum_{n=1}^\infty M_n < \infty$.
Ex.5.11 Recall that $\mathrm{C}[a,b]$ is complete with respect to the supremum norm.
Ex.5.12

The following is a kind of converse to Theorem 5.16. However, we shall not use it in the rest of the book and state it merely for the sake of completeness.

***Theorem 5.18.** *Let $(E, \|\cdot\|)$ be a normed vector space such that every absolutely convergent series converges in E. Then E is a Banach space.*

Proof. Let $(f_n)_{n\in\mathbb{N}} \subseteq E$ be a Cauchy sequence in E. By Lemma 5.2.c) it suffices to find a subsequence that converges. Pick successively $(f_{n_k})_{k\in\mathbb{N}}$ such that

$$\left\|f_{n_k} - f_{n_{k+1}}\right\| \le 2^{-k} \qquad (k \in \mathbb{N}).$$

(This is possible since $(f_n)_{n\in\mathbb{N}}$ is Cauchy!) For $g_k := f_{n_k} - f_{n_{k+1}}$ one therefore has $\sum_{k=1}^\infty \|g_k\| < \infty$. By assumption,

$$g := \lim_{N\to\infty} \sum\nolimits_{k=1}^N g_k = \lim_{N\to\infty} \sum\nolimits_{k=1}^N f_{n_k} - f_{n_{k+1}} = \lim_{N\to\infty} f_{n_1} - f_{n_{N+1}}$$

exists in E. But this implies that $f_{n_N} \to f_{n_1} - g$ as $N \to \infty$. $\square$

Exercises 5A

Exercise 5.1. Let $(E, \|\cdot\|_E)$ and $(F, \|\cdot\|_F)$ be normed spaces and let $T : E \to F$ be a bounded linear mapping. Show that T maps Cauchy sequences in E to Cauchy sequences in F.

Exercise 5.2. Let $\|\cdot\|_1$ and $\|\cdot\|_2$ be two equivalent norms on a vector space E. Show that a sequence $(x_n)_{n\in\mathbb{N}} \subseteq E$ is $\|\cdot\|_1$-Cauchy if and only if it is $\|\cdot\|_2$-Cauchy.

Exercise 5.3. Show that every discrete metric space is complete.

Exercise 5.4. Prove assertions b) and c) from Lemma 5.5.

Exercise 5.5. Prove that the space c_{00} of finite sequences is not a Hilbert space with respect to the standard inner product.

Exercise 5.6. Show that c_{00} (the space of finite sequences) is not a Banach space with respect to the supremum norm.

Exercise 5.7. Show that $\mathrm{C}[a,b]$ is not a Banach space with respect to $\|\cdot\|_1$.

Exercise 5.8. Prove the assertions from Example 5.14.

Exercise 5.9. Show that the set of scalar null sequences

$$\mathrm{c}_0 = \{(x_n)_{n\in\mathbb{N}} \subseteq \mathbb{K} \mid \lim_{n\to\infty} x_n = 0\}$$

is a $\|\cdot\|_\infty$-closed subspace of ℓ^∞, and hence a Banach space with respect to the supremum norm.

Exercise 5.10. Show that ℓ^1 is a Banach space with respect to $\|\cdot\|_1$, but c_{00} is not.

Exercise 5.11. Let $(E, \|\cdot\|)$ be a normed space, and let $(f_n)_{n\in\mathbb{N}}$ be a sequence in E such that $\sum_{n=1}^{\infty} f_n$ converges. Show that $\lim_{n\to\infty} f_n = 0$.

Exercise 5.12. Let $\alpha = (\alpha_n)_{n\in\mathbb{N}_0}$ such that $\sum_{n=0}^{\infty} |\alpha_n| < \infty$. Consider the trigonometric series

$$\sum_{n=0}^{\infty} \alpha_n e^{2\pi i n \cdot t}$$

and show that it converges uniformly in $t \in \mathbb{R}$ to a continuous and 1-periodic function on $\mathbb{R}$.

Exercises 5B

Exercise 5.13. The space of **convergent sequences** is

$$\mathrm{c} := \{(x_j)_{j\in\mathbb{N}} \subseteq \mathbb{K} \mid \lim_{j\to\infty} x_j \text{ exists in } \mathbb{K}\}.$$

Show that c is $\|\cdot\|$-closed subpace of ℓ^∞, and hence a Banach space.

Exercise 5.14 (Uniformly continuous functions). Let (Ω, d) be a metric space, and let

$$\mathrm{UC_b}(\Omega) := \{f : \Omega \longrightarrow \mathbb{K} \mid f \text{ is bounded and uniformly continuous}\}$$

(see Exercise 4.24). Show that $\mathrm{UC_b}(\Omega)$ is a closed subspace of $\mathrm{C_b}(\Omega)$ with respect to $\|\cdot\|_\infty$. Conclude that it is a Banach space.

Exercise 5.15. Let $(E, \|\cdot\|_E)$ and $(F, \|\cdot\|_F)$ be Banach spaces and let $T : E \to F$ be a bounded linear mapping. Suppose that there is $c \geq 0$ such that

$$\|f\|_E \leq c \|Tf\|_F$$

for all $f \in E$. Show that $\ker T = \{0\}$ and $\operatorname{ran}(T)$ is closed.

Exercise 5.16. Let $(E, \|\cdot\|_E)$ and $(F, \|\cdot\|_F)$ be normed spaces. A linear mapping $T : E \to F$ is called **invertible** or a **topological isomorphism** if T is bounded and bijective, and T^{-1} is bounded too.

Suppose that $T : E \to F$ is invertible and E is a Banach space. Show that F is a Banach space, too.

Exercise 5.17. Let $d_1(x, y) := |x - y|$ and

$$d_2(x, y) := |\arctan(x) - \arctan(y)|$$

for $x, y \in \mathbb{R}$. Show that d_1 and d_2 are equivalent metrics on $\mathbb{R}$. Then show that $(\mathbb{R}; d_2)$ is not complete.

Exercise 5.18 (Cauchy Double Sequences, cf. Exercise 3.17). A double sequence $(x_{n,m})_{n,m\in\mathbb{N}}$ in a metric space (Ω, d) is called **Cauchy** if for each $\epsilon > 0$ there is $N \in \mathbb{N}$ such that $d(x_{n,m}, x_{k,l}) \leq \epsilon$ for all $n, m, k, l \geq N$.

a) Prove that if $\lim_{n,m\to\infty} x_{n,m} = x$ in Ω, then the double sequence $(x_{n,m})_{n,m\in\mathbb{N}}$ is Cauchy.

b) Prove that if $(x_{n,m})_{n,m\in\mathbb{N}}$ is Cauchy and $\lim\limits_{n\to\infty} x_{n,n}$ exists, then $\lim\limits_{n,m\to\infty} x_{n,m}$ exists, too.

c) Prove that if (Ω, d) is complete, then each Cauchy double sequence has a limit.

Exercise 5.19. Let $(x_{n,m})_{n,m\in\mathbb{N}}$ be a double sequence in a metric space (Ω, d) such that $x = \lim_{n,m\to\infty} x_{n,m}$ exists in Ω.

a) Suppose that for each $n \in \mathbb{N}$ the limit $a_n := \lim_{m\to\infty} x_{n,m}$ exists in Ω. Show that $a_n \to x$.

b) Show that if (Ω, d) is complete, then for each $n \in \mathbb{N}$ the limit $a_n := \lim\limits_{m\to\infty} x_{n,m}$ exists in Ω.

Exercise 5.20 (The Fundamental Principle of Analysis). Let (Ω, d) be a metric space, and let $(x_{n,m})_{n,m\in\mathbb{N}}$ be a double sequence in Ω. Suppose that

$$x_{n,m} \to a_n \quad (m \to \infty) \quad \text{and} \quad x_{n,m} \to b_m \quad (n \to \infty)$$

for certain $a_n, b_m \in \Omega$.

$$\begin{array}{ccc} x_{n,m} & \xrightarrow{\ n\ } & b_m \\ {\scriptstyle m}\big\downarrow & & \\ a_n & & \end{array}$$

Suppose further that the convergence $x_{n,m} \to b_m$ is *uniform* in $m \in \mathbb{N}$, i.e.,

$$\sup_{m\in\mathbb{N}} d(x_{n,m}, b_m) \to 0 \qquad \text{as } n \to \infty.$$

Show that the following assertions:

(i) $a := \lim_{n\to\infty} a_n$ exists in Ω,

(ii) $b := \lim_{m\to\infty} b_m$ exists in Ω,

(iii) $x := \lim_{n,m\to\infty} x_{n,m}$ exists in Ω,

are equivalent and one has $a = b = x$.

Exercise 5.21 (The Fundamental Principle of Analysis II). Let (Ω, d) be a *complete* metric space, and let $(x_{n,m})_{n,m\in\mathbb{N}}$ be a double sequence in Ω. Suppose that for each n the sequence $(x_{n,m})_{m\in\mathbb{N}}$ is Cauchy and that the sequence $(x_{n,m})_{n\in\mathbb{N}}$ is Cauchy *uniformly in* $m \in \mathbb{N}$. By this we mean that

$$\sup_{m\in\mathbb{N}} d(x_{k,m}, x_{l,m}) \to 0 \quad (k, l \to \infty).$$

Show that under these hypotheses there are elements $a_n, b_m, a \in \Omega$ such that

$$\begin{array}{ccc} x_{n,m} & \xrightarrow{\ n\ } & b_m \\ {\scriptstyle m}\big\downarrow & & \big\downarrow{\scriptstyle m} \\ a_n & \xrightarrow{\ n\ } & a \end{array}$$

Exercise 5.22 (Double Series). Let E be a normed space. Each two-sided sequence $(f_j)_{j\in\mathbb{Z}}$ in E determines a **double series**

$$\sum_{j=-\infty}^{\infty} f_j := (s_{n,m})_{n,m\in\mathbb{N}}, \qquad s_{n,m} := \sum_{j=-n}^{m} f_j.$$

That is, the double series is just the double sequence of its partial sums. If this double sequence converges, we also write

$$\sum_{j=-\infty}^{\infty} f_j := \lim_{n,m\to\infty} \sum_{j=-n}^{m} f_j$$

for this limit. Formulate and prove the analogues of Theorems 5.16 and 8.13 for double series.

Exercises 5C

Exercise 5.23. Show that for each $1 \le p \le \infty$ the space ℓ^p of p-summable scalar sequences is a Banach space with respect to $\|\cdot\|_p$; cf. Section 2.5.

Exercise 5.24. Let E be a Banach space and let Ω be a nonempty set. Show that the space $\mathcal{B}(\Omega; E)$ of bounded E-valued functions on Ω is a Banach space with respect to $\|\cdot\|_\infty$; cf. Exercise 2.33.

Exercise 5.25. Let E be a Banach space. Show that the space $\mathrm{BV}([a, b]; E)$ of E-valued functions of bounded variation on $[a, b] \subseteq \mathbb{R}$ is a Banach space with respect to the norm $\|f(a)\|_E + \|f\|_v$; cf. Exercise 2.34.

Exercise 5.26. Let E be a Banach space and $\alpha \in (0, 1]$. Show that the space $\mathrm{C}^\alpha([a, b]; E)$ of E-valued Hölder-continuous functions is a Banach space with respect to the norm $\|f\|_{\mathrm{C}^\alpha} = \|f(a)\|_E + \|f\|_{(\alpha)}$; cf. Exercise 2.35.

Exercise 5.27. Let E be a Banach space. The closure of the space $\mathrm{St}([a, b]; E)$ of E-valued step functions (cf. Exercise 2.36) in $\mathcal{B}([a, b]; E)$ with respect to $\|\cdot\|_\infty$ is the space $\mathrm{Reg}([a, b]; E)$ of **regulated functions**. Show that this is a Banach space and for each regulated function f and each point $x_0 \in [a, b]$ the one-sided limits $\lim_{x\nearrow x_0} f(x)$ and $\lim_{x\searrow x_0} f(x)$ exist. (One can show that, conversely, each function with one-sided limits is regulated; cf. Exercise 16.41.) Show that a regulated function can have at most countably many points of discontinuity.

Exercise 5.28. By Lemmas 4.39 and 5.5, each (sequentially) compact metric space is precompact and complete. Prove the converse: *Every precompact and complete metric space is (sequentially) compact.*

Exercise 5.29 (Quotient Spaces). Let E be a normed vector space and $F \subseteq E$ a closed linear subspace. Let $s : E \to E/F$ be the canonical surjection onto the factor space. We define the **quotient norm** on E/F by

$$\|g + F\| := \inf\{\|h\| \mid h \in g + F\} = \inf\{\|g - f\| \mid f \in F\}.$$

The space E/F endowed with the quotient norm is called the **quotient space**. Prove the following assertions:

a) The quotient norm is a norm, and $s : E \to E/F$ is bounded. Moreover, $\|s\| = 1$ if $E \neq F$.

b) A linear mapping $T : E/F \to G$ into another normed space is bounded if and only if the linear mapping $T \circ s : E \to G$ is bounded.

c) If E is a Banach space, so is E/F.

Exercise 5.30. Recall from Exercise 3.25 that

$$d(x,y) := \sum_{j=1}^{\infty} 2^{-j} \frac{|x_j - y_j|}{1 + |x_j - y_j|}$$

is a metric on the space

$$\mathrm{s} := \mathcal{F}(\mathbb{N}) = \{x = (x_j)_{j\in\mathbb{N}} \mid x_j \in \mathbb{K} \ \forall j \in \mathbb{N}\}$$

of all scalar sequences. Show that this metric is complete.

Chapter 6

*The Contraction Principle

Many problems arising in concrete (engineering) situations can be reduced to solving a single equation

$$F(x) = 0 \tag{6.1}$$

where x is an unknown quantity. In undergraduate courses, this quantity is often a scalar or a vector in a finite-dimensional real or complex vector space. However, if the equation is a differential equation, the unknown quantity is a function, and so it seems reasonable to ensure great flexibility and look at the problem from an abstract perspective.

Are there general principles to find a solution of an equation? Well, if $F : \mathbb{R}^n \to \mathbb{R}^m$ is linear, then finding x in (6.1) is just the same as determining the kernel of F. That can be done by standard *Gaussian elimination*[1] using a matrix representation of F and is treated in elementary linear algebra classes. If F is a real function, then you may know of *Newton's method*[2] to solve it. Whereas Gaussian elimination is a finite step procedure which ultimately yields a solution (in fact, all of them), Newton's method is recursive: it begins with a (well-chosen) start value x_0 and then proceeds by defining "approximate solutions" x_n recursively by

$$x_{n+1} = x_n - \frac{F(x_n)}{F'(x_n)} \qquad (n \geq 0).$$

[1] Carl Friedrich Gauß (1777–1855), German mathematician and astronomer.
[2] Isaac Newton (1643–1727), English mathematician and physicist.

If the sequence $(x_n)_{n\in\mathbb{N}}$ converges to a limit x_∞, and if F and F' are continuous and F' does not have a zero in a neighborhood of x_∞, then one has

$$x_\infty = x_\infty - F(x_\infty)/F'(x_\infty) \quad \text{whence} \quad F(x_\infty) = 0.$$

If we write

$$Tx := x - F(x)/F'(x)$$

then the original equation $F(x) = 0$ is (under suitable hypotheses on F') equivalent to the *fixed point equation*

$$Tx = x. \tag{6.2}$$

With this notation the sequence of approximate solutions $(x_n)_{n\in\mathbb{N}}$ is computed as $x_{n+1} = Tx_n$, $n \in \mathbb{N}$, i.e.,

$$x_n = T^n x_0 \qquad (n \in \mathbb{N}).$$

Of course, one has to ensure that the sequence $(x_n)_{n\in\mathbb{N}}$ is well-defined and has a limit x_∞. (See [**GL06**, Section 5.4] or [**Che01**, Section 3.3] for some suitable sets of hypotheses for Newton's method.)

6.1. Banach's Contraction Principle

Newton's method is a special case of an *iterative procedure* to approximate a solution of the fixed point equation (6.2). The following general result in this direction, due to Banach, yields a simple setting in which such an iterative procedure always works.

Theorem 6.1 (Banach's Contraction Principle). *Let (Ω, d) be a nonempty complete metric space, and let $T : \Omega \to \Omega$ be a continuous mapping such that there is a real number $0 \le q < 1$ satisfying*

$$d(Tx, Ty) \le q\, d(x, y) \qquad (x, y \in \Omega).$$

Then there is a unique $z \in \Omega$ such that $Tz = z$. Moreover, if $x_0 \in \Omega$ is arbitrary, then the sequence $(x_n)_{n\in\mathbb{N}}$ defined by

$$x_{n+1} := Tx_n \qquad (n \in \mathbb{N})$$

converges to z, and one has the error estimate

$$d(x_n, z) \le \frac{q^n}{1-q} d(x_0, Tx_0) \qquad (n \in \mathbb{N}).$$

Proof. Uniqueness: Suppose that z, y are fixed points of T, then

$$d(y, z) = d(Ty, Tz) \le q d(y, z)$$

which implies by $q < 1$ that $d(y, z) = 0$, i.e., $y = z$.

Existence: Fix $x_0 \in \Omega$ and define the sequence $(x_n)_{n\in\mathbb{N}}$ recursively by $x_{n+1} := Tx_n$, $n \geq 0$. Suppose that $z := \lim_{n\to\infty} x_n$ exists. Then by continuity of T,

$$Tz = T(\lim_{n\to\infty} x_n) = \lim_{n\to\infty} Tx_n = \lim_{n\to\infty} x_{n+1} = z$$

and hence z is a fixed point of T.

To show that the sequence $(x_n)_{n\in\mathbb{N}}$ converges, by completeness of (Ω, d) it suffices to show that it is a Cauchy sequence. To this end, note first that, by definition,

$$T^j x_i = x_{i+j} \qquad \text{for all } i, j \geq 0.$$

Now take $n, l \in \mathbb{N}$. Then, by the triangle inequality,

$$\begin{aligned} d(x_0, x_l) &\leq d(x_0, x_1) + d(x_1, x_2) + \cdots + d(x_{l-1}, x_l) \\ &= d(x_0, x_1) + d(Tx_0, Tx_1) + \cdots + d(T^{l-1}x_0, T^{l-1}x_1) \\ &\leq (1 + q + \cdots + q^{l-1}) d(x_0, x_1) = \frac{1 - q^l}{1 - q} d(x_0, x_1) \leq \frac{d(x_0, x_1)}{1 - q}. \end{aligned}$$

Hence, if $n, l \in \mathbb{N}$, we obtain

$$d(x_n, x_{n+l}) = d(T^n x_0, T^n x_l) \leq q^n d(x_0, x_l) \leq \frac{q^n}{1 - q} d(x_0, x_1). \tag{6.3}$$

Since $0 \leq q < 1$, we have $q^n \to 0$. This shows that the sequence $(x_n)_{n\in\mathbb{N}}$ is indeed Cauchy. By completeness, $z := \lim_{n\to\infty} x_n$ exists.

Finally, let $l \to \infty$ in (6.3). Since the function $x \mapsto d(x_n, x)$ is continuous (Corollary 4.10) we obtain $d(x_n, z) \leq ({q^n}/{1-q})\, d(x_0, x_1)$ as claimed. □ Ex.6.1

The contraction principle is a simple and everyday tool for finding solutions to equations. However, this does not mean that it renders the problem trivial. It just tells you what to look for, namely a strict contraction on a complete metric space. To find those objects may still be a highly nontrivial matter.

6.2. Application: Ordinary Differential Equations

The standard application of the contraction principle is to the existence and uniqueness of solutions for the **initial value problem**

$$x'(t) = f(t, x(t)) \qquad x(0) = x_0 \in E \tag{6.4}$$

where $E = \mathbb{R}^d$ and $f : U \to E$ is defined on an open subset U of $\mathbb{R} \times E$ containing $(0, x_0)$. A **solution** of (6.4) is a differentiable function

$$x : I \longrightarrow E$$

such that I is an open interval containing 0, and $\{(t, x(t)) : t \in I\} \subseteq U$ such that (6.4) holds.

If f is continuous, then — by the fundamental theorem of calculus — (6.4) is equivalent to the integral equation

$$x(t) = x_0 + \int_0^t f(s, x(s))\, \mathrm{d}s \qquad (t \in I). \tag{6.5}$$

This can be seen formally as a fixed point equation $Tx = x$ by writing

$$(Tx)(t) = x_0 + \int_0^t f(s, x(s))\, \mathrm{d}s.$$

In order to find a solution by the contraction principle one has to devise a complete metric space of continuous functions on which T acts as a strict contraction. Since there are situations where the initial value problem (6.4) is not uniquely solvable, this will work only under certain hypotheses on f. Fortunately, these hypotheses are so general as to grant a wide applicability.

For the following we fix a norm $\|\cdot\|_E$ on E, so that E is a Banach space.

Definition 6.2. Let $U \subseteq \mathbb{R} \times E$ be an open subset. A continuous function $f : U \to E$ is said to satisfy a **Lipschitz condition** in the second argument if there is $L \ge 0$ such that

$$\|f(t, x) - f(t, y)\|_E \le L \|x - y\|_E \tag{6.6}$$

for all $x, y \in E$ and $t \in \mathbb{R}$ such that $(t, x), (t, y) \in U$.

Let us now suppose that $f : U \to E$ satisfies such a Lipschitz condition, with constant L as in (6.6). Since f is continuous, and continuous functions are bounded on compact sets, we can make U smaller in order to get a bound

$$\|f(s, x)\|_E \le K \quad \text{for all } (s, x) \in U. \tag{6.7}$$

Since U is open and contains $(0, x_0)$, we find $a, b > 0$ such that

$$|t| \le a, \quad \|x - x_0\|_E \le b \quad \Longrightarrow \quad (t, x) \in U.$$

Let $0 < c \le \min(a, {}^b\!/\!_K)$, consider $I := [-c, c]$ and

$$\Omega := \{x \in \mathrm{C}([-c, c]; E) \mid \|x(s) - x_0\|_E \le b \text{ for all } s \in [-c, c]\}.$$

Then by construction we have $(s, x(s)) \in U$ for every $x \in \Omega$ and $s \in I$.

Lemma 6.3. *In the construction from above, if $x \in \Omega$ and*

$$(Tx)(t) := x_0 + \int_0^t f(s, x(s))\, \mathrm{d}s \qquad (-c \le t \le c),$$

then $Tx \in \Omega$, too.

Proof. By virtue of Exercise 6.2 this follows from

$$\|(Tx)(t) - x_0\|_E = \left\| \int_0^t f(s, x(s))\, \mathrm{d}s \right\|_E \le \operatorname{sgn}(t) \int_0^t \|f(s, x(s))\|_E \,\mathrm{d}s$$
$$\le \int_0^{|t|} K \,\mathrm{d}s \le cK \le b.$$ □

The space $\mathrm{C}([-c, c]; E)$ of continuous E-valued functions on $I = [-c, c]$ clearly is a vector space with respect to the usual pointwise operations. Furthermore, with respect to the supremum norm

$$\|g\|_\infty := \sup\{\|g(s)\|_E \mid -c \le s \le c\} \qquad (g \in \mathrm{C}([-c, c]; E))$$

it is a Banach space. This is proved in the same way as for scalar functions (Examples 2.8, 5.11 and 5.13): one just replaces absolute values everywhere by E-norm signs. Since uniform convergence implies pointwise convergence, Ex.6.3
the set Ω is closed in the Banach space $\mathrm{C}([-c, c]; E)$ and hence, with respect to the induced metric, is a complete metric space.

Consequently, we have all the necessary conditions at hand to take the final step.

Lemma 6.4. *In the situation above,*

$$|\!|\!| g |\!|\!|_\infty := \sup_{s \in [-c,c]} e^{-L|s|} \|g(s)\|_E$$

is an equivalent norm for the Banach space $\mathrm{C}([-c, c]; E)$, *and*

$$|\!|\!| Tx - Ty |\!|\!|_\infty \le q\, |\!|\!| x - y |\!|\!|_\infty \qquad (x, y \in \Omega) \tag{6.8}$$

with $q := 1 - e^{-Lc} < 1$.

Proof. The equivalence of the norms follows by taking the supremum with respect to $t \in [-c, c]$ in the inequalities

$$e^{-L|t|} \|g(t)\|_E \le \|g(t)\|_E \le e^{cL} e^{-L|t|} \|g(t)\|_E .$$

For the contraction property, we estimate

$$e^{-L|t|} \|(Tx)(t) - (Ty)(t)\|_E = e^{-L|t|} \left\| \int_0^t f(s, x(s)) - f(s, y(s)) \,\mathrm{d}s \right\|_E$$
$$\le e^{-L|t|} \operatorname{sgn}(t) \int_0^t \|f(s, x(s)) - f(s, y(s))\|_E \,\mathrm{d}s$$
$$\le e^{-L|t|} \operatorname{sgn}(t) \int_0^t L \|x(s) - y(s)\|_E \,\mathrm{d}s$$
$$\le e^{-L|t|} \int_0^{|t|} L e^{L|s|}\, |\!|\!| x - y |\!|\!|_\infty \,\mathrm{d}s$$
$$= e^{-L|t|}(e^{L|t|} - 1)\, |\!|\!| x - y |\!|\!|_\infty = (1 - e^{-L|t|})\, |\!|\!| x - y |\!|\!|_\infty \le q\, |\!|\!| x - y |\!|\!|_\infty.$$

Taking the supremum with respect to $t \in [-c, c]$ we obtain (6.8). □

Lemma 6.4 shows that T is a strict contraction for the metric on Ω coming from the norm $\|\cdot\|_\infty$. Since this norm is equivalent to the usual supremum norm, $\mathrm{C}([-c,c];E)$ is complete with respect to this norm, and Ω is closed, whence also complete. Hence the Banach contraction principle applies and yields a unique $x \in \Omega$ such that $Tx = x$. This is clearly a solution for (6.4).

Remark 6.5 (Local Uniqueness). Note that our definition of Ω in the proof above is actually dependent on the choice of c, so we should better write Ω_c instead of just Ω. Now, given any two solutions x_1, x_2 of (6.4), by continuity we have $x_1, x_2 \in \Omega_c$ for some $c > 0$, and since the fixed point guaranteed by the contraction principle is unique, we must have $x_1 = x_2$ on $[-c, c]$. This is called the **local uniqueness** of solutions.

6.3. Application: Google's PageRank

We sketch the original idea behind the web-search engine Google. The world wide web can be pictured as a huge graph, where the nodes are the individual webpages and the edges are the links that can be found on these pages. In order to rank these pages without external information, one uses this link-structure and regards it as a democratic voting system: each page has one vote, and by pointing to n other pages, it attributes one n-th of its own importance to each of these.

More formally, let W be the collection of all pages. For each $w \in W$ let $n(w)$ be the number of links that point from page w to other pages, and let $I(w)$ be the collection of webpages w' that point to w. Then the required **importance vector** $x : W \to \mathbb{R}$ should satisfy $x \geq 0$ and

$$x(w) = \sum_{w' \in I(w)} \frac{1}{n(w')} x(w') \qquad (w \in W). \tag{6.9}$$

That means, each page w' pointing to w contributes with $1/n(w')$ of its own importance $x(w')$ to the importance $x(w)$ of w.

Let $d := \#W$ be the number of all websites, so we may identify $W = \{1, \dots, d\}$. Then $x \in \mathbb{R}^d$ (viewed as a column vector) and (6.9) can be written as

$$x = Ax$$

where A is a gigantic matrix in which all entries of the w'-th column are the same, namely $1/n(w')$.

Of course, nothing so far tells us that such an importance vector x must exist. The ingenious idea of the fathers of Google, Brin[3] and Page,[4] was to *modify* the matrix A slightly in order to ensure the solvability of the problem, and at the same time preserve the essential features of A so that the obtained solution x yields a meaningful ranking of the webpages. The modification is in two steps, and in the end the importance vector x, which is a fixed point of A, will be found by an application of the Banach contraction principle.

The *first step* deals with the so-called **dangling nodes**, i.e., webpages without any outgoing links. These correspond to zero-columns in the web-graph matrix A. Motivated by the voting metaphor, we replace each of these columns by the column $(1/d)\mathbf{1}$, i.e., we consider a nonvoter as somebody who values every candidate equally. The effect of this step is that A becomes a **column stochastic** matrix, i.e., a positive matrix such that the sum of the entries of each column is equal to 1. (In other words: every column of A is a **probability vector**.)

In the *second step* one chooses a number $0 < q < 1$ and defines

$$B := qA + (1-q)\,(1/d)\mathbf{1}\cdot\mathbf{1}^\top.$$

Note that $\mathbf{1}$ is the column vector that has a 1 in each component, so $(1/d)\mathbf{1}\cdot\mathbf{1}^\top$ is the $d\times d$ matrix that has $1/d$ in each component. Consequently, B is also column stochastic, i.e.,

$$B \geq 0 \quad \text{and} \quad \mathbf{1}^\top B = \mathbf{1}^\top.$$

We now claim that B *has a unique fixed vector within the set*

$$\Omega := \{x \in \mathbb{R}^d \mid x \geq 0,\ \mathbf{1}^\top x = \mathbf{1}^\top\}$$

of probability vectors.

Lemma 6.6. *Let Q be any column stochastic $d\times d$-matrix. Then Q is a contraction for the 1-norm. Moreover, if $x \in \Omega$, then $Qx \in \Omega$ as well.*

Proof. Let $x = (x_1,\dots,x_d)^\top \in \mathbb{R}^d$. We denote by $|x|$ the vector $|x| = (|x_1|,\dots,|x_d|)^\top$. Then $\|x\|_1 = \mathbf{1}^\top|x|$. Since $Q \geq 0$, i.e., all entries of Q are ≥ 0, it follows easily that $|Qx| \leq Q\,|x|$, where the ordering $\leq$ is understood componentwise. Hence

$$\|Qx\|_1 = \mathbf{1}^\top|Qx| \leq \mathbf{1}^\top Q\,|x| = \mathbf{1}^\top|x| = \|x\|_1,$$

since $\mathbf{1}^\top Q = \mathbf{1}^\top$ by hypothesis. This proves the first assertion. If $x \in \Omega$, then $x \geq 0$ and hence $Qx \geq 0$ as well. Moreover, $\mathbf{1}^\top(Qx) = (\mathbf{1}^\top Q)x = \mathbf{1}^\top x = 1$, whence $Qx \in \Omega$ as claimed. □

[3]Sergey Brin (1973–), American computer scientist and Internet entrepreneur.

[4]Lawrence "Larry" Page (1973–), American computer scientist and Internet entrepreneur.

By Lemma 6.6 the set Ω is invariant under the action of B. It is closed in $\mathbb{R}^d$ and hence complete as a metric space. If $x, y \in \Omega$, then

$$\|Bx - By\|_1 = \|qAx - qAy\|_1 = q\,\|A(x-y)\|_1 \le q\,\|x-y\|_1$$

by Lemma 6.6, since A is column stochastic as well. Hence, the Banach contraction principle applies: for any probability vector $x_0 \in \Omega$, the limit

$$x := \lim_{n\to\infty} B^n x_0$$

exists and is the unique fixed point of B in Ω. The rate of convergence is by powers of q.

This is basically, what "PageRank", the algorithm governing the search engine, does. Given that d is very large, one is interested to make q small, in order to avoid many iterations. On the other hand, if q gets smaller, the matrix B will lose the characteristic features of A, and become more and more similar to the uniform matrix $(1/d)\mathbf{1}\cdot\mathbf{1}^\top$. In practice, the actual choice of q is a matter of trial and error; it is reported that $q = 0.8$ was used for a long time as a good compromise. A very good one it seems, regarded Google's tremendous success.

For more about PageRank and the mathematics of web-search engines see [**BL06**] and [**LM06**].

6.4. Application: The Inverse Mapping Theorem

Our last application of the contraction principle is to *many-variable calculus*. We begin with recalling the basic notions and notations.

Let us fix some finite-dimensional real spaces E and F together with any norms on them; cf. Theorem 4.29. On the space $\mathcal{L}(E;F)$ of linear mappings we consider the associated operator norm. (Via a choice of bases one may of course identify $E = \mathbb{R}^d$ and $F = \mathbb{R}^e$ and $\mathcal{L}(E;F) = \mathbb{R}^{d\times e}$.)

A mapping $f : U \to F$ defined on some open subset $U \subseteq E$ is called *(totally) differentiable* at $x_0 \in U$ if there is a linear mapping $A : E \to F$ such that

$$f(x) = f(x_0) + A(x - x_0) + \eta(x) \quad \text{for all } x \in U, \qquad \lim_{x\to x_0} \frac{\eta(x)}{\|x - x_0\|} = 0.$$

In this case, A is unique with this property. It is then denoted by $\mathrm{D}f(x_0)$ and called the *derivative* of f at x_0. (After a choice of bases for E and F, the derivative is represented by the matrix $\mathrm{D}f(x_0) \sim \left(\frac{\partial f_j}{\partial x_i}(x_0)\right)_{i,j}$ of partial derivatives.) Note that the derivative is a bounded linear map; cf. Example 2.19.

The mapping f is called a C^1*-map* if f is differentiable at *each* $x_0 \in U$, and the new map

$$\mathrm{D}f : U \to \mathcal{L}(E; F)$$

is continuous. (In matrix representation, this just means that all the component functions, i.e., partial derivatives, are continuous. Conversely, by a classical result from many-variable calculus, if f has continuous partial derivatives then it is totally differentiable at each point, hence a C^1-map.)

Lemma 6.7 (Mean Value Theorem). *Let $f : U \to F$ be a C^1-map, and let $x, x' \in U$ be such that the segment*

$$[x, x'] := \{x + t(x' - x) \mid t \in [0, 1]\}$$

is contained in U. Then

$$\|f(x) - f(x')\| \le \Big(\sup_{z \in [x,x']} \|\mathrm{D}f(z)\| \Big) \|x - x'\| .$$

Proof. Let $\varphi(t) := f(x' + t(x - x'))$ for $t \in [0, 1]$. Then $\varphi : [0, 1] \to F$ is differentiable, and by the chain rule

$$\varphi'(t) = \mathrm{D}f(x' + t(x - x')) \cdot (x - x') \qquad (t \in [0, 1]).$$

Since the function φ is continuous, the fundamental theorem of calculus yields

$$f(x) - f(x') = \varphi(1) - \varphi(0) = \int_0^1 \varphi'(t)\,\mathrm{d}t = \int_0^1 \mathrm{D}f(x' + t(x - x')) \cdot (x - x')\,\mathrm{d}t.$$

Taking norms we estimate (cf. Exercise 6.2)

$$\begin{aligned} \|f(x) - f(x')\| &\le \int_0^1 \|\mathrm{D}f(x' + t(x - x')) \cdot (x - x')\|\,\mathrm{d}t \\ &\le \int_0^1 \|\mathrm{D}f(x' + t(x - x'))\| \cdot \|x - x'\|\,\mathrm{d}t \\ &\le \Big(\sup_{t \in [0,1]} \|\mathrm{D}f(x' + t(x - x'))\| \Big) \|x - x'\| . \end{aligned}$$ □

We now come to the promised application of the contraction principle. A C^1-map $f : U \to F$ is called a *local* C^1*-diffeomorphism* at $x_0 \in U$, if f maps an open neighborhood U_0 of x_0 bijectively onto an open neighborhood V_0 of $f(x_0)$, and the inverse mapping $f^{-1} : V_0 \to U_0$ is again a C^1-map. (Note that this can only happen if $\dim E = \dim F$.)

Theorem 6.8 (Inverse Mapping Theorem). *Let $f : U \to F$ be a C^1-map, and let $x_0 \in U$ be such that $\mathrm{D}f(x_0) : E \to F$ is invertible. Then f is a local C^1-diffeomorphism at x_0.*

Proof. By passing to the function $\tilde{f}(x) := [\mathrm{D}f(x_0)]^{-1}(f(x + x_0) - f(x_0))$ we may suppose that $E = F$, $x_0 = f(x_0) = 0$, and $\mathrm{D}f(0) = \mathrm{I}$. Define $g(x) := x - f(x)$. Then $\mathrm{D}g(x) = \mathrm{I} - \mathrm{D}f(x)$. Hence $\mathrm{D}g(0) = 0$, and by continuity there is $\delta > 0$ such that

$$\|x\| \le \delta \quad \Longrightarrow \quad x \in U \quad \text{and} \quad \|\mathrm{D}g(x)\| \le 1/2. \tag{6.10}$$

The mean value theorem (Lemma 6.7) with $x' = 0$ now yields that

$$\|x\| \le \delta \quad \Longrightarrow \quad \|g(x)\| \le 1/2\,\|x\| \le \delta/2,$$

which means that $g : \mathrm{B}[0, \delta] \to \mathrm{B}[0, \delta/2]$. After these preparatory considerations, we can now make a decisive step.

Claim: *For each $y \in \mathrm{B}[0, \delta/2]$ there is a unique $x \in \mathrm{B}[0, \delta]$ with $f(x) = y$.*

To prove the claim we fix $y \in \mathrm{B}[0, \delta/2]$ and consider the function

$$h(x) := y + x - f(x) = y + g(x) \qquad (x \in U).$$

If $\|x\| \le \delta$, then

$$\|h(x)\| \le \|y\| + \|g(x)\| \le \delta/2 + \delta/2 = \delta.$$

Hence $h : \mathrm{B}[0, \delta] \to \mathrm{B}[0, \delta]$. Moreover, if $x, x' \in \mathrm{B}[0, \delta]$, then

$$\|h(x) - h(x')\| = \|g(x) - g(x')\| \le 1/2\,\|x - x'\|$$

by (6.10) and the mean value theorem. Hence, h is a strict contraction on the complete(!) metric space $\mathrm{B}[0, \delta]$. By the contraction principle, there is a unique $x \in \mathrm{B}[0, \delta]$ with $h(x) = x$, i.e., $f(x) = y$. This proves the claim.

We can now invert f on the set $\mathrm{B}[0, \delta/2]$, i.e., we can form a function $\varphi : \mathrm{B}[0, \delta/2] \to \mathrm{B}[0, \delta]$ with $f \circ \varphi = \mathrm{Id}$. The function φ is continuous. Indeed, writing $x = g(x) + f(x)$,

$$\|x - x'\| \le \|g(x) - g(x')\| + \|f(x) - f(x')\| \le \|f(x) - f(x')\| + 1/2\,\|x - x'\|$$

which amounts to $\|x - x'\| \le 2\,\|f(x) - f(x')\|$ for $x, x' \in \mathrm{B}[0, \delta]$. Replacing x, x' by $\varphi(y), \varphi(y')$ we hence obtain

$$\|\varphi(y) - \varphi(y')\| \le 2\,\|y - y'\| \qquad (y, y' \in \mathrm{B}[0, \delta/2]). \tag{6.11}$$

Now let $0 < \epsilon < \delta/2$ be so small that with $V_0 := \mathrm{B}(0, \epsilon)$ one has $f^{-1}(V_0) \subseteq \mathrm{B}(0, \delta)$. (This is possible since f is continuous and $f(0) = 0$.) Then

$$U_0 := \varphi(V_0) = f^{-1}(V_0) \cap \mathrm{B}[0, \delta] = f^{-1}(V_0)$$

is open, since V_0 is open. It follows that $f : U_0 \to V_0$ is bijective with (continuous) inverse $\varphi : V_0 \to U_0$.

Claim: *The mapping $\varphi : V_0 \to E$ is a C^1-map.*

Let $y, y_0 \in V_0$ and define $x := \varphi(y), x_0 := \varphi(x_0)$. We define $\eta(x)$ through the identity

$$f(x) - f(x_0) = \mathrm{D}f(x_0) \cdot (x - x_0) + \eta(x).$$

Then

$$\varphi(y) - \varphi(y_0) = [\mathrm{D}f(x_0)]^{-1}(y - y_0) + \rho(y)$$

with $\rho(y) := -[\mathrm{D}f(x_0)]^{-1}\eta(\varphi(y))$, and we want to show that $\|\rho(y)\|/\|y-y_0\| \to 0$ as $y \to y_0$. To this end, we estimate with (6.11)

$$\begin{aligned} \frac{\|\rho(y)\|}{\|y - y_0\|} &\le \left\|[\mathrm{D}f(x_0)]^{-1}\right\| \frac{\|\eta(\varphi(y))\|}{\|\varphi(y) - \varphi(y_0)\|} \frac{\|\varphi(y) - \varphi(y_0)\|}{\|y - y_0\|} \\ &\le 2\left\|[\mathrm{D}f(x_0)]^{-1}\right\| \frac{\|\eta(\varphi(y))\|}{\|\varphi(y) - \varphi(y_0)\|} \\ &= 2\left\|[\mathrm{D}f(x_0)]^{-1}\right\| \frac{\|\eta(x)\|}{\|x - x_0\|}. \end{aligned}$$

If $y \to y_0$ then, by the continuity of φ, $x \to x_0$ and hence $\|\eta(x)\|/\|x-x_0\| \to 0$ by the differentiability of f at x_0.

It follows that φ is differentiable on V_0 with $\mathrm{D}\varphi(y) = [\mathrm{D}f(\varphi(y))]^{-1}$ for each $y \in V_0$. This is continuous in y since $\mathrm{D}\varphi : V_0 \to \mathcal{L}(F; E)$ is a composition of three continuous operations (first φ, then $\mathrm{D}f$, then taking the inverse); see Exercise 6.10. □

The finite-dimensionality of E and F has not been used in an essential way. That is to say, if we make the boundedness of the derivative a requirement for differentiablity, and require that all involved normed spaces are complete, then everything carries over to infinite dimensions. See [**Lan93**, XIII & XIV] or [**Che01**, Chapter 3].

By using the inverse mapping theorem one obtains a quick proof (cf. Exercise 6.11) of the so-called **implicit function theorem**. This result states that if E, F, G are (finite-dimensional) Banach spaces, U and V are open subsets of E and F, respectively, $(x_0, y_0) \in U \times V$ and

$$f : U \times V \to G$$

is a C^1-map such that $f(x_0, y_0) = 0$ and the partial derivative $\mathrm{D}_2 f(x_0, y_0) : F \to G$ is invertible, then one can "solve for x" in the equation

$$f(x, y) = 0$$

locally around x_0, i.e., there is a C^1-map g defined on a open neighborhood U_0 of x_0 with $g(x_0) = y_0$ and such that

$$f(x, g(x)) = 0 \quad \text{for all } x \in U_0.$$

This theorem is the basis for the important theorem on *Lagrange multipliers*[5]; see [**Che01**, Section 3.5].

Exercises 6A

Exercise 6.1. In the setting of Theorem 6.1 show that

$$d(x_{n+1}, z) \le q d(x_n, z) \quad \text{for all } n \in \mathbb{N}.$$

Hence in each step of the iterative procedure, the error decreases at least by a factor q.

Exercise 6.2. Let $\|\cdot\|$ be any norm on $\mathbb{K}^d$, and let $\gamma : [a, b] \to \mathbb{K}^d$ be a continuous function. Then the integral $\int_a^b \gamma(s)\, \mathrm{d}s$ is defined componentwise. Show that

$$\left\| \int_a^b \gamma(s)\, \mathrm{d}s \right\| \le \int_a^b \|\gamma(s)\| \, \mathrm{d}s,$$

approximating the integral by Riemann sums. For the special case of the p-norms $\|\cdot\|_p$ with $p = 1, 2, \infty$ also give a direct argument.

Exercise 6.3. Let E be any normed vector space. Show that $C([a, b]; E)$, the space of E-valued continuous functions on $[a, b]$ is a normed space with respect to the pointwise operations and the supremum norm

$$\|f\|_\infty := \sup_{s \in [a,b]} \|f(s)\|_E \, .$$

Show that $C([a, b]; E)$ is complete if E is.

Exercises 6B

Exercise 6.4. Consider $\Omega := \mathbb{R}_+$ and $Tx := x + \frac{1}{x+1}$ for $x \ge 1$. Show that for each $x, y \ge 0$ one has $|Tx - Ty| < |x - y|$, but T does not have a fixed point.

Exercise 6.5. Prove the following generalization of Theorem 6.1: *Let (Ω, d) be a complete metric space, and let $T : \Omega \to \Omega$ be a continuous mapping such that for some $0 < q < 1$ and some $m \in \mathbb{N}$,*

$$d(T^m x, T^m y) \le q d(x, y) \qquad \textit{for all } x, y \in \Omega.$$

Then T has a unique fixed point.

Exercise 6.6. Let E be a Banach space, $r > 0$ and suppose that $f : B[0, r] \to E$ is such that there is $0 < q < 1$ such that the following properties hold:

1) $\|f(x) - f(y)\| \le q \|x - y\|$ for all $x, y \in B[0, r]$;
2) $\|f(0)\| \le r(1 - q)$.

[5] Joseph-Louis de Lagrange (1736–1813), Italian-French mathematician and physicist.

Show that there is a unique $x \in \mathrm{B}[0,r]$ such that $f(x) = x$.

Exercise 6.7. Consider the subset

$$F := \{f \in \mathrm{C}[0,1] \mid 0 \le f \le 1,\ f(0) = 0,\ f(1) = 1\}$$

of $\mathrm{C}[0,1]$, endowed with the metric coming from the supremum norm. Furthermore, let T be the operator on $\mathrm{C}[0,1]$ defined by

$$(Tf)(t) := tf(t) \qquad (t \in [0,1]).$$

Show that

a) F is bounded and closed, whence complete.

b) T maps F into F and satisfies

$$\|Tf - Tg\|_\infty < \|f - g\|_\infty \quad (f, g \in F,\ f \ne g).$$

c) T does not have a fixed point in F.

Exercise 6.8. Let (Ω, d) be a *compact* metric space, and let $T : \Omega \to \Omega$ be a mapping satisfying

$$d(Tx, Ty) < d(x, y) \quad \text{for all } x, y \in \Omega,\ x \ne y.$$

Show that T has a unique fixed point. [Hint: Consider the continuous(!) function $f(x) := d(x, Tx)$.]

Exercises 6C

Exercise 6.9. Let E, F be finite-dimensional normed spaces, and let $U \subseteq \mathbb{R} \times E$ be an open set. Suppose that $f : U \to F$ is a C^1-map. Show that *locally*, f satisfies a Lipschitz condition in the second argument (cf. Definition 6.2). More precisely, show that for each $(t_0, x_0) \in U$ there is an open set $V \subseteq U$ that contains (t_0, x_0) and such that $f|_V$ satisfies a Lipschitz condition in the second argument. [Hint: Mean value theorem]

Exercise 6.10. Show that the mapping $A \mapsto A^{-1}$ is continuous on the set of invertible $d \times d$-matrices. (See also Exercise 11.32.)

Exercise 6.11. Prove the *implicit function theorem* as stated at the end of Section 6.4; cf. [**Lan93**, XIV, §2].

Chapter 7

The Lebesgue Spaces

We have seen in Chapter 5 that the space $\mathrm{C}[a, b]$ is not complete with respect to the p-norm ($p = 1, 2$). By Appendix B, for each of these norms $\mathrm{C}[a, b]$ can be embedded isometrically onto a dense subspace of a certain Banach space, the respective "completion". However, such a completion is an abstract object, and the major challenge is to give a representation of it as a *function space* over $[a, b]$. In order to achieve this, we have to employ the theory of measure and integration, rooting in the fundamental works of Lebesgue[1] from the beginning of the twentieth century.

If Lebesgue theory is new to you, please read first the "Advice for the Reader" on page 125 below.

Evidently, the 1-norm and the 2-norm both use the notion of *integral* of a continuous function, and this integral is to be understood in the Riemann[2] sense. To construct a "completion" of $\mathrm{C}[a, b]$ we (at least) must assign to every $\|\cdot\|_1$-Cauchy sequence in $\mathrm{C}[a, b]$ a natural limit function. In the example used in the proof of Theorem 5.8 it is not hard to see what this limit would be in this case, namely the function f that is 0 on $[-1, 0]$ and 1 in $(0, 1]$. So we are still in the domain of Riemann integrable functions and we could try to consider the space

$$\mathrm{R}[a, b] := \{f : [a, b] \longrightarrow \mathbb{K} \mid f \text{ is Riemann-integrable}\}$$

endowed with the "norm" $\|f\|_1 = \int_a^b |f(x)| \, \mathrm{d}x$.

[1] Henri Lebesgue (1875–1941), French mathematician.

[2] Georg Friedrich Bernhard Riemann (1826–1866), German mathematician.

A first difficulty arises here: the "norm" is actually *not a norm*, since there are many nonzero positive Riemann-integrable functions with zero integral. Below we shall encounter a very elegant mathematical method which shall allow us to ignore this fact most of the time.

The more urgent problem is that the space $\mathrm{R}[a, b]$ is still not complete with respect to $\|\cdot\|_1$, a fact that is by no means obvious. (The proof uses so-called generalized Cantor sets.) So we are forced to go beyond Riemann-integrable functions, and this means that a new notion of integral has to be found that is defined on a larger class of functions and has better properties. This is the so-called **Lebesgue integral**.

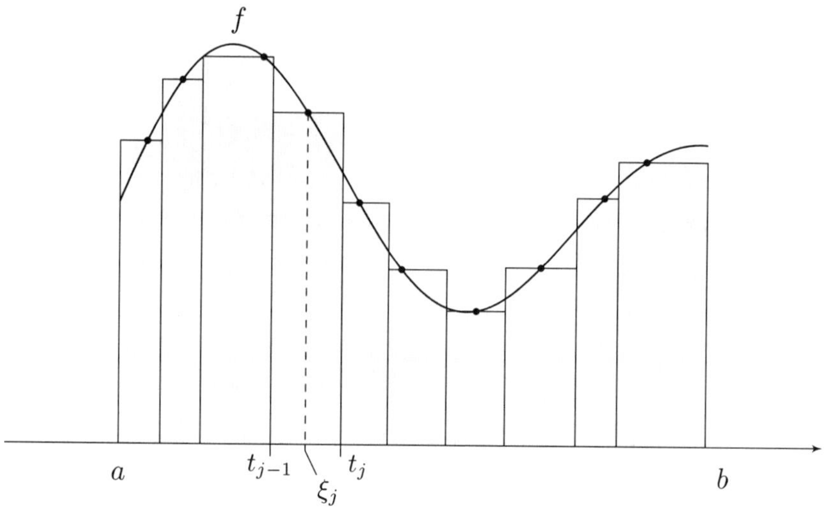

Figure 9. Riemann's approximation of a function and its integral.

To understand the major shift in approach from Riemann to Lebesgue, recall the use of "Riemann sums"

$$\sum_{j=1}^{n} f(\xi_j)(t_j - t_{j-1}) \quad \to \quad \int_a^b f(x)\, \mathrm{d}x$$

as approximations of the integral. The quantity $(t_j - t_{j-1})$ is simply the **length** $|A_j|$ of the interval $A_j := (t_{j-1}, t_j]$. Underlying this approximation of the integral is an approximation (in mean) of *functions*, namely

$$\sum_{j=1}^{n} f(\xi_j)\mathbf{1}_{A_j}(x) \quad \to \quad f(x) \qquad (x \in [a, b]).$$

Here we use the notation $\mathbf{1}_A$ to denote the **characteristic function**

$$\mathbf{1}_A(x) := \begin{cases} 1 & \text{if } x \in A, \\ 0 & \text{if } x \notin A \end{cases}$$

of the set A. So in the Riemann approach, the *domain* of the function is partitioned into intervals and this partion is then used for an approximation of f and its integral; see Figure 9.

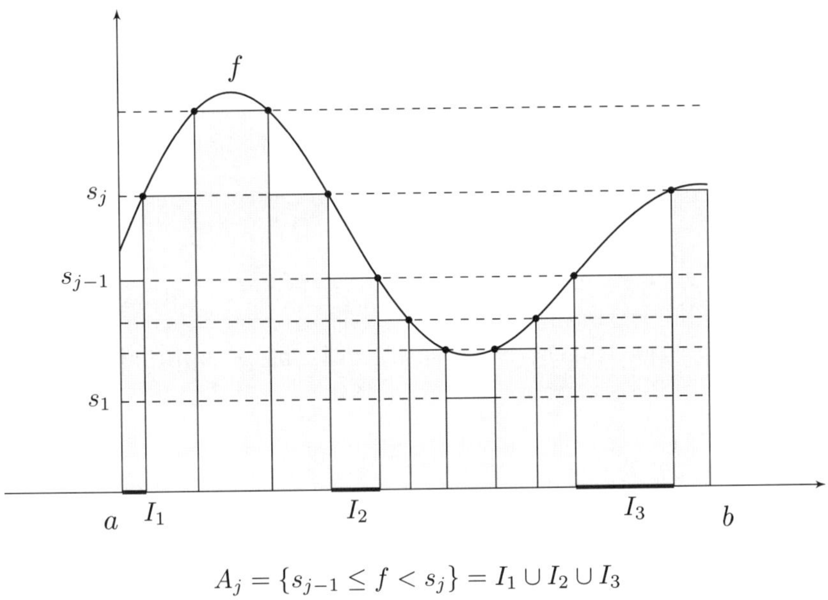

Figure 10. Lebesgue's approximation of a function and its integral.

In Lebesgue's approach, however, it is the *range* of the function that is partitioned into intervals. Suppose for simplicity that we are dealing with a positive function f on $[a, b]$. Then the range of f is contained in $[0, \infty)$. Every partition of $[0, \infty)$,

$$0 = s_0 < s_1 < \cdots < s_n = \infty$$

induces a partition of the domain $[a, b]$ of f into the sets

$$A_j = \{s_{j-1} \leq f < s_j\} := f^{-1}[s_{j-1}, s_j) \quad (j = 1, \ldots, n).$$

This partition is used to approximate f from below. The function

$$g := s_0 \mathbf{1}_{A_1} + s_1 \mathbf{1}_{A_2} + \cdots + s_{n-1} \mathbf{1}_{A_n} = \sum_{j=1}^{n} s_{j-1} \mathbf{1}_{A_j}$$

satisfies $0 \le g \le f$ and is a "step function" in that it is constant (to s_{j-1}) on each set A_j. The difference $f - g$ is at most $s_j - s_{j-1}$ on A_j, hence as small as one wishes. (Suppose for simplicity that f is a bounded function).

Now, *if* all the sets A_j are intervals or — as in Figure 10 — disjoint unions of finitely many intervals, then we can integrate g and obtain

$$\int_a^b \sum_{j=1}^n s_{j-1}\mathbf{1}_{A_j}(x)\,\mathrm{d}x = \sum_{j=1}^n s_{j-1}\,|A_j|$$

as an approximation of $\int_a^b f(x)\,\mathrm{d}x$. Here $|A_j|$ is the sum of the lengths of the intervals whose disjoint union is A_j. In general, however, the sets A_j may not be as nice, and so we are led to search for a natural extension of the length (= 1-dimensional volume) of an interval to more general sets. This is the notion of the *Lebesgue measure* of the set.

7.1. The Lebesgue Measure

The basic idea of measuring a set $A \subseteq \mathbb{R}$ is to *cover* it with intervals, and take the sum of the lengths of these intervals as an approximation of the measure of A. Here is the precise definition.

Definition 7.1. The **Lebesgue outer measure** of a set $A \subseteq \mathbb{R}$ is

$$\lambda^*(A) := \inf \sum\nolimits_{n=1}^{\infty} |Q_n|$$

where the infimum is taken over all sequences of intervals $(Q_n)_{n\in\mathbb{N}}$ such that $A \subseteq \bigcup_{n\in\mathbb{N}} Q_n$. (Such a sequence is called a **cover** of A.)

The new feature here is that *infinite* covers are used. In Riemann's theory and the volume theories based on it the approximations use only *finite* covers or sums.

It is true — but not at all trivial to prove — that $\lambda^*(Q) = |Q|$ if Q is an interval. So the set function λ^* extends the notion of 'length' from intervals to the whole power set of $\mathbb{R}$,

$$\lambda^* : \mathcal{P}(\mathbb{R}) \longrightarrow [0, \infty].$$

Unfortunately, its properties are not quite as good as one wishes. For example, as a notion of (1-dimensional) volume, we would like λ^* to be **finitely additive**, i.e.,

$$A \cap B = \emptyset \quad \Longrightarrow \quad \lambda^*(A \cup B) = \lambda^*(A) + \lambda^*(B); \tag{7.1}$$

it turns out that λ^* *fails* this if one allows arbitrary sets A, B here. However, by a famous result of Caratheodory[3] there is a quite rich class of subsets of $\mathbb{R}$ on which λ^* behaves as it should.

Theorem 7.2. *There is a set $\Sigma \subseteq \mathcal{P}(\mathbb{R})$ of subsets of $\mathbb{R}$ such that the following statements hold:*

a) *Every interval is contained in Σ.*

b) *The set system Σ satisfies* Ex.7.1

$$\emptyset \in \Sigma; \quad A \in \Sigma \Rightarrow A^c \in \Sigma; \quad A_n \in \Sigma \ (n \in \mathbb{N}) \Rightarrow \bigcup_{n \in \mathbb{N}} A_n \in \Sigma.$$

c) *The outer measure λ^* satisfies*

$$A_n \in \Sigma \ (n \in \mathbb{N}), \ A_n \cap A_m = \emptyset \ (n \neq m) \implies \lambda^*\Big(\bigcup_{n=1}^{\infty} A_n\Big) = \sum_{n=1}^{\infty} \lambda^*(A_n).$$

d) *If $\lambda^*(A) = 0$, then $A \in \Sigma$.*

In measure theoretic terminology, b) says that Σ is a **σ-algebra**, and c) says that the restriction of the Lebesgue outer measure to Σ is **countably additive**, i.e., a **measure**.

An element A of the set Σ from Theorem 7.2 is called **Lebesgue measurable**, and the restriction $\lambda := \lambda^*|_\Sigma$ of the Lebesgue outer measure to Σ is called the **Lebesgue measure**. It is very important to keep in mind that $\Sigma \neq \mathcal{P}(\mathbb{R})$ is *not* the whole power set, i.e., there exist subsets of $\mathbb{R}$ that are not Lebesgue measurable.

The difference between the outer measure λ^* and the Lebesgue measure λ lies only in the domain of definition. Every subset $A \subseteq \mathbb{R}$ has an outer measure $\lambda^*(A)$, but only if $A \in \Sigma$ we may write $\lambda(A)$ for this number.

Lemma 7.3. *Every open and every closed subset of $\mathbb{R}$ is Lebesgue measurable.*

Proof. Let $O \subseteq \mathbb{R}$ be open. For each $x \in O$ there are rational numbers $a_x, b_x \in \mathbb{Q}$ such that $x \in (a_x, b_x) \subseteq O$. So we can write $O = \bigcup_{x \in O}(a_x, b_x)$. But there are only countably many pairs of rational numbers, so the union is effectively one of countably many intervals. Since each interval is measurable, Theorem 7.2.b) ensures that O is measurable, too. $\square$ Ex.7.2

Based on the notion of a measurable set, we define measurable functions.

[3]Constantin Carathéodory (1873–1950), German mathematician of Greek origin.

Definition 7.4. Let X be any interval. A function $f : X \to [-\infty, \infty]$ is called **(Lebesgue) measurable** if

$$\{a \le f < b\} \in \Sigma$$

for all $a, b \in \mathbb{R}$. A complex-valued function f is measurable if $\operatorname{Re} f$ and $\operatorname{Im} f$ are both measurable.

Ex.7.3 **Lemma 7.5.** *If $A \in \Sigma$, then its characteristic function $\mathbf{1}_A$ is measurable. Every continuous function is measurable.*

Proof. The first assertion is Exercise 7.3. For the second, let $f : X \to \mathbb{R}$ be continuous and $a, b \in \mathbb{R}$. Then

$$\{a \le f < b\} = \bigcap\nolimits_{n \in \mathbb{N}} \{a - \tfrac{1}{n} < f < b\}$$

and each set $\{a - 1/n < f < b\} = f^{-1}(a - 1/n, b)$ is open. Hence $\{a \le f < b\} \in \Sigma$ by Lemma 7.3 and Exercise 7.1. □

One can even prove that *all Riemann-integrable functions are measurable.* The class of measurable functions has nice closure properties. As a rule of thumb, all pointwise defined operations involving at most countably many measurable functions produce again measurable functions. Instead of making this principle precise, we give some examples.

Lemma 7.6. *Let $f, g, f_n : X \to [-\infty, \infty]$ be measurable and $\lambda \in \mathbb{R}$. Then fg, λf, $\inf_{n \in \mathbb{N}} f_n$ and $\sup_{n \in \mathbb{N}} f_n$ are measurable. If $f_n(x) \to h(x)$ as $n \to \infty$ for every $x \in X$, then h is also measurable. If $f(x) + g(x)$ is defined for*
Ex.7.4 *every $x \in X$, then $f + g$ is measurable.*

Since we are dealing with functions that can assume the values $\pm\infty$, we need conventions for computing with these values. The most important conventions are that $0 \cdot \pm\infty = \pm\infty \cdot 0 = 0$ and the expressions $\infty - \infty$, $\infty + (-\infty)$ as well as $(-\infty) + \infty$ are not defined.

As a consequence, we obtain that

$$\mathcal{M}(X) = \mathcal{M}(X; \mathbb{K}) := \{f : X \longrightarrow \mathbb{K} \mid f \text{ measurable}\}$$

is a vector space with respect to the pointwise operations. Also, the set

$$\mathcal{M}_+(X) := \{f : X \longrightarrow [0, \infty] \mid f \text{ measurable}\}$$

is a *cone*, i.e., closed under addition and multiplication with positive scalars.

7.2. The Lebesgue Integral and the Space $\mathrm{L}^1(X)$

Following the original idea of Lebesgue one can define an integral

$$\mathcal{M}_+(X) \longrightarrow [0,\infty] \qquad f \longmapsto \int_X f \,\mathrm{d}\lambda$$

on the cone $\mathcal{M}_+(X)$ of positive measurable functions in such a way that $\int_X \mathbf{1}_A \,\mathrm{d}\lambda = \lambda(A)$ for every $A \in \Sigma$ and that the integral is **additive** and **positively-homogeneous**:

$$\int_X (f + \alpha g) \,\mathrm{d}\lambda = \int_X f \,\mathrm{d}\lambda + \alpha \int_X g \,\mathrm{d}\lambda \qquad (f, g \in \mathcal{M}_+(X),\ \alpha \ge 0). \tag{7.2}$$

Furthermore, the integral is **monotonic**:

$$f \le g \quad \Longrightarrow \quad \int_X f \,\mathrm{d}\lambda \le \int_X g \,\mathrm{d}\lambda, \tag{7.3}$$

and one has the so-called **monotone convergence theorem**:

If $f_n \in \mathcal{M}_+(X)$, $n \in \mathbb{N}$, and $f_1 \le f_2 \le \dots f_n \nearrow f$ pointwise, then

$$f \in \mathcal{M}_+(X) \quad \textit{and} \quad \int_X f \,\mathrm{d}\lambda = \lim_{n\to\infty} \int_X f_n \,\mathrm{d}\lambda.$$

Note that this integral is only defined for positive functions and it may take the value infinity. A (not necessarily positive) measurable function $f \in \mathcal{M}(X)$ is called **(Lebesgue) integrable** if

$$\|f\|_1 := \|f\|_{\mathrm{L}^1} := \int_X |f| \,\mathrm{d}\lambda < \infty.$$

We denote by

$$\mathcal{L}^1(X) := \{ f \in \mathcal{M}(X) \mid \|f\|_1 < \infty \}$$

the space of integrable functions.

Theorem 7.7. *The space $\mathcal{L}^1(X)$ is a vector space, and one has*

$$\|f + g\|_1 \le \|f\|_1 + \|g\|_1, \quad \|\alpha f\|_1 = |\alpha| \, \|f\|_1$$

for all $f, g \in \mathcal{L}^1(X), \alpha \in \mathbb{K}$.

Proof. This is a straightforward consequence of (7.2) and (7.3). □

For a real-valued $f \in \mathcal{L}^1(X)$ we define its integral by

$$\int_X f \,\mathrm{d}\lambda := \int_X f^+ \,\mathrm{d}\lambda - \int_X f^- \,\mathrm{d}\lambda \tag{7.4}$$

where

$$f^+ := \frac{1}{2}(|f| + f) \qquad \text{and} \qquad f^- := \frac{1}{2}(|f| - f)$$

are the **positive part** and the **negative part** of f, respectively. For a $\mathbb{C}$-valued f we define

$$\int_X f \,\mathrm{d}\lambda := \int_X \mathrm{Re}\, f \,\mathrm{d}\lambda + \mathrm{i} \int_X \mathrm{Im}\, f \,\mathrm{d}\lambda. \tag{7.5}$$

Then we arrive at the following.

Lemma 7.8. *Let $X \subseteq \mathbb{R}$ be an interval. The integral*

$$\mathcal{L}^1(X) \longrightarrow \mathbb{C}, \qquad f \longmapsto \int_X f \,\mathrm{d}\lambda$$

is a linear mapping satisfying

$$\left| \int_X f \,\mathrm{d}\lambda \right| \le \int_X |f| \,\mathrm{d}\lambda = \|f\|_1$$

for every $f \in \mathcal{L}^1(X)$.

Ex.7.5 **Proof.** Showing linearity is tedious but routine, involving (7.4) and (7.5). Let us prove the second statement. Find $c \in \mathbb{C}$ with $|c| = 1$ such that

$$c\Big(\int_X f \,\mathrm{d}\lambda \Big) = \Big| \int_X f \,\mathrm{d}\lambda \Big|.$$

Taking real parts we obtain

$$\Big| \int_X f \,\mathrm{d}\lambda \Big| = \mathrm{Re} \Big(c \int_X f \,\mathrm{d}\lambda \Big) = \int_X \mathrm{Re}(cf) \,\mathrm{d}\lambda \le \int_X |f| \,\mathrm{d}\lambda.$$

Here we used first linearity, then (7.5) and finally monotonicity. □

Connection with the Riemann Integral. The Lebesgue integral coincides with the Riemann integral for Riemann integrable functions defined on an interval $[a, b]$. (This fact is not obvious and its proof requires the dominated convergence theorem from below.) Therefore we may write

$$\int_a^b f(x) \,\mathrm{d}x$$

in place of $\int_{[a,b]} f \,\mathrm{d}\lambda$, and we shall usually do this.

Change of Variables. As for the Riemann integral there is a change-of-variables formula for the Lebesgue integral. For $A \subseteq \mathbb{R}$, $c \in \mathbb{R}$ and $\alpha > 0$ we define

$$-A := \{-x \mid x \in A\}, \quad c + A := \{c + x \mid x \in A\}, \quad \alpha A := \{\alpha x \mid x \in A\}.$$

Then it easily follows from the definition of the outer measure that

$$\lambda^*(A) = \lambda^*(-A) = \lambda^*(c + A) = \alpha^{-1} \lambda^*(\alpha A). \tag{7.6}$$

Furthermore, it can be shown that if A is measurable, then also $-A, c+A, \alpha A$ are measurable; it follows (how?) that if f is a measurable function, then the functions

$$x \longmapsto f(-x), \quad x \longmapsto f(c+x), \quad x \longmapsto f(\alpha x)$$

are again measurable (on the respective new domain interval). In combination with (7.6) this leads to the formulae

$$\int_a^b f(x)\,\mathrm{d}x = \int_{-b}^{-a} f(-x)\,\mathrm{d}x = \int_{a-c}^{b-c} f(x+c)\,\mathrm{d}x = \alpha \int_{a/\alpha}^{b/\alpha} f(\alpha x)\,\mathrm{d}x \tag{7.7}$$

for $-\infty \le a < b \le \infty$ and f a measurable function on (a,b) which is either positive or integrable. In particular, if $(a,b) = \mathbb{R}$, we have

$$\int_{\mathbb{R}} f(x)\,\mathrm{d}x = \int_{\mathbb{R}} f(-x)\,\mathrm{d}x = \int_{\mathbb{R}} f(c+x)\,\mathrm{d}x = \alpha \int_{\mathbb{R}} f(\alpha x)\,\mathrm{d}x. \tag{7.8}$$

Measurability Issues. It is a (sometimes annoying) fact that $\Sigma \neq \mathcal{P}(\mathbb{R})$, i.e., there exist nonmeasurable subsets; see [**Ran02**, §4.6]. Consequently, there also exist nonmeasurable functions, and hence measurability of functions is an issue in proofs and constructions.

For example, a function f is integrable if it is measurable and satisfies $\int |f|\,\mathrm{d}\lambda < \infty$. Note that the second condition does not make sense without the first, because the integral is only defined for measurable functions.

This is typical: often one needs to show measurability plus a certain estimate for integrals. However, for the sake of simplicity and since this is a book on functional analysis, we shall usually ignore the measurability issues and concentrate on the estimates.

7.3. Null Sets

We call a set $A \subseteq \mathbb{R}$ a (Lebesgue) **null set** if $\lambda^*(A) = 0$. By Theorem 7.2.d), each null set is Lebesgue measurable.

Example 7.9. If $x \in \mathbb{R}$ is a single point and $\epsilon > 0$, then setting

$$Q_1 := (x - \epsilon/2, x + \epsilon/2), \quad Q_n := \emptyset \quad (n \ge 2)$$

shows that $\{x\}$ is a null set.

To generate more null sets, one may use the following lemma.

Lemma 7.10. a) *Every subset of a null set is a null set.*

b) *If $(A_n)_{n\in\mathbb{N}}$ is a sequence of null sets, then $\bigcup_{n\in\mathbb{N}} A_n$ is a null set, too.*

Proof. The first assertion is trivial. To prove the second, fix $\epsilon > 0$ and find for A_k a cover $(Q_{kn})_{n\in\mathbb{N}}$ such that $\sum_{n=1}^{\infty} |Q_{kn}| < \epsilon/2^k$. Then

$$A = \bigcup_{k\in\mathbb{N}} A_k \subseteq \bigcup_{k,n\in\mathbb{N}} Q_{kn}$$

and $\sum_{k,n=1}^{\infty} |Q_{kn}| < \sum_{k=1}^{\infty} \epsilon/2^k = \epsilon$. Note that since $\mathbb{N}\times\mathbb{N}$ is countable we may arrange the Q_{nk} into a single sequence. □

Example 7.11. *Each countable subset of* $\mathbb{R}$ *is a null set.* Indeed, if $A = \{a_n \mid n \in \mathbb{N}\} \subseteq \mathbb{R}$ is countable, then $A = \bigcup_{n\in\mathbb{N}}\{a_n\}$ and as each $\{a_n\}$ is a null set (already seen), A is a null set, too.

It is tempting to believe that each null set is countable. However, this is far from true. A nice example of an uncountable null set is the so-called "Cantor middle thirds" set. It is constructed by removing from the interval $[0, 1]$ the open middle third interval $(1/3, 2/3)$, then doing the same for the two remaining intervals and proceed iteratively. What remains (i.e., the intersection of all the constructed sets) is clearly a null set. But it contains exactly those real numbers $r \in [0, 1]$ which can be written in a tryadic notation as

$$r = \sum_{j=1}^{\infty} d_j/3^j$$

with $d_j \in \{0, 2\}$ for all $j \in \mathbb{N}$, and there are uncountably many such numbers.

Definition 7.12. We say that a property P of points of an interval X holds **almost everywhere** (a.e.) if the set

$$\{x \in X \mid P(x) \text{ is not true}\}$$

is a null set. For example, if $f, g : X \to \mathbb{K}$ are two functions then we say that

$$f = g \quad \text{almost everywhere}$$

Ex.7.6 if the set $\{f \neq g\}$ is a null set. In this case we write $f \sim_\lambda g$.

Ex.7.7

Ex.7.8

Example 7.13. A sequence of functions $(f_n)_{n\in\mathbb{N}}$ on X converges to a function f almost everywhere if $f_n(x) \to f(x)$ except for x from a set of measure zero. For instance, if $f_n(x) := x^n$, $x \in [0, 1]$, then $f_n \to 0$ almost everywhere. (Note that this is false if we consider the f_n as functions on $\mathbb{R}$.)

The role of null sets in integration theory stems from the following fact.

Lemma 7.14. *Let $f \in \mathcal{M}_+(X)$ such that $\int_X f \, \mathrm{d}\lambda < \infty$. Then $f(x) < \infty$ for almost every $x \in X$. Furthermore,*

$$\int_X f \, \mathrm{d}\lambda = 0 \quad \Longleftrightarrow \quad f = 0 \text{ almost everywhere.}$$

Proof. Let $A = \{f = \infty\}$. Then, for each $\alpha > 0$, $\alpha \mathbf{1}_A \leq f$, and hence

$$\alpha\lambda(A) = \int_X \alpha \mathbf{1}_A \, \mathrm{d}\lambda \leq \int_X f \, \mathrm{d}\lambda$$

by monotonicity of the integral. Hence $\int_X f \, \mathrm{d}\lambda < \infty$ only if $\lambda(A) = 0$.

Suppose that $\int_X f \, \mathrm{d}\lambda = 0$, and let $n \in \mathbb{N}$. Then $\frac{1}{n}\mathbf{1}_{\{f \geq 1/n\}} \leq f$ and hence

$$\tfrac{1}{n}\lambda\{f \geq 1/n\} = \int_X \tfrac{1}{n}\mathbf{1}_{\{f \geq 1/n\}} \, \mathrm{d}\lambda \leq \int_X f \, \mathrm{d}\lambda = 0.$$

Hence $\{f \geq 1/n\}$ is a null set, and so is $\{f > 0\} = \bigcup_{n \in \mathbb{N}} \{f \geq 1/n\}$.

For the converse implication, we suppose that $f = 0$ almost everywhere. Let $f_n(x) := \min\{f(x), n\}$ for $x \in X$ and $n \in \mathbb{N}$. Then $f_n \leq n\mathbf{1}_{\{f \neq 0\}}$ and hence, by monotonicity,

$$\int_X f_n \, \mathrm{d}\lambda \leq \int_X n\mathbf{1}_{\{f \neq 0\}} = n\lambda\{f \neq 0\} = 0.$$

Hence $\int_X f \, \mathrm{d}\lambda = \lim_{n\to\infty} \int_X f_n \, \mathrm{d}\lambda = 0$ by the monotone convergence theorem. □

Lemma 7.15. *The relation $\sim_\lambda$ ("is equal almost everywhere to") is an equivalence relation on $\mathcal{F}(X)$. Moreover, the following statements hold.*

a) *If $f = \tilde{f}$ a.e. and $g = \tilde{g}$ a.e., then*

$$|f| = |\tilde{f}|, \quad \lambda f = \lambda \tilde{f}, \quad f + g = \tilde{f} + \tilde{g}, \quad fg = \tilde{f}\tilde{g}$$

almost everywhere.

b) *If $f_n = g_n$ almost everywhere, for all $n \in \mathbb{N}$ and if $\lim_n f_n = f$ a.e. and $\lim_n g_n = g$ a.e., then $f = g$ almost everywhere.*

Proof. Obviously one has $f \sim_\lambda f$ for every f since $\{f \neq f\} = \emptyset$ is a null set. *Symmetry* is trivial, so let us show *transitivity*. Suppose that $f = g$ almost everywhere and $g = h$ almost everywhere. Now $\{f \neq h\} \subseteq \{f \neq g\} \cup \{g \neq h\}$, and so this is a null set by Lemma 7.10.

The proof of a) is left as exercise, we prove b). Set $A_n := \{f_n \neq g_n\}$ and

$$A = \{x \mid f_n(x) \not\to f(x)\}, \quad B = \{x \mid g_n(x) \not\to g(x)\}.$$

Then $\{f \neq g\} \subseteq A \cup B \cup \bigcup_n A_n$, because if $x \notin A_n$ for each n and $x \notin A$ and $x \notin B$, then $f_n(x) = g_n(x)$ and $f_n(x) \to f(x)$ and $g_n(x) \to g(x)$. But

that implies that $f(x) = g(x)$. Now, by Lemma 7.10 the set $\{f \neq g\}$ must
Ex.7.9 be a null set. □

Recall that $\mathcal{L}^1(X)$ misses a decisive property of a norm: definiteness. Indeed, by Lemma 7.14,

$$\|f-g\|_1 = 0 \quad \Leftrightarrow \quad |f-g| = 0 \quad \text{a.e.} \quad \Leftrightarrow \quad f = g \quad \text{a.e.}$$

To remedy this defect, we pass to *equivalence classes* modulo equality almost everywhere and define

$$\mathrm{L}^1(X) := \mathcal{L}^1(X)/\sim_\lambda.$$

Another way to view this is as a factor space of $\mathcal{L}^1(X)$ with respect to the linear subspace(!) $\mathcal{N} := \{f \in \mathcal{L}^1(X) \mid \|f\|_1 = 0\}$. The computations (addition, scalar multiplication, taking $\|\cdot\|_1$, taking the integral) are done by using *representatives* for the classes. Of course one must show that these
Ex.7.10 operations are well-defined.

We shall write $f \in \mathrm{L}^1(X)$ and work with f as if it was a function. This turns out to be very convenient, and after some time one tends to forget about the fact that these objects *are* not really functions but are only *represented* by functions.

The most annoying consequence of this is that for $f \in \mathrm{L}^1(X)$ the "value" $f(x_0)$ of f at a point $x_0 \in \mathbb{R}$ is meaningless! Indeed, if we alter f on the single point x_0 then we remain still in the same equivalence class, since $\{x_0\}$ is a null set.

Because finite sets are null sets, in the definition of $\mathrm{L}^1(X)$ it is inessential whether one starts with closed or open intervals. For instance,

$$\mathrm{L}^1[a,b] = \mathrm{L}^1(a,b) \quad \text{and} \quad \mathrm{L}^1[0,\infty) = \mathrm{L}^1(0,\infty).$$

7.4. The Dominated Convergence Theorem

The advantage of the Lebesgue integral lies in its flexibility and especially its convergence results. The following is the most important instance.

Theorem 7.16 (Dominated Convergence Theorem). *Let $(f_n)_{n\in\mathbb{N}}$ be a sequence in $\mathrm{L}^1(X)$ such that $f := \lim_{n\to\infty} f_n$ exists pointwise almost everywhere. If there is $0 \le g \in \mathrm{L}^1(X)$ such that $|f_n| \le g$ almost everywhere, for each $n \in \mathbb{N}$, then $f \in \mathrm{L}^1(X)$, $\|f_n - f\|_1 \to 0$ and*

$$\int_X f_n \,\mathrm{d}\lambda \to \int_X f \,\mathrm{d}\lambda.$$

Proof. Note that the *function* f here is defined only almost everywhere. But as such it determines a unique equivalence class modulo equality almost everywhere. It is actually easy to see that $f \in \mathrm{L}^1(X)$: since $f_n \to f$ almost everywhere and $|f_n| \le g$ almost everywhere, for every $n \in \mathbb{N}$, by "throwing away" countably many null sets we see that $|f| \le g$ almost everywhere, and hence

$$\int_X |f| \,\mathrm{d}\lambda \le \int_X g \,\mathrm{d}\lambda < \infty$$

since $g \in \mathrm{L}^1(X)$. So, indeed, $f \in \mathrm{L}^1(X)$.

Second, if we know already that $\|f_n - f\|_1 \to 0$, then the convergence of the integrals is clear from

$$\left| \int_X f_n \,\mathrm{d}\lambda - \int_X f \,\mathrm{d}\lambda \right| = \left| \int_X f_n - f \,\mathrm{d}\lambda \right| \le \|f_n - f\|_1 \to 0$$

(Lemma 7.8). In other words, the integral is a bounded linear mapping from $\mathrm{L}^1(X)$ to $\mathbb{K}$.

So the real step in the dominated convergence theorem is the assertion that $\|f - f_n\|_1 \to 0$. A proof is in Exercise 7.28. □

The dominated convergence theorem has a vast number of applications. Examples are the continuity of the Laplace transform and the Fourier transform of an L^1-function; see Exercises 7.19 and 7.21. Here is a simple model of how this works.

Example 7.17 (Integration Operator). For $f \in \mathrm{L}^1(a,b)$ one defines the function Jf by

$$(Jf)(t) := \int_a^b \mathbf{1}_{[a,t]} f \,\mathrm{d}\lambda = \int_a^t f(x)\,\mathrm{d}x \qquad (t \in [a,b]).$$

Then Jf *is continuous*: indeed, if $t_n \to t$ in $[a,b]$, then $\mathbf{1}_{[a,t_n]} \to \mathbf{1}_{[a,t]}$ pointwise, except for the point t itself. So $\mathbf{1}_{[a,t_n]} f \to \mathbf{1}_{[a,t]} f$ almost everywhere, and since

$$\left|\mathbf{1}_{[a,t_n]} f\right| \le |f| \in \mathrm{L}^1(a,b)$$

one can apply dominated convergence to conclude that $Jf(t_n) \to Jf(t)$.

Hence $J : \mathrm{L}^1(a,b) \to \mathrm{C}[a,b]$ is a linear operator. It is also bounded, since

$$|Jf(t)| = \left| \int_a^t f(s)\,\mathrm{d}s \right| \le \int_a^t |f(s)|\,\mathrm{d}s \le \int_a^b |f(s)|\,\mathrm{d}s = \|f\|_1$$

for all $t \in [a,b]$. This yields $\|Jf\|_\infty \le \|f\|_1$ for all $f \in \mathrm{L}^1(a,b)$.

Using the dominated convergence theorem one can prove the completeness of $\mathrm{L}^1(X)$.

Theorem 7.18 (Completeness of L^1). *The space $L^1(X)$ is a Banach space. More precisely, let $(f_n)_{n\in\mathbb{N}}$ be a Cauchy sequence in $L^1(X)$. Then there are functions $f, g \in L^1(X)$ and a subsequence $(f_{n_k})_{k\in\mathbb{N}}$ such that*

$$|f_{n_k}| \le g \quad a.e. \quad and \quad f_{n_k} \to f \quad a.e..$$

Furthermore, $\|f_n - f\|_1 \to 0$.

Proof. Note first that if we have found f, g and the subsequence with the stated properties, then $\|f_{n_k} - f\|_1 \to 0$ by dominated convergence, and hence $\|f_n - f\|_1 \to 0$ since the sequence $(f_n)_{n\in\mathbb{N}}$ is $\|\cdot\|_1$-Cauchy.

We find the subsequence in the following way. By using the Cauchy property we may pick $n_k < n_{k+1}$, $k \in \mathbb{N}$, such that $\|f_{n_k} - f_{n_{k+1}}\|_1 < \frac{1}{2^k}$. To facilitate notation let $g_k := f_{n_k}$. Then for every $N \in \mathbb{N}$,

$$\int_X \sum_{k=0}^{N} |g_k - g_{k+1}| \, d\lambda = \sum_{k=0}^{N} \|g_k - g_{k+1}\|_1 \le \sum_{k=0}^{\infty} \frac{1}{2^k} = 2.$$

Define $G := \sum_{k=0}^{\infty} |g_k - g_{k+1}|$ pointwise. Then by the monotone convergence theorem

$$\int_X G \, d\lambda = \lim_{N\to\infty} \int_X \sum_{k=0}^{N} |g_k - g_{k+1}| \, d\lambda \le 2,$$

and hence $G \in L^1(X)$. By Lemma 7.14, $\sum_{k=0}^{\infty} |g_k - g_{k+1}| = G < \infty$ a.e. Hence for almost all $x \in X$ the limit

$$h(x) := \sum_{k=0}^{\infty} g_k(x) - g_{k+1}(x) = g_0(x) - \lim_{n\to\infty} g_{n+1}(x)$$

exists. Hence $g_k \to f := g_0 - h$ almost everywhere, and

$$|g_k| \le |g_0| + \sum_{j=0}^{k-1} |g_j - g_{j+1}| \le |g_0| + G \quad \text{a.e.}$$

Ex.7.11 Thus, if we set $g := |g_0| + G$, the theorem is completely proved. □

We have already discussed the relation between pointwise convergence and convergence in various norms in Section 3.3. In general, pointwise convergence does not imply convergence in $\|\cdot\|_1$; cf. also Exercise 7.11.

Conversely, $\|\cdot\|_1$-convergence does not imply pointwise convergence, not even convergence almost everywhere. Indeed, consider the sequence $(f_k)_{k\in\mathbb{N}}$ given by

$$\mathbf{1}_{[0,1/2]}, \ \mathbf{1}_{[1/2,1]}, \ \mathbf{1}_{[0,1/3]}, \ \mathbf{1}_{[2/3,1]}, \ \mathbf{1}_{[0,1/4]}, \ \ldots$$

This sequence $\|\cdot\|_1$-converges to zero since $\left\|\mathbf{1}_{[j/n, j+1/n]}\right\|_1 = 1/n$. On the other hand, for every $x \in [0,1]$, the sequence $(f_k(x))_{k\in\mathbb{N}}$ is a $\{0,1\}$-sequence with both values occurring infinitely often. Hence $(f_k(x))_{k\in\mathbb{N}}$ does not converge at any point in $[0,1]$.

This failure is remedied by Theorem 7.18 which tells us that one must at least have a *subsequence* that converges almost everywhere.

7.5. The Spaces $L^p(X)$ with $1 \le p \le \infty$

Apart from the space $L^1(X)$ of all (equivalence classes of) integrable functions, integration theory provides other important and useful Banach spaces. Most important is the space $L^2(X)$, because it is a Hilbert space.

The Space $L^2(X)$. A function $f \in \mathcal{M}(X)$ is called **square integrable** if

$$\|f\|_2 := \|f\|_{L^2} := \left(\int_X |f|^2 \, d\lambda\right)^{1/2} < \infty.$$

One defines

$$\mathcal{L}^2(X) := \{f \in \mathcal{M}(X) \mid \|f\|_2 < \infty\}, \qquad L^2(X) := \mathcal{L}^2(X)/\sim_\lambda.$$

Note the following elementary inequalities for real numbers $a, b \ge 0$:

$$(a+b)^2 \le 2a^2 + 2b^2, \tag{7.9}$$

$$ab \le 1/2\, a^2 + 1/2\, b^2. \tag{7.10}$$

Given $f, g \in L^2(X)$ we hence obtain for all $x \in X$

$$|f(x) + g(x)|^2 \le (|f(x)| + |g(x)|)^2 \le 2\,|f(x)|^2 + 2\,|g(x)|^2$$

and integrating this shows that $f + g \in L^2(X)$ again. Since obviously $0 \in L^2(X)$ and $\lambda f \in L^2(X)$ for each $\lambda \in \mathbb{C}$, we see that $L^2(X)$ is a vector space. Moreover, from (7.10) we obtain

$$\left|f(x)\overline{g(x)}\right| = |f(x)|\,|g(x)| \le 1/2\,|f(x)|^2 + 1/2\,|g(x)|^2$$

for all $x \in X$. Integrating yields $f\overline{g} \in L^1(X)$ and hence we may define the **standard inner product** by

$$\langle f, g\rangle_2 := \langle f, g\rangle_{L^2} := \int_X f\,\overline{g}\, d\lambda.$$

It is routine to check the properties of an inner product; for the definiteness, note that $\langle f, f\rangle_2 = \|f\|_2 = 0$ implies that $|f|^2 = 0$ almost everywhere, whence $f = 0$ in $L^2(X)$.

From the general theory of inner product spaces, it now follows that $\|\cdot\|_2$ is a norm on $\mathrm{L}^2(X)$ and we have the Cauchy–Schwarz inequality

$$\left|\int_X f\overline{g}\,\mathrm{d}\lambda\right| \le \int_X |fg|\,\mathrm{d}\lambda \le \|f\|_2\,\|g\|_2\,.$$

Observe that $g \in \mathrm{L}^2(X)$ if and only if $\overline{g} \in \mathrm{L}^2(X)$. Replacing g by $\overline{g}$ in the Cauchy–Schwarz inequality yields the 2/2-case of **Hölder's inequality**

$$\left|\int_X fg\,\mathrm{d}\lambda\right| \le \|f\|_2\,\|g\|_2\,.$$

One can easily derive an L^2-version of the dominated convergence theorem from Theorem 7.16; cf. Exercise 7.22. Using this, one arrives at completeness of $\mathrm{L}^2(X)$ in a similar fashion as before.

Theorem 7.19 (Completeness of L^2). *The space $\mathrm{L}^2(X)$ is a Hilbert space. More precisely, let $(f_n)_{n\in\mathbb{N}}$ be a Cauchy sequence in $\mathrm{L}^2(X)$. Then there are functions $f, F \in \mathrm{L}^2(X)$ and a subsequence $(f_{n_k})_{k\in\mathbb{N}}$ such that*

$$|f_{n_k}| \le F \quad \text{a.e.} \quad \text{and} \quad f_{n_k} \to f \quad \text{a.e.}$$

Furthermore, $\|f_n - f\|_2 \to 0$.

Proof. We only sketch the proof and leave the details as Exercise 7.23. Find a subsequence $g_k = f_{n_k}$ such that $\|g_k - g_{k+1}\|_2 \le 2^{-k}$. Then estimate

$$\begin{aligned}\int_X \Big(\sum_{k=0}^N |g_k - g_{k+1}|\Big)^2 \mathrm{d}\lambda &= \Big\|\sum_{k=0}^N |g_k - g_{k+1}|\Big\|_2^2 \le \Big(\sum_{k=0}^N \|g_k - g_{k+1}\|_2\Big)^2 \\ &\le \Big(\sum_{k=0}^\infty \frac{1}{2^k}\Big)^2 = 2^2 = 4.\end{aligned}$$

Define $g := \sum_{k=0}^\infty |g_k - g_{k+1}|$. Then $g \in \mathrm{L}^2(X)$ by the monotone convergence theorem. The remaining part is as in the proof of Theorem 7.18. □

The space $\mathrm{L}^\infty(X)$. For a measurable function $f \in \mathcal{M}(X)$ we define its L^∞-norm by

$$\|f\|_{\mathrm{L}^\infty} := \inf\{c > 0 \mid |f| \le c \text{ almost everywhere}\}$$

(with $\|f\|_{\mathrm{L}^\infty} = \infty$ if the set on the right-hand side is empty). If $\|f\|_{\mathrm{L}^\infty} < \infty$ we call the function f **essentially bounded**. Thus, a function is essentially bounded if it is almost everywhere equal to a bounded function. We further define

$$\mathcal{L}^\infty(X) := \{f \in \mathcal{M}(X) \mid f \text{ is essentially bounded}\}$$

and $\mathrm{L}^\infty(X) := \mathcal{L}^\infty(X)/{\sim_\lambda}$.

Lemma 7.20. *One has* $|f| \le \|f\|_{L^\infty}$ *almost everywhere, and* $\|f\|_{L^\infty} = 0$ *if and only if* $f = 0$ *almost everywhere. The* L^∞*-norm turns* $L^\infty(X)$ *into a Banach space.*

Proof. This is not very difficult, and is left as Exercises 7.12 and 7.24. □ Ex.7.12

It is clear that each bounded function f is also essentially bounded, with $\|f\|_{L^\infty} \le \|f\|_\infty$. The function

$$f = \sum_{n=1}^{\infty} n\mathbf{1}_{\{n\}}$$

is not bounded, but essentially bounded. (In fact, $f = 0$ almost everywhere.) However, *if f is continuous, then* $\|f\|_\infty = \|f\|_{L^\infty}$. If $f \in L^\infty(X)$ and $g \in$ Ex.7.13
$L^1(X)$ we have $|fg| \le \|f\|_{L^\infty} |g|$ almost everywhere. It follows that $fg \in L^1(X)$. Integrating yields

$$\left| \int_X fg \,\mathrm{d}\lambda \right| \le \|f\|_{L^\infty} \|g\|_1 ,$$

which is the $1/\infty$-case of **Hölder's inequality**.

***The Spaces $L^p(X)$ for $1 \le p < \infty$.** In this (optional) section we generalize the construction of the spaces of integrable ($p = 1$) and square integrable ($p = 2$) functions towards all exponents $p \in [1, \infty)$.

Let $1 \le p < \infty$. A measurable function $f \in \mathcal{M}(X)$ is called **p-integrable** if

$$\|f\|_p := \|f\|_{L^p} := \left(\int_X |f|^p \,\mathrm{d}\lambda \right)^{1/p} < \infty.$$

One defines

$$\mathcal{L}^p(X) := \{f \in \mathcal{M}(X) \mid \|f\|_p < \infty\}, \qquad L^p(X) := \mathcal{L}^p(X)/\sim_\lambda.$$

In order to see that $L^p(X)$ is a vector space and $\|\cdot\|_p$ is a norm on it, we may employ Lemma 2.29 as in the proof of Theorem 2.28. This is left as Exercise 7.29.a).

As in the special cases $p = 1, 2$ there is a dominated convergence theorem for $L^p(X)$ (Exercise 7.29.b)). Also the completeness theorem for the space $L^p(X)$ is a direct generalization of the special cases $p = 1, 2$.

Theorem 7.21 (Completeness of L^p). *The space* $L^p(X)$ *is a Banach space. More precisely, let* $(f_n)_{n\in\mathbb{N}}$ *be a Cauchy sequence in* $L^p(X)$*. Then there are functions* $f, F \in L^p(X)$ *and a subsequence* $(f_{n_k})_{k\in\mathbb{N}}$ *such that*

$$|f_{n_k}| \le F \quad \text{a.e.} \quad \text{and} \quad f_{n_k} \to f \quad \text{a.e.}$$

Furthermore, $\|f_n - f\|_p \to 0$.

Proof. The proof is analogous to the case $p = 2$ (Theorem 7.21) and left as Exercise 7.29.c). □

We now treat one of the most important inequalities in analysis.

Theorem 7.22 (Hölder's Inequality). *Let q be the dual exponent defined by $1/p + 1/q = 1$. If $f \in L^p(X)$ and $g \in L^q(X)$, then $fg \in L^1(X)$ and*

$$\left| \int_X fg \, d\lambda \right| \le \|f\|_p \, \|g\|_q .$$

Proof. The case $p, q \in \{1, \infty\}$ has been treated above, so we shall suppose $1 < p, q < \infty$ in the following. The proof proceeds as the proof of Theorem 2.30. Recall from Lemma 2.31 the identity

$$ab = \inf_{t>0} \frac{t^p}{p} a^p + \frac{t^{-q}}{q} b^q$$

for real numbers $a, b \ge 0$. Inserting $a = |f(x)|, b = |g(x)|$ we obtain

$$|f(x)g(x)| \le \frac{t^p}{p} |f(x)|^p + \frac{t^{-q}}{q} |g(x)|^q$$

for all $t > 0$ and all $x \in X$. Integrating yields

$$\int_X |fg| \, d\lambda \le \frac{t^p}{p} \int_X |f|^p \, d\lambda + \frac{t^{-q}}{q} \int_X |g|^q \, d\lambda$$

for all $t > 0$. Taking the infimum over $t > 0$ again yields

$$\begin{aligned} \int_X |fg| \, d\lambda &\le \inf_{t>0} \left[\frac{t^p}{p} \int_X |f|^p \, d\lambda + \frac{t^{-q}}{q} \int_X |g|^q \, d\lambda \right] \\ &= \left(\int_X |f|^p \, d\lambda \right)^{1/p} \left(\int_X |g|^q \, d\lambda \right)^{1/q} = \|f\|_p \, \|g\|_q . \end{aligned}$$

This shows that $fg \in L^1(X)$ and concludes the proof. □

Finally, we note that multiplying an L^p-function with an L^∞-function yields again an L^p-function.

Lemma 7.23. *Let $1 \le p \le \infty$, $f \in L^\infty(X)$ and $g \in L^p(X)$. Then $fg \in L^p(X)$ and $\|fg\|_{L^p} \le \|f\|_{L^\infty} \|g\|_{L^p}$.*

Proof. We suppose first that $p < \infty$. The function fg is measurable and $|fg| = |f| \, |g| \le \|f\|_{L^\infty} |g|$ almost everywhere, by Lemma 7.20. Taking the p-th power and integrating yields

$$\|fg\|_p^p = \int_X |fg|^p \le \int_X \|f\|_{L^\infty}^p |g|^p = \|f\|_{L^\infty}^p \|g\|_p^p < \infty.$$

This proves the claim for the case $p < \infty$. The case $p = \infty$ are left as Exercise 7.14. □ Ex.7.14

In the remainder of the book we shall not make substantial use of the spaces $\mathrm{L}^p(X)$ for $p \neq 1, 2, \infty$. On the other hand, a restriction to these special cases in the formulation of lemmas or theorems would often look awkward or strange. Therefore we have chosen to allow for general exponents p when that seems natural, but you are encouraged to confine to $p \in \{1, 2, \infty\}$ whenever it pleases you.

Density. Finally, we return to our starting point, namely the question of a natural "completion" of $\mathrm{C}[a,b]$ with respect to $\|\cdot\|_1$ or $\|\cdot\|_2$. If $X = [a,b]$ is a finite interval, $1 < p < \infty$ and $1/p + 1/q = 1$, then one has

$$\mathrm{C}[a,b] \subseteq \mathrm{C}_b(a,b) \subseteq \mathrm{L}^\infty(a,b) \subseteq \mathrm{L}^p(a,b) \subseteq \mathrm{L}^1(a,b)$$

with

$$\begin{aligned} \|f\|_1 &\le (b-a)^{1/q} \, \|f\|_p && \text{for all } f \in \mathrm{L}^p(a,b), \\ \|f\|_p &\le (b-a)^{1/p} \, \|f\|_{\mathrm{L}^\infty} && \text{for all } f \in \mathrm{L}^\infty(a,b), \\ \|f\|_\infty &= \qquad \|f\|_{\mathrm{L}^\infty} && \text{for all } f \in \mathrm{C}_b(a,b). \end{aligned}$$

(The proof is an exercise.) The following result gives the desired answer to our question, but once again, we can do nothing but quote the result without being able to provide a proof here. Ex.7.15

Theorem 7.24. *The space* $\mathrm{C}[a,b]$ *is* $\|\cdot\|_p$*-dense in* $\mathrm{L}^p(a,b)$ *for* $1 \le p < \infty$.

Note: The space $\mathrm{C}_b(a,b)$ is *not* $\|\cdot\|_{\mathrm{L}^\infty}$-dense in $\mathrm{L}^\infty(a,b)$. Ex.7.16

Advice for the Reader

Sections 7.1–7.5 contain a survey of the basic definitions, constructions and results of Lebesgue integration theory. Most proofs are omitted since a thorough treatment is beyond the scope of this book. Details can be found in the relevant literature, e.g., [**Bar01, Bar95, Bau01, Sch05, Ran02**].

If you see Lebesgue integration for the first time now, you might feel a little overwhelmed. So it may console you that for the remainder of this book only a rather superficial understanding of Lebesgue measure (Section 7.1) and Lebesgue integral (Section 7.2) is really needed. From the other sections we shall need

- the definition of a null set, the notion "almost everywhere" and Lemma 7.14;
- the dominated convergence theorem (Theorem 7.16);
- the definition of the spaces $L^1(X)$ and $L^2(X)$ and their completeness (as a fact, i.e., without the proof);
- Hölder's inequality and Lemma 7.23 for the cases $p = 1, 2, \infty$;
- the density of $C[a, b]$ in $L^1(a, b)$ and $L^2(a, b)$ (Theorem 7.24).

One may always restrict to the cases $p = 1, 2, \infty$ when we speak of $L^p(X)$, the case $p = \infty$ playing only a marginal role.

Exercises 7A

Exercise 7.1. Suppose $\Sigma \subseteq \mathcal{P}(\mathbb{R})$ is as in Theorem 7.2.b). Show that if $(A_n)_{n\in\mathbb{N}}$ is a sequence in Σ, then $\bigcap_{n\in\mathbb{N}} A_n \in \Sigma$. Show that Σ is closed under the usual set-theoretic operations $\cup$ (union), $\cap$ (intersection) and $\setminus$ (set difference).

Exercise 7.2. Show that every closed set $A \subseteq \mathbb{R}$ is Lebesgue measurable.

Exercise 7.3. Let $A \in \Sigma$. Show that $\mathbf{1}_A$ is a measurable function.

Exercise 7.4. Use Lemma 7.6 to prove the following. If $f, g : X \to \mathbb{C}$ are measurable and $\lambda, \mu \in \mathbb{C}$, then $\alpha f + \mu g$ and $|f|$ is measurable. [Hint: $|z| = \sup_{t\in\mathbb{Q}} \operatorname{Re}(e^{it}z)$ for $z \in \mathbb{C}$.]

Exercise 7.5. Show that the integral, defined in (7.4) and (7.5), is a linear functional on $\mathcal{L}^1(X)$.

Exercise 7.6. Show that if $f : X \to \mathbb{K}$ is such that $f = 0$ a.e., then $f \in \mathcal{M}(X)$. Conclude that if $f \in \mathcal{M}(X)$ and $g = f$ a.e., then $g \in \mathcal{M}(X)$, too.

Exercise 7.7. Let $f, g \in C[a, b]$ and $f = g$ almost everywhere. Show that then $f = g$, i.e., $f(x) = g(x)$ for *all* $x \in X$.

Exercise 7.8. Show that A is a null set if and only if $\mathbf{1}_A = 0$ almost everywhere.

Exercise 7.9. Let X be an interval and let $f, g, \tilde{f}, \tilde{g}$ be functions on X. Show that $f = \tilde{f}$ a.e. and $g = \tilde{g}$ a.e., then

$$|f| = |\tilde{f}|, \quad \lambda f = \lambda \tilde{f}, \quad f + g = \tilde{f} + \tilde{g}, \quad fg = \tilde{f}\tilde{g}, \quad \operatorname{Re} f = \operatorname{Re} \tilde{f}$$

almost everywhere.

Exercise 7.10. Let $f, g \in \mathcal{L}^1(X)$ such that $f = g$ almost everywhere. Show that $\|f\|_1 = \|g\|_1$ and

$$\int_X f \, d\lambda = \int_X g \, d\lambda.$$

Exercise 7.11. Let $X = \mathbb{R}$, and let $f_n := \mathbf{1}_{[n,n+1]}$, $n \in \mathbb{N}$. Show that $f_n \to 0$ everywhere, but $(f_n)_n$ is not a $\|\cdot\|_1$-Cauchy sequence.

Exercise 7.12. For a measurable function $f \in \mathcal{M}(X)$ show that $|f| \le \|f\|_{L^\infty}$ almost everywhere. Use this to show that $\|f\|_{L^\infty} = 0$ if and only if $f = 0$ almost everywhere. Then show that $L^\infty(X)$ is a vector space and $\|\cdot\|_{L^\infty}$ is a norm on it.

Exercise 7.13. Show that $C_b(a,b) \subseteq L^\infty(a,b)$ and $\|f\|_\infty = \|f\|_{L^\infty}$ for each $f \in C_b(a,b)$.

Exercise 7.14. Complete the proof of Lemma 7.23.

Exercise 7.15. a) Let $[a,b] \subseteq \mathbb{R}$ be a finite interval, let $1 < p < \infty$ and $1/p + 1/q = 1$. Show that

$$L^\infty(a,b) \subseteq L^p(a,b) \subseteq L^1(a,b)$$

with $\|f\|_1 \le (b-a)^{1/q} \|f\|_p$ for all $f \in L^p(X)$ and $\|f\|_p \le (b-a)^{1/p} \|f\|_{L^\infty}$ for all $f \in L^\infty(a,b)$. Then show that these inclusions are proper.

b) Show that none of the spaces $L^\infty(\mathbb{R})$, $L^p(\mathbb{R})$ and $L^1(\mathbb{R})$ is included in any of the others.

[Restrict to the case $p = 2$ if you skipped the section on general L^p-spaces.]

Exercise 7.16. Show that the space $C_b(a,b)$ is not $\|\cdot\|_{L^\infty}$-dense in $L^\infty(a,b)$. [Hint: Exercise 7.13.]

Exercises 7B

Exercise 7.17. Let $\alpha \in \mathbb{R}$ and consider $f_n := n^\alpha \mathbf{1}_{[0,1/n]}$ for $n \in \mathbb{N}$. Compute $\|f_n\|_p$ for $p = 1, 2$. What is the a.e. behaviour of the sequence $(f_n)_{n\in\mathbb{N}}$?

Exercise 7.18. Let $(A_n)_{n\in\mathbb{N}}$ be an *increasing* sequence of measurable subsets of X and let $A := \bigcup_{n\in\mathbb{N}} A_n$. Show that if $f \in L^p(X)$ for $p = 1$ or $p = 2$, then $\mathbf{1}_{A_n} f \to \mathbf{1}_A f$ pointwise and in $\|\cdot\|_p$.

Exercise 7.19 (Laplace Transform[4]). For $f \in L^1(\mathbb{R}_+)$ define its **Laplace transform**

$$(\mathcal{L}f)(t) := \int_0^\infty e^{-ts} f(s)\,ds \qquad (t \ge 0).$$

Show that $\|\mathcal{L}f\|_\infty \le \|f\|_1$, that $\mathcal{L}f : \mathbb{R}_+ \to \mathbb{K}$ is continuous, and $\lim_{t\to\infty}(\mathcal{L}f)(t) = 0$.

Exercise 7.20 (Laplace Transform (2)). Similar to Exercise 7.19 we define the *Laplace transform* of $f \in L^2(\mathbb{R}_+)$

$$(\mathcal{L}f)(t) := \int_0^\infty e^{-ts} f(s)\,ds \qquad (t > 0).$$

(Note that $\mathcal{L}f(0)$ is not defined in general.) Show that $\mathcal{L}f : (0,\infty) \to \mathbb{K}$ is continuous with $\lim_{t\to\infty}(\mathcal{L}f)(t) = 0$. (Cf. Example 11.7.)

[4]Pierre-Simon Marquis de Laplace (1749–1827), French mathematician and politician.

Exercise 7.21 (Fourier Transform). We define the **Fourier transform** of $f \in \mathrm{L}^1(\mathbb{R})$

$$(\mathcal{F}f)(t) := \int_{-\infty}^{\infty} \mathrm{e}^{-its} f(s)\, \mathrm{d}s \qquad (t \in \mathbb{R}).$$

Show that $\|\mathcal{F}f\|_\infty \le \|f\|_1$ and that $\mathcal{F}f : \mathbb{R} \to \mathbb{K}$ is continuous.

Exercise 7.22. Formulate and prove a L^2-version of the dominated convergence theorem, e.g., by using Theorem 7.16.

Exercise 7.23. Complete the proof of Theorem 7.19.

Exercise 7.24. Show that if $(f_n)_{n\in\mathbb{N}} \subseteq \mathcal{L}^\infty(X)$ is a $\|\cdot\|_{\mathrm{L}^\infty}$-Cauchy sequence, then there is a null set $N \subseteq \mathbb{R}$ such that $(f_n)_n$ converges *uniformly* on $X \setminus N$. Conclude that $(\mathrm{L}^\infty(X), \|\cdot\|_{\mathrm{L}^\infty})$ is a Banach space.

Exercise 7.25. Let $X \subseteq \mathbb{R}$ be any interval and $m \in \mathrm{L}^\infty(X)$ and $p \in \{1, 2\}$. By Lemma 7.23 the **multiplication operator**

$$T : \mathrm{L}^p(X) \longrightarrow \mathrm{L}^p(X), \qquad Tf := m \cdot f$$

is bounded with operator norm $\|T\| \le \|m\|_{\mathrm{L}^\infty}$. Show that actually $\|T\| = \|m\|_{\mathrm{L}^\infty}$. [Hint: For $t < \|m\|_{\mathrm{L}^\infty}$ consider $A := \{|m| > t\}$. Show that $\lambda(A) > 0$, then consider $f = \mathbf{1}_A$.]

Exercise 7.26. Let $[a, b]$ be a bounded interval. Show that

$$\|f\|_{\mathrm{L}^2} \le \sqrt{\|f\|_{\mathrm{L}^1} \cdot \|f\|_{\mathrm{L}^\infty}} \qquad (f \in \mathrm{L}^2(a, b)).$$

Show that on the set $\Omega := \{f \in \mathrm{L}^\infty(a, b) \mid \|f\|_\infty \le 1\}$, i.e., the unit ball in $\mathrm{L}^\infty(a, b)$, the metrics associated with $\|\cdot\|_1$ and $\|\cdot\|_2$ are equivalent (cf. Exercise 4.26).

Exercises 7C

Exercise 7.27. Let $Q \subseteq \mathbb{R}$ be any interval. Show that that the outer measure of Q equals its length, i.e., $\lambda^*(Q) = |Q|$. [Hint: Only $|Q| \le \lambda^*(Q)$ is not obvious; reduce the problem to the case that Q is compact; then show that it suffices to use open intervals in the covers, and use the compactness to obtain a finite cover.]

Exercise 7.28 (Proof of the Dominated Convergence Theorem). Complete the proof of Theorem 7.16. [Hint: Define $g_n := \sup_{k\ge n} |f_k - f|$. Then $0 \le g_n \le 2g$ and $g_n \searrow 0$ pointwise. Apply the monotone convergence theorem to the sequence $2g - g_n$.)]

Exercise 7.29 (L^p-spaces for $1 \le p < \infty$). a) Show that $\mathrm{L}^p(X)$ is a vector space and $\|\cdot\|_p$ is a norm on it. [Hint: Use Lemma 2.29 as in the proof of Theorem 2.28.]

b) Establish a dominated convergence theorem for $\mathrm{L}^p(X)$; cf. Exercise 7.22.]

c) Show that $\mathrm{L}^p(X)$ is a Banach space with respect to the norm $\|\cdot\|_p$. [Hint: Go along the proof of Theorem 7.19.]

d) Show that if $f \in \mathrm{L}^1(X) \cap \mathrm{L}^\infty(X)$, then $f \in \mathrm{L}^p(X)$ with $\|f\|_p \le \|f\|_1^{1/p} \, \|f\|_\infty^{1-1/p}$.

Chapter 8

Hilbert Space Fundamentals

Hilbert spaces, i.e., inner product spaces that are complete with respect to the associated norm, have been introduced in Chapter 5. In this chapter we investigate their basic properties, which make them the favourite tool in the study of partial differential equations.

8.1. Best Approximations

The most fundamental and yet far-reaching property of Hilbert spaces has to do with a certain approximation problem, conveniently formulated in the framework of metric spaces.

Let (Ω, d) be a metric space, let $A \subseteq \Omega$ be a subset and $x \in \Omega$. Recall from Definition 3.15 that

$$d(x, A) = \inf\{d(x, y) \mid y \in A\}$$

is the **distance** of x to the set A. The function $d(\cdot, A)$ is continuous (Exercise 8.9). Any element $a \in A$ which *realizes* this distance, i.e., such that

$$d(x, A) = d(x, a),$$

is called a **best approximation** to x in A.

A special case of this concept occurs when A is a closed subset of a normed space and $x = 0$ is the zero vector. Then a best approximation is just an element of A with *minimal norm*. This concept has a very applied side: The actual state of a physical system is usually that with the smallest energy. In many cases, the energy is a certain norm (adapted to the problem one is considering) and so "minimal energy" becomes "minimal norm", hence a best approximation problem. See also Section 10.4 and Exercise 10.18.

A best approximation a to x in A is a *minimizer* of the function

$$(a \longmapsto d(x,a)) : A \longrightarrow \mathbb{R}_+.$$

In general situations such minimizers do not necessarily exist and when they exist, they need not be unique.

Example 8.1. By Lemma 3.16, $d(x,A) = 0$ if and only if $x \in \overline{A}$. Hence, if A is not closed, then to $x \in \overline{A} \setminus A$ there cannot be a best approximation in A: since $d(x,A) = 0$, a best approximation $a \in A$ would satisfy $d(x,a) = 0$, and hence $x = a \in A$, which is false by choice of x.

A special case of this is $A = c_{00}$ the space of finite sequences and $\Omega = \ell^2$ and $x = (1/n)_{n \in \mathbb{N}}$.

If the set A is closed and we are in a finite-dimensional setting, then a
Ex.8.1 best approximation always exists (Exercise 8.1). This is not true in infinite-dimensional situations.

Example 8.2 (Nonexistence). Let $E := \{f \in \mathrm{C}[0,1] \mid f(0) = 0\}$ with the supremum norm. (This is a Banach space!). Let

$$A := \Big\{f \in E \mid \int_0^1 f(t)\,\mathrm{d}t = 0\Big\}.$$

Then A is a closed subspace of E. Let $f(t) := t$, $t \in [0,1]$. Then $f \in E \setminus A$, since $f(0) = 0$ but $\int_0^1 f(t)\,\mathrm{d}t = 1/2 \neq 0$. Hence

$$\frac{1}{2} = \int_0^1 f(t) - g(t)\,\mathrm{d}t \le \int_0^1 |f(t) - g(t)|\,\mathrm{d}t \le \|f - g\|_\infty$$

for every $g \in A$. One can show that $d(f,A) = 1/2$ but there exists no best approximation to f in A. (Exercise 8.14)

So existence can fail. On the other hand, the following example shows that in some cases there are *several different* best approximations.

Example 8.3 (Nonuniqueness). Consider $E = \mathbb{R}^2$ with the norm $\|\cdot\|_1$ and $A := \mathbb{R}(1,-1)$ the straight line through the points $(-1,1), (0,0), (1,-1)$. If

$x := (1, 1)$, then $d(x, A) = 2$ and every $a = (\lambda, -\lambda)$ with $\lambda \in [-1, 1]$ is a distance minimizer since

$$\|(\lambda, -\lambda) - (1, 1)\|_1 = |1 - \lambda| + |1 + \lambda| = 2$$

for $-1 \leq \lambda \leq 1$. See Figure 11.

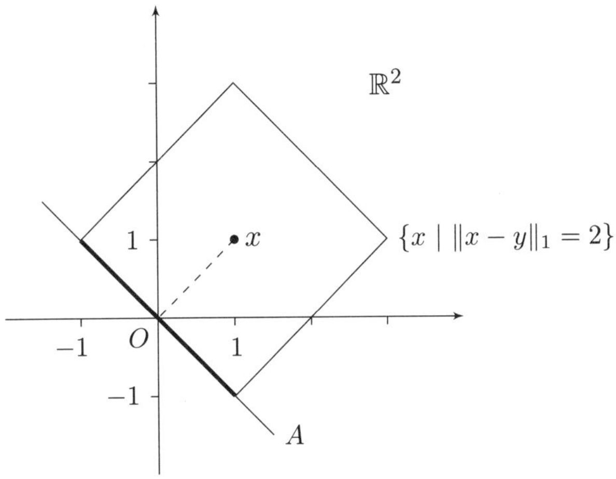

Figure 11. Nonuniqueness of best approximation.

We shall show that in Hilbert spaces, unique best approximations exist under a (relatively weak) condition for the set A, namely *convexity.*

Definition 8.4. A subset A of a normed vector space E is called **convex** if

$$f, g \in A,\ t \in [0, 1] \quad \Rightarrow \quad tf + (1 - t)g \in A.$$

Hence a set A is convex if it contains with any two points also the whole straight line segment joining them. We can now formulate and prove the main result of this chapter.

Theorem 8.5. *Let H be an inner product space, and let $A \neq \emptyset$ be a complete convex subset of H. Furthermore, let $f \in H$. Then there is a unique vector $P_A f := g \in A$ with $\|f - g\| = d(f, A)$.*

Proof. Let us define $d := d(f, A) = \inf\{\|f - g\| \mid g \in A\}$. For $g, h \in A$ we have $\frac{1}{2}(g + h) \in A$ as A is convex. If both $h, g \in A$ minimize $\|\cdot - f\|$ we see from Figure 12 that $g = h$. Algebraically, we use the parallelogram identity

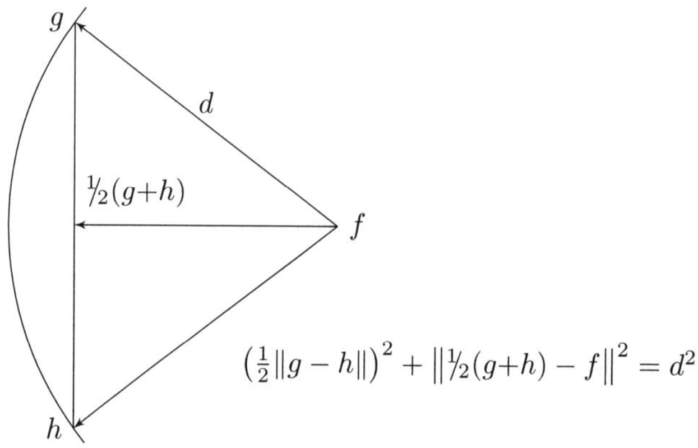

Figure 12. Uniqueness of best approximation in a convex subset.

and compute

$$\begin{aligned}\|g-h\|^2 &= \|(g-f)-(h-f)\|^2 \\ &= 2\|g-f\|^2 + 2\|h-f\|^2 - 4\left\|\tfrac{1}{2}(g+h)-f\right\|^2 \\ &\le 2\|g-f\|^2 + 2\|h-f\|^2 - 4d^2.\end{aligned}$$

Hence, if $\|g-f\|^2 = d^2 = \|h-f\|^2$ and we obtain

$$\|g-h\|^2 \le 2d^2 + 2d^2 - 4d^2 = 0.$$

To show existence, let $(g_n)_{n\in\mathbb{N}}$ be a minimizing sequence in A, i.e., $g_n \in A$ and $d_n := \|f - g_n\| \searrow d$. For $m \ge n$ we replace g, h by g_n, g_m in the estimation above and obtain

$$\|g_n - g_m\|^2 \le 2\|g_n - f\|^2 + 2\|g_m - f\|^2 - 4d^2 \le 4(d_n^2 - d^2).$$

Since $d_n \to d$, also $d_n^2 \to d^2$. Therefore, $(g_n)_{n\in\mathbb{N}}$ is a Cauchy sequence in A, and since A is complete, there is a limit $g := \lim_{n\to\infty} g_n \in A$. But the norm is continuous, and so

$$\|f-g\| = \lim_{n\to\infty} \|f-g_n\| = \lim_{n\to\infty} d_n = d,$$

and we have found our desired minimizer. □

Note that the proof shows actually that *every minimizing sequence* converges to the best approximation! The conditions of Theorem 8.5 are, in particular, satisfied if A is a closed convex subset of a Hilbert space. It is, in general, not easy to compute the best approximation explicitly. (See
Ex.8.2 Exercise 8.2 for an instructive example.)

8.2. Orthogonal Projections

Let H be a Hilbert space, and let $F \subseteq H$ be a closed linear subspace of H. Then F is, in particular, convex and complete, so for every $f \in H$ there exists the best approximation $P_F f \in F$ to f in F. Here is a second characterization of the vector $P_F f$.

Lemma 8.6. *Let H be a Hilbert space, let $F \subseteq H$ be a closed linear subspace of H, and let $f, g \in H$. Then the following assertions are equivalent:*

(i) $g = P_F f$.

(ii) $g \in F \quad \text{and} \quad f - g \perp F$.

Proof. (ii) $\Rightarrow$ (i): Take $h \in F$. Since $f - g \perp F$ and $g \in F$, one has $f - g \perp g - h$, and hence Pythagoras yields

$$\|f - h\|^2 = \|f - g\|^2 + \|g - h\|^2 \geq \|f - g\|^2 .$$

Taking square roots and the infimum over $h \in F$ yields $d(f, F) \geq d(f, g)$, and this shows that g is a best approximation to f in F, i.e., $g = P_F f$.
(i) $\Rightarrow$ (ii): Suppose $g = P_F f$. Then $\|f - g\|^2 \leq \|f - h\|^2$ for all $h \in F$. Since F is a linear subspace, we may replace h by $g - h$ in this inequality, i.e.,

$$\|f - g\|^2 \leq \|(f - g) + h\|^2 = \|f - g\|^2 + 2\,\mathrm{Re}\,\langle f - g, h\rangle + \|h\|^2$$

for all $h \in F$. Now we replace h by th with $t > 0$ and divide by t. We obtain

$$0 \leq 2\,\mathrm{Re}\,\langle f - g, h\rangle + t\,\|h\|^2 \qquad (h \in F).$$

Since this is true for all $t > 0$, we can let $t \searrow 0$ to get

$$0 \leq \mathrm{Re}\,\langle f - g, h\rangle \qquad (h \in F).$$

Finally, we can replace h by $-h$ to see that $\mathrm{Re}\,\langle f - g, h\rangle = 0$ for all $h \in F$. So if $\mathbb{K} = \mathbb{R}$ we are done; in the complex case we replace h by ih to finally obtain (ii). □

Lemma 8.6 facilitates the computation of best approximations. We now have a closer look at the mapping P_F if F is a closed subspace of a Hilbert space. Ex.8.3

Definition 8.7. If F is a closed subspace of a Hilbert space, then the mapping

$$P_F : H \longrightarrow F$$

is called the **orthogonal projection** onto F.

The following theorem summarizes the properties of the orthogonal projection.

Theorem 8.8. *Let F be a closed subspace of a Hilbert space H. Then the orthogonal projection P_F has the following properties:*

a) $P_F f \in F$ *and* $f - P_F f \perp F$ *for all* $f \in H$.

b) $P_F f \in F$ *and* $\|f - P_F f\| = d(f, F)$ *for all* $f \in H$.

c) $P_F : H \to H$ *is a bounded linear mapping satisfying* $(P_F)^2 = P_F$ *and*

$$\|P_F f\| \le \|f\| \qquad (f \in H).$$

In particular, either $F = \{0\}$ *or* $\|P_F\| = 1$.

d) $\operatorname{ran}(P_F) = F$ *and* $\ker(P_F) = F^\perp$.

e) $\mathrm{I} - P_F = P_{F^\perp}$, *the orthogonal projection onto* $F^\perp$.

Proof. b) is true by definition, a) by Lemma 8.6. By a), $\operatorname{ran}(P_F) \subseteq F$. Furthermore, if $f \in F$, then $d(f, F) = 0$, so f is the best approximation to f in F. This shows that $P_F f = f$ for $f \in F$; in particular, $F \subseteq \operatorname{ran}(P_F)$. Again by a), $\ker(P_F) \subseteq F^\perp$. On the other hand, if $f \perp F$ then $g := 0$ satisfies (ii) of Lemma 8.6, so $P_F f = 0$, i.e., $f \in \ker(P_F)$. Hence d) is proved.

To prove c), fix $f \in H$ and note first that $P_F f \in F$. But P_F acts as the identity on F, which means that $P_F^2 f = P_F(P_F f) = P_F f$. Since $f - P_F f \perp P_F f$, Pythagoras yields

$$\|f\|^2 = \|f - P_F f\|^2 + \|P_F f\|^2 \ge \|P_F f\|^2 .$$

This yields $\|P_F\| \le 1$. If $0 \ne f \in F$ one has $\|P_F f\| = \|f\| \ne 0$, whence $\|P_F\| = 1$. To show that P_F is a linear mapping, let $f, g \in H$ and $\alpha \in \mathbb{K}$. Let $h := P_F f + \alpha P_F g$. Then $h \in F$ and

$$(f + \alpha g) - h = (f - P_F f) + \alpha(g - P_F g) \perp F;$$

cf. Lemma 1.8. By Lemma 8.6 we obtain $h = P_F(f + \alpha g)$, and this is linearity.

Finally, to prove e) let $f \in H$ and $g := (\mathrm{I} - P_F) f$. Then $g = f - P_F f \in F^\perp$ and $f - g = P_F f \in F \perp F^\perp$. Hence, by Lemma 8.6 with F replaced by $F^\perp$,
Ex.8.4 $g = P_{F^\perp} f$. □

Remark 8.9. It follows from Theorem 8.8.a) that our new concept of orthogonal projection coincides with the one for finite-dimensional F, introduced in Chapter 1.

A prominent example of an orthogonal projection appears in probability theory. If $(\Omega, \Sigma, \mathrm{P})$ is a probability space and $\mathcal{F} \subseteq \Sigma$ is a sub-σ-algebra, then $\mathrm{L}^2(\Omega, \mathcal{F}, \mathrm{P})$ is in a natural way a closed subspace of $\mathrm{L}^2(\Omega, \Sigma, \mathrm{P})$. Then the orthogonal projection $P : \mathrm{L}^2(\Omega, \Sigma, \mathrm{P}) \to \mathrm{L}^2(\Omega, \mathcal{F}, \mathrm{P})$ is just the *conditional expectation* operator $\mathbb{E}(\,\cdot\,|\,\mathcal{F})$; see e.g. [**Bob05**, Chap. 3].

Corollary 8.10 (Orthogonal Decomposition). *Let H be a Hilbert space, and let $F \subseteq H$ be a closed linear subspace. Then every vector $f \in H$ can be written in a unique way as $f = u + v$ where $u \in F$ and $v \in F^\perp$.*

Proof. Uniqueness: if $f = u + v = u' + v'$ with $u, u' \in F$ and $v, v' \in F^\perp$, then

$$u - u' = v' - v \in F \cap F^\perp = \{0\}$$

by the definiteness of the scalar product. Hence $u = u', v = v'$ as claimed. Existence: Simply set $u = P_F f$ and $v = f - P_F f$. □

Employing terminology from linear algebra (see Appendix A.7), we may say that H is the *direct sum*

$$H = F \oplus F^\perp$$

of the subspaces $F, F^\perp$. Ex.8.5

Corollary 8.11. *Let F be a subspace of a Hilbert space H. Then $F^{\perp\perp} = \overline{F}$. Moreover,*

$$\overline{F} = H \quad \textit{if and only if} \quad F^\perp = \{0\}.$$

Proof. Suppose first that F is closed. The inclusion $F \subseteq F^{\perp\perp}$ is trivial. To prove the converse, let $f \in F^{\perp\perp}$ and $g := P_F f$. Then $f - g \in F^{\perp\perp} \cap F^\perp = \{0\}$, i.e., $f = g$. So we have proved that $F = F^{\perp\perp}$ if F is closed.

If F is not closed we note that $F^\perp = \overline{F}^\perp$ by Corollary 4.14b). Hence by the previous considerations $F^{\perp\perp} = \overline{F}^{\perp\perp} = \overline{F}$.

Finally, suppose that $\overline{F} = H$. Then $\{0\} = H^\perp = \overline{F}^\perp = F^\perp$. Conversely, if $F^\perp = \{0\}$, then $H = \{0\}^\perp = F^{\perp\perp} = \overline{F}$. □

8.3. The Riesz–Fréchet Theorem

Let H be a Hilbert space. If we fix $g \in H$ as the second component in the inner product, we obtain a linear functional

$$\varphi_g : H \longrightarrow \mathbb{K}, \qquad f \longmapsto \varphi_g(f) := \langle f, g \rangle$$

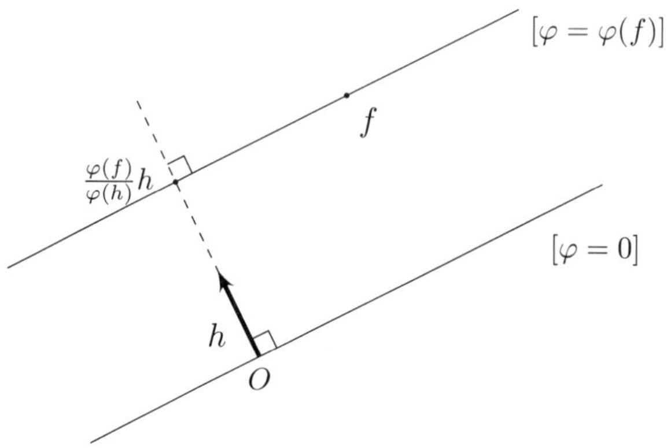

Figure 13. The construction of the vector g.

By Cauchy–Schwarz, one has

$$|\varphi_g(f)| = |\langle f, g\rangle| \leq \|f\| \, \|g\|$$

for all $f \in H$, hence φ_g is bounded; see Example 2.23. The Riesz–Fréchet theorem asserts that *every* bounded linear functional on H is of this form.

Theorem 8.12 (Riesz–Fréchet[1]). *Let H be a Hilbert space and let $\varphi : H \to \mathbb{K}$ be a bounded linear functional on H. Then there exists a unique $g \in H$ such that*

$$\varphi(f) = \langle f, g\rangle \qquad \textit{for all } f \in H.$$

Proof. Uniqueness: If $g, h \in H$ are such that $\langle f, g\rangle = \varphi(f) = \langle f, h\rangle$ for all $f \in H$, then

$$\langle f, g - h\rangle = \langle f, g\rangle - \langle f, h\rangle = \varphi(f) - \varphi(f) = 0 \qquad (f \in H).$$

Hence $g - h \perp H$ which is only possible if $g = h$.

Existence: If $\varphi = 0$, we can take $g := 0$, so we may suppose that $\varphi \neq 0$. In this case the closed linear subspace $\ker(\varphi)$ is not the whole space, so we may pick an orthogonal vector $h \perp \ker(\varphi)$ with $\|h\| = 1$. In particular, $\varphi(h) \neq 0$. Given $f \in H$ we hence have

$$h \perp f - \tfrac{\varphi(f)}{\varphi(h)} h, \quad \text{i.e,} \quad \langle f, h\rangle = \tfrac{\varphi(f)}{\varphi(h)}.$$

It follows that $\varphi(f) = \left\langle f, \overline{\varphi(h)} h\right\rangle$. Since f was arbitrary, we may take $g := \overline{\varphi(h)} h$ and are done. □

[1]Maurice René Fréchet (1878–1973), French mathematician.

We shall see a typical application of the Riesz–Fréchet theorem to differential equations in Chapter 10. Exercise 8.21 covers the connection of the Riesz–Fréchet theorem with **variational problems**, more precisely with **quadratic minimization**.

8.4. Orthogonal Series and Abstract Fourier Expansions

One of the central themes of analysis is the summation of infinite series. As introduced in Section 5.4, we have the definition of the series $\sum_{j=1}^{\infty} f_j$ as the limit of partial sums

$$\sum_{j=1}^{\infty} f_j := \lim_{n\to\infty} \sum_{j=1}^{n} f_j$$

if this limit exists. If the normed space is complete, one can of course use the Cauchy criterion, stating that

$$\left\| \sum_{j=n}^{m} f_j \right\| \to 0 \quad \text{as } n, m \to \infty,\ n \le m.$$

In many situations, however, this would be too tedious and one asks for conditions which are more easily verified.

Probably the most convenient criterion here is *absolute summability*, already treated in Theorem 5.16. But, of course, there are many convergent, but not absolutely convergent series, e.g., the alternating harmonic series $\sum_{n=1}^{\infty} (-1)^n/n$.

While for scalar series there is a wealth of convergence criteria (e.g., the Leibniz criterion), the situation is much worse for general Banach spaces, where Theorem 5.16 is in many cases the only way to prove convergence. However, for **orthogonal series** in a *Hilbert* space we have the following important result.

Theorem 8.13. *Let $(H, \langle\cdot,\cdot\rangle)$ be an inner product space and let $(f_n)_{n\in\mathbb{N}}$ be a sequence of pairwise orthogonal elements of H. Consider the statements*

(i) *The series $f := \sum_{n=1}^{\infty} f_n$ converges in H.*

(ii) $\sum_{n=1}^{\infty} \|f_n\|^2 < \infty$.

Then (i) *implies* (ii) *and one has* **Parseval's identity**[2]

$$\|f\|^2 = \sum_{n=1}^{\infty} \|f_n\|^2 . \tag{8.1}$$

If H is a Hilbert space, then (ii) *implies* (i).

[2]Marc-Antoine Parseval des Chênes (1755–1836), French mathematician.

Proof. Write $s_n := \sum_{j=1}^n f_j$ for the partial sums. If $f = \lim_{n\to\infty} s_n$ exists in H, then by the continuity of the norm and Pythagoras one obtains

$$\begin{aligned}\|f\|^2 &= \Big\|\lim_{m\to\infty} s_m\Big\|^2 = \lim_{m\to\infty} \|s_m\|^2 = \lim_{m\to\infty} \Big\|\sum\nolimits_{j=1}^m f_j\Big\|^2 \\ &= \lim_{m\to\infty} \sum\nolimits_{j=1}^m \|f_j\|^2 = \sum\nolimits_{j=1}^\infty \|f_j\|^2 .\end{aligned}$$

Since $\|f\| < \infty$, this implies (ii).

Conversely, suppose that (ii) holds and that H is a Hilbert space. Hence (i) holds if and only if the partial sums $(s_n)_{n\in\mathbb{N}}$ form a Cauchy sequence. If $m > n$, then by Pythagoras' theorem

$$\|s_m - s_n\|^2 = \Big\|\sum\nolimits_{j=n+1}^m f_j\Big\|^2 = \sum\nolimits_{j=n+1}^m \|f_j\|^2 \le \sum\nolimits_{j=n+1}^\infty \|f_j\|^2 \to 0$$

as $n \to \infty$ by (ii), and this concludes the proof. □

Example 8.14. Let $(e_n)_{n\in\mathbb{N}}$ be the sequence of standard unit vectors in ℓ^2. Then

$$\sum_{n=1}^\infty \frac{1}{n} e_n$$

converges in $\|\cdot\|_2$ to $(1, 1/2, 1/3, \dots) \in \ell^2$. Note that this series does *not* converge absolutely, since $\sum_{n=1}^\infty \|(1/n)e_n\|_2 = \sum_{n=1}^\infty 1/n = \infty$.

Example 8.14 is an instance of an abstract Fourier series. These are special orthogonal series and will be treated next.

Let H be a Hilbert space and let $(e_j)_{j\in\mathbb{N}}$ be an orthonormal system in H. Analogous to the finite-dimensional situation considered in Chapter 1 we study now the *infinite* abstract Fourier series

$$Pf := \sum\nolimits_{j=1}^\infty \langle f, e_j\rangle\, e_j$$

for given $f \in H$. Of course, there is an issue of convergence here.

Theorem 8.15. *Let H be a Hilbert space, let $(e_j)_{j\in\mathbb{N}}$ be an orthonormal system in H, and let $f \in H$. Then one has* **Bessel's inequality**

$$\sum\nolimits_{j=1}^\infty |\langle f, e_j\rangle|^2 \le \|f\|^2 < \infty. \tag{8.2}$$

Moreover, the series

$$Pf := \sum\nolimits_{j=1}^\infty \langle f, e_j\rangle\, e_j$$

is convergent in H, and $Pf = P_F f$ is the orthogonal projection of f onto the closed subspace

$$F := \overline{\operatorname{span}}\{e_j \mid j \in \mathbb{N}\}.$$

Finally, one has **Parseval's identity** $\|Pf\|^2 = \sum_{j=1}^\infty |\langle f, e_j\rangle|^2$.

Proof. For Bessel's inequality it suffices to establish the estimate

$$\sum_{j=1}^{n} |\langle f, e_j\rangle|^2 \le \|f\|^2$$

for arbitrary $n \in \mathbb{N}$. This is immediate from Lemma 1.10; see (1.1). By Bessel's inequality and the fact that H is complete (by assumption) Theorem 8.13 yields that the sum $Pf := \sum_{j=1}^{\infty} \langle f, e_j\rangle e_j$ is indeed convergent in H with Parseval's identity being true.

To see that $Pf = P_F f$ we only need to show that $Pf \in F$ and $f - Pf \perp F$. Since Pf is a limit of sums of vectors in F, and F is closed, $Pf \in F$. For the second condition, note that

$$\langle f - Pf, e_k\rangle = \langle f, e_k\rangle - \sum_{j=1}^{\infty} \langle f, e_j\rangle \langle e_j, e_k\rangle = \langle f, e_k\rangle - \langle f, e_k\rangle = 0$$

for every k. Hence $f - Pf \perp F$ by Corollary 4.14. □

Corollary 8.16. *Let H be a Hilbert space, let $(e_j)_{j\in\mathbb{N}}$ be an orthonormal system in H. Then the following assertions are equivalent:*

(i) $\{e_j \mid j \in \mathbb{N}\}^{\perp} = \{0\}$.

(ii) $\operatorname{span}\{e_j \mid j \in \mathbb{N}\}$ *is dense in H.*

(iii) $f = \sum_{j=1}^{\infty} \langle f, e_j\rangle e_j$ *for all $f \in H$.*

(iv) $\|f\|^2 = \sum_{j=1}^{\infty} |\langle f, e_j\rangle|^2$ *for all $f \in H$.*

(v) $\langle f, g\rangle_H = \sum_{j=1}^{\infty} \langle f, e_j\rangle \overline{\langle g, e_j\rangle}$ *for all $f, g \in H$.*

Proof. We use the notation from above. Then (i) just says that $F^{\perp} = \{0\}$ which is (by orthogonal decomposition) equivalent to $F = H$, i.e., (ii). Now, (iii) simply expresses that $f = Pf$ for all $f \in H$, and since $P = P_F$ is the orthogonal projection onto F, this is equivalent to $F = H$. If (iii) holds , then (iv) is also true, by Parseval's identity. On the other hand, by Pythagoras' lemma and since P is an orthogonal projection,

$$\|f\|^2 = \|Pf\|^2 + \|f - Pf\|^2$$

which implies that $\|f\|^2 = \|Pf\|^2$ if and only if $f = Pf$. This proves the equivalence (iii)⇔(iv). The equivalence of (iv) and (v) is established in Exercise 8.6. □ Ex.8.6

Definition 8.17. An orthonormal system $(e_j)_{j\in\mathbb{N}}$ in the inner product space H is called **maximal** or an **orthonormal basis** of H, if it satisfies the equivalent conditions of Corollary 8.16.

Attention: In algebraic terminology, a *basis* of a vector space is a linearly independent subset such that every vector can be represented as a *finite* linear combination of basis vectors. Hence an *orthonormal basis* in our sense is usually *not* an (algebraic) basis. To distinguish the two notions of bases, in analytic contexts one sometimes uses the term *Hamel basis*[3] for an algebraic basis.

Sometimes an orthonormal basis is called a "complete" orthonormal system in the literature. We shall not use this term since it is outdated.

Example 8.18. Let us apply these results to the space $H = \ell^2$ with its orthonormal system $(e_n)_{n\in\mathbb{N}}$ of standard unit vectors. In Example 3.18 it was shown that their linear span, c_{00}, is dense in ℓ^2. Hence $(e_n)_{n\in\mathbb{N}}$ is indeed an orthonormal basis of ℓ^2.

The following result shows that — in a sense — ℓ^2 is the prototype of a Hilbert space.

Theorem 8.19. *Let H be a Hilbert space, and let $(e_j)_{j\in\mathbb{N}}$ be an orthonormal basis of H. Then the "coordinatization map"*

$$T : H \longrightarrow \ell^2, \qquad f \longmapsto (\langle f, e_j\rangle)_{j\in\mathbb{N}}$$

is an isometric isomorphism.

Proof. It follows directly from Theorem 8.15 and Corollary 8.16 that T is a
Ex.8.7 well-defined linear isometry. The surjectivity is left as an (easy) exercise. □

Theorem 8.19 says that — in a sense — ℓ^2 is the *only* Hilbert space with a countable orthonormal basis. However, the choice of basis is not canonical in most cases.

Finally and at last, we turn to the question of existence of orthonormal bases. By employing the Gram–Schmidt procedure (Lemma 1.11), we obtain the following.

Lemma 8.20. *A Hilbert space has a countable orthonormal basis if and only if there is a sequence $(f_n)_{n\in\mathbb{N}}$ in H such that* $\operatorname{span}\{f_n \mid n \in \mathbb{N}\}$ *is dense in H.*

[3]Georg Hamel (1877–1954), German mathematician.

Proof. By succesively discarding linearly dependent vectors from the sequence, one ends up with a linearly independent one. Then one can apply literally the Gram–Schmidt Lemma 1.11 to find a countable orthonormal basis. □

A Hilbert space satisfying the condition of Lemma 8.20 is called **separable**; cf. Definition 4.35 and Theorem 4.37. One can ask what happens for nonseparable Hilbert spaces. Then one encounters uncountable orthonormal systems. With a little effort one can extend the theory to that case, and with the help of Zorn's lemma one can prove that every Hilbert space has a (possibly uncountable) orthonormal basis. An account of these facts can be found in Appendix F. Ex.8.8

Exercises 8A

Exercise 8.1. Let $A \subseteq E$ be closed subset of a normed space E, and let $x \in E$. Show that if A is (sequentially) compact then a best approximation to x exists in A. Conclude that if E is finite-dimensional then there is a best approximation to x in A even if A is not compact. [Hint: Consider the set $A \cap \mathrm{B}[x,r]$ for sufficiently large $r > 0$.]

Exercise 8.2. Let $X \subseteq \mathbb{R}$ be an interval and let $H := \mathrm{L}^2(X;\mathbb{R})$. Let $\mathrm{L}^2_+ := \{f \in \mathrm{L}^2(X;\mathbb{R}) \mid f \geq 0 \,\text{a.e.}\}$ be the *positive cone.* Show that L^2_+ is closed and convex. Then show that for $f, g \in \mathrm{L}^2(X)$ with $g \geq 0$ one has

$$|f - g| \geq f^- = \left|f - f^+\right| \quad \text{a.e.}$$

Conclude that f^+ is the best approximation to f in L^2_+.

Exercise 8.3. Let $H = \mathrm{L}^2(0,1)$ and $f(t) = e^t$, $t \in [0,1]$. Find best approximations to f within F in the following cases:

a) F is the space of polynomials of degree at most 1.

b) F is the space $\{at + bt^2 \mid a, b \in \mathbb{C}\}$.

c) F is the space $\{g \in \mathrm{L}^2(0,1) \mid \int_0^1 g = 0\}$.

Exercise 8.4 (Characterization of Orthogonal Projections). Let H be a Hilbert space, and let $P : H \to H$ be a linear mapping satisfying $P^2 = P$. Show that $Q := \mathrm{I} - P$ satisfies $Q^2 = Q$, and that $\ker(P) = \operatorname{ran}(Q)$. Then show that the following assertions are equivalent:

(i) $\operatorname{ran}(P) \perp \ker(P)$.

(ii) $\langle Pf, g\rangle = \langle f, Pg\rangle$ for all $f, g \in H$.

(iii) $\|Pf\| \leq \|f\|$ for all $f \in H$.

(iv) $F := \operatorname{ran}(P)$ is closed and $P = P_F$.

[Hint for the implication (iii) $\Rightarrow$ (i): If (iii) holds, then $\|P(f+cg)\|^2 \le \|f+cg\|^2$ for all $c \in \mathbb{K}$ and $f, g \in H$; fix $f \in \operatorname{ran}(P)$, $g \in \ker(P)$, use Lemma 1.5 and vary c to conclude that $\langle f, g\rangle = 0$.]

Exercise 8.5. Let $E := \mathrm{L}^1(-1,1)$ and consider the mapping

$$(Tf)(t) := f(-t) \quad f \in \mathrm{L}^1(-1,1).$$

A function $f \in \mathrm{L}^1(-1,1)$ is called *even* if $f = Tf$ almost everywhere.

a) Show that if $f \in \mathrm{L}^1(-1,1)$, then $Tf \in \mathrm{L}^1(-1,1)$ as well and $\int_{-1}^1 Tf\,\mathrm{d}\lambda = \int_{-1}^1 f\,\mathrm{d}\lambda$. [Hint: Show first that it is true for $f \in C[-1,1]$. Then use Theorem 7.24.]

b) Show that $Tf \in \mathrm{L}^2(-1,1)$ and $\|Tf\|_2 = \|f\|_2$ for all $f \in \mathrm{L}^2(-1,1)$. [Hint: Use a)]

c) Let $H = \mathrm{L}^2(-1,1)$. Show that the space

$$F := \{f \in \mathrm{L}^2(-1,1) \mid f \text{ is even}\}$$

is a closed linear subspace of H and show that $P_F = \frac{1}{2}(\mathrm{I}+T)$ is the orthogonal projection onto F.

d) Describe $F^\perp$ and the orthogonal projection onto $F^\perp$.

Exercise 8.6. Let H be an inner product space with an ONS $(e_n)_{n\in\mathbb{N}}$ and vectors $f, g \in H$. Show that the series

$$\sum_{j=1}^{\infty} \langle f, e_j\rangle \overline{\langle g, e_j\rangle}$$

converges *absolutely*. Then prove the equivalence of (iii) and (iv) in Corollary 8.16. [Hint: See Exercise 1.9.]

Exercise 8.7. Let $(e_j)_{j\in\mathbb{N}}$ an ONS in a Hilbert space H. Show that for every $\alpha \in \ell^2$ there is $f \in H$ such that $\langle f, e_j\rangle = \alpha_j$ for all $j \in \mathbb{N}$.

Exercise 8.8. Let H be a Hilbert space with a countable orthonormal basis. Let $F \subseteq H$ be a closed subspace. Show that F also has a countable orthonormal basis. [Hint: Use the orthogonal projection P_F.]

Exercises 8B

Exercise 8.9. Let (Ω, d) be a metric space and let $A \subseteq \Omega$ be any subset. Show that

$$|d(x,A) - d(y,A)| \le d(x,y)$$

for all $x, y \in \Omega$, and conclude that $d(\cdot, A)$ is a continuous function on Ω.

Exercise 8.10 (Minimal Norm Problems)**.** Let E be a vector space. Each subset of E of the form $h + F$, where F is a linear subspace and $h \in E$, is called an **affine subspace** of E. If $G = h + F$ is an affine subspace, then $F = G - g$ for *each* $g \in G$ (why?). In particular, F is uniquely determined by G. (Make a little sketch to see what's going on!)

Let H be a Hilbert space, and let $G \subseteq H$ be a closed affine subspace with associated linear subspace F. Show that for $g \in G$ the following assertions are equivalent:

(i) g is the (unique) element of G with minimal norm.

(ii) $g \perp F$.

Then, in each of the cases

1) $H = \mathrm{L}^2(0,1)$, $G := \{f \in \mathrm{L}^2(0,1) \mid \int_0^1 tf(t)\,\mathrm{d}t = 1\}$,

2) $H = \mathrm{L}^2(0,1)$, $G := \{f \in \mathrm{L}^2(0,1) \mid \int_0^1 f(t)\,\mathrm{d}t = 1/3,\quad \int_0^1 t^2 f(t)\,\mathrm{d}t = 1/15\}$,

determine an element of minimal norm in G.

Exercise 8.11. Define the space of square-summable *two-sided* sequences

$$\ell^2(\mathbb{Z}) := \bigl\{(x_n)_{n\in\mathbb{Z}} \mid x_n \in \mathbb{C},\ \sum\nolimits_{n=-\infty}^{\infty} |x_n|^2 < \infty\bigr\}.$$

Show, e.g., as in Example 2.4 that this is a Hilbert space with norm

$$\|x\| = \Bigl(\sum\nolimits_{n=-\infty}^{\infty} |x_n|^2\Bigr)^{1/2}, \qquad x = (x_n)_{n\in\mathbb{Z}} \in \ell^2(\mathbb{Z}).$$

(See Exercise 5.22 for the definition of double series.)

Exercise 8.12 ($\mathbb{Z}$-indexed Orthonormal Systems)**.** Let H be a Hilbert space and $(e_j)_{j\in\mathbb{Z}}$ an ONS in H. Prove the analogue of Theorem 8.15 in this context. (See Exercise 5.22.) More precisely, show **Bessel's inequality**

$$\sum\nolimits_{j=-\infty}^{\infty} |\langle f, e_j\rangle|^2 \le \|f\|^2 < \infty. \tag{8.3}$$

Then show that the double series

$$Pf := \sum\nolimits_{j=-\infty}^{\infty} \langle f, e_j\rangle\, e_j$$

is convergent in H, and $Pf = P_F f$ is the orthogonal projection of f onto the closed subspace

$$F := \overline{\mathrm{span}}\{e_j \mid j \in \mathbb{Z}\}.$$

Finally, show **Parseval's identity** $\|Pf\|^2 = \sum\nolimits_{j=-\infty}^{\infty} |\langle f, e_j\rangle|^2$.

Exercise 8.13. Let H be a Hilbert space and $(e_j)_{j\in\mathbb{Z}}$ an ONS in H. Prove the analogue of Corollary 8.16 in this context. In the case that the ONS $(e_j)_{j\in\mathbb{Z}}$ is maximal, prove that

$$T : H \longrightarrow \ell^2(\mathbb{Z}), \qquad Tf := (\langle f, e_j\rangle)_{j\in\mathbb{Z}}$$

is an isometric isomorphism.

Exercise 8.14. Let $E := \{f \in \mathrm{C}[0,1] \mid f(0) = 0\}$ with the supremum norm. Let

$$A := \{f \in E \mid \int_0^1 f(t)\,\mathrm{d}t = 0\},$$

and let $f(t) := t$, $t \in [0,1]$.

a) Show that E is a closed subspace of $C[0,1]$ and that A is a closed subspace of E.

b) For given $\epsilon > 0$, find $g \in A$ such that $\|g - f\|_\infty \le 1/2 + \epsilon$. [Hint: Modify the function $f - 1/2$ appropriately.]

c) Conclude from this and Example 8.2 that $d(f, A) = 1/2$.

d) Show that for each $g \in A$ one must have $\|f - g\|_\infty > 1/2$.

Exercise 8.15. Consider the series of functions

$$\sum_{n=1}^{\infty} \frac{\cos \pi n t}{n^\alpha} \qquad (t \in [0,1]).$$

Determine for which values of $\alpha \ge 0$ the series converges

a) in $C[0,1]$ with respect to the supremum norm;

b) in $L^2(0,1)$ with respect to $\|\cdot\|_2$.

Justify your answers.

Exercises 8C

Exercise 8.16. Let $H = \ell^2$ with canonical unit vectors $(e_n)_{n\in\mathbb{N}}$ and consider the closed linear subspaces

$$F := \overline{\operatorname{span}}\{e_{2n-1} \mid n \in \mathbb{N}\} \quad \text{and} \qquad G := \overline{\operatorname{span}}\{e_{2n-1} + (1/n)e_{2n} \mid n \in \mathbb{N}\}.$$

Give formulae for P_F and P_G. Show that $F \cap G = \{0\}$. Show that the subspace $F + G$ is dense in H, but not closed.

Exercise 8.17. Let $(e_j)_{j\in\mathbb{N}}$ be an orthonormal system in a Hilbert space H, and let $a_j \ge 0$ be scalars with $\sum_j a_j^2 < \infty$. Show that the set

$$C := \{\sum_{j=1}^{\infty} \lambda_j e_j \mid \lambda_j \in \mathbb{K},\ |\lambda_j| \le a_j\}$$

is compact in H.

Exercise 8.18. Let A be a closed convex subset of a Hilbert space H. Show that $P_A : H \to A$, the best approximation map onto A, is continuous.

Exercise 8.19. Let H be an inner product space, and let $\emptyset \ne A \subseteq H$ be a complete, convex subset of H. Furthermore, let $f \in H$. Show that for $g \in H$ the following assertions are equivalent:

(i) $g = P_A f$.

(ii) $g \in A$ and $\operatorname{Re}\langle f - g, h - g\rangle \le 0$ for all $h \in A$.

Give a geometric interpretation of the condition (ii) in case $\mathbb{K} = \mathbb{R}$. [Hint: Imitate the proof of Lemma 8.6.]

Exercise 8.20 (Geometric Hahn–Banach Theorem for Hilbert Spaces). For a real vector space E, a linear functional $\varphi : E \to \mathbb{R}$ and $c \in \mathbb{R}$ we abbreviate

$$[\varphi \le c] := \{f \in E \mid \varphi(f) \le c\}$$

and likewise $[\varphi = c]$ and $[\varphi \ge c]$.

If $\varphi = 0$ then these sets are either empty or the whole space E. In case $\varphi \neq 0$ we call $[\varphi = c]$ an (affine) **hyperplane**. We say that a set $A \subseteq E$ is **separated** by the hyperplane $[\varphi = c]$ from a point $f \in E$, if $A \subseteq [\varphi \le c]$ and $\varphi(f) > c$.

Use Exercise 8.19 to show that if $E = H$ is a real Hilbert space, $A \subseteq H$ is a nonempty closed convex subset of H and $f \in H \setminus A$, then A can be separated from f by a *closed* hyperplane. [Hint: Closed hyperplanes correspond to bounded linear functionals.]

Exercise 8.21 (Quadratic Minimization Problem). Let H be an inner product space, let $\varphi : H \to \mathbb{K}$ be a bounded linear functional on H, and let $A \subseteq H$ be a nonempty subset of H. The associated **quadratic minimization problem** on A consists in finding $g \in A$ such that the real function

$$\Phi : H \longrightarrow \mathbb{R}, \quad \Phi(u) = \tfrac{1}{2} \|u\|^2 - \operatorname{Re} \varphi(u)$$

has a minimum on A at $u = g$.

a) Suppose that $f \in H$ is such that $\varphi(h) = \langle h, f \rangle$ for all $h \in H$. Show that $g \in A$ solves this quadratic minimization problem if and only if g is a best approximation to f in A.

b) Suppose again that $f \in H$ is such that $\varphi(h) = \langle h, f \rangle$ for all $h \in H$. Show that Φ takes an absolute minimum on H precisely in the vector f.

c) Conclude, employing the Riesz–Fréchet theorem, that if A is convex and complete then any quadratic minimization problem on A has a unique solution. Next, give an alternative proof of this statement without using the Riesz–Fréchet theorem. [Hint: Prove first that $\inf_{u \in A} \Phi(u) > -\infty$, then establish the formula

$$\Phi(u) + \Phi(v) = \left\| \tfrac{u-v}{2} \right\|^2 + 2\Phi\left(\tfrac{u+v}{2}\right)$$

and proceed as in the proof of Theorem 8.5.]

d) Employ c) to give an alternative proof of the Riesz-Fréchet theorem. [Hint: show first that an absolute minimizer f of Φ must satisfy

$$\operatorname{Re} \langle f - h, f + h \rangle \le 2 \operatorname{Re} \varphi(f - h) \quad \text{for all } h \in H,$$

then proceed as in the proof of Lemma 8.6.]

Exercise 8.22 (Semi-inner product spaces). A **semi-inner product** on a $\mathbb{K}$-vector space E is a positive, symmetric sesquilinear form $\langle \cdot, \cdot \rangle : E \times E \to \mathbb{K}$; cf. Definition 1.2. That means, $\langle \cdot, \cdot \rangle$ has all the properties of an inner product apart from definiteness. As with inner products, we write

$$\|f\| := \sqrt{\langle f, f \rangle} \qquad (f \in E)$$

and $f \perp g$ as an abbreviation for $\langle f, g \rangle = 0$. Show that the following results about inner products remain true for semi-inner products:

a) Lemma 1.5, in particular, the polarization identity and the parallelogram law.

b) Pythagoras' Lemma 1.9.

c) The Cauchy–Schwarz inequality $|\langle f, g\rangle| \le \|f\| \, \|g\|$.

d) Homogeneity and triangle inequality for $\|\cdot\|$ (cf. Definition 2.5).

In the following let $(E, \langle\cdot,\cdot\rangle)$ be a semi-inner product and let $E_0 := \{f \in E \mid \|f\| = 0\}$. Prove that

$$\|f\| = 0 \quad \Longleftrightarrow \quad f \perp E.$$

Show that Theorem 8.5 about the existence of best approximations remains literally true if one assumes that

$$f, g \in A, \quad \|f - g\| = 0 \quad \to \quad f = g.$$

(This means that the "pseudo-metric" $d(f, g) := \|f - g\|$ when restricted to A is a proper metric.) Then show that Lemma 8.6 remains valid for semi-inner products, if the restriction of $\langle\cdot,\cdot\rangle$ to F is a proper inner product.

Exercise 8.23 (Lax–Milgram Theorem, Operator Form). Let H be a Hilbert space, and let $T \in \mathcal{L}(H)$ such that for some $\delta > 0$,

$$|\langle Tf, f\rangle| \ge \delta \, \|f\|^2 \qquad \text{for all } f \in H.$$

Show that T is bijective and its inverse is bounded with $\|T^{-1}\| \le 1/\delta$. [Hint: Exercise 5.15.]

Chapter 9

Approximation Theory and Fourier Analysis

In this chapter we apply the functional analytic concepts obtained so far to problems from classical Fourier analysis. Our focus lies, however, not on the results in that field but on the underlying principles, in particular the role of density and approximation arguments. Consequently, the first part of this chapter is mainly devoted to studying techniques of approximation and enlarging our stocks of approximation results.

In abstract terms, an approximation problem has the following form: given $f \in \Omega$ and $A \subseteq \Omega$ for a certain metric space (Ω, d), prove that $f \in \overline{A}$, i.e., for each $\epsilon > 0$ there is $g \in A$ such that $d(f, g) < \epsilon$. In the situation of Weierstrass' theorem, $\Omega = \mathrm{C}[a, b]$ with the supemum norm, $A = \mathrm{P}[a, b]$ is the space of polynomials, and $f \in \mathrm{C}[a, b]$ is arbitrary. Note that the set A of possible approximants is easily described as the linear span of the set $\{x^n \mid n \geq 0\}$ of monomials. As such a situation occurs quite frequently, we give it a special name.

Definition 9.1. A subset $A \subseteq E$ of a normed space E is called **fundamental** in E if $\overline{\mathrm{span}}(A) = E$.

In this terminology, Weierstrass' theorem simply states that the set of monomials is fundamental in $\mathrm{C}[a, b]$ with respect to the supremum norm. And an orthonormal system $(e_n)_n$ in a Hilbert space H is maximal if and only if it is fundamental in H.

Approximation and Permanence Principles. There exists a handful of general techniques, we call them *approximation principles*, that help in solving approximation problems. We formulate them in the framework of normed spaces; but see also Exercise 9.1.

Theorem 9.2 (Approximation Principles).

a) ("dense in dense is dense") *Let E be a normed space and $A, B \subseteq E$. If $B \subseteq \overline{A}$ ($B \subseteq \overline{\operatorname{span}}(A)$) and B is dense (fundamental) in E, then so is A.*

b) ("image of dense is dense in the image") *Let E, F be normed space and $T : E \to F$ be a bounded linear operator. If $A \subseteq E$ is dense (fundamental) in E, then $T(A)$ is dense (fundamental) in $T(E)$.*

c) ("dense is dense in a weaker norm") *Let $\|\cdot\|_w$, $\|\cdot\|_s$ be two norms on a vector space E, such that there is a constant $c > 0$ with*

$$\|f\|_w \le c\,\|f\|_s \quad \text{for all } f \in E.$$

If $A \subseteq E$ is $\|\cdot\|_s$-dense (fundamental) in E, then it is also $\|\cdot\|_w$-dense (fundamental) in E.

Proof. a) is a simple consequence of the triangle inequality and the fact that $\overline{\operatorname{span}}(A)$ is a subspace of E (Corollary 4.14). b) is a simple consequence of
Ex.9.1 the continuity and linearity of T, and c) is a special case of b). □
Ex.9.2

Statement a) of Theorem 9.2 is the most important, for it allows a *stepwise* approach in solving an approximation problem. That means, to approximate f from within the "small" set A, one first approximates it with h from something bigger than A, say B, and then approximates h from within A. Since in each step one can make the distance arbitrarily small, by the triangle inequality the overall distance is arbitrarily small as well. A typical example of such a stepwise approach is Lebesgue's proof of Weierstrass' theorem, to be treated below.

Whereas an approximation principle helps to establish that certain sets are dense or fundamental, a *permanence princple* is used to exploit such information. More precisely, a permanence principle asserts that a property is shared by all elements of a space in case that it is shared by the elements of a dense (or fundamental) subset. A simple example, again stated in the
Ex.9.3 framework of normed spaces, is the following; see also Exercise 9.3.

Lemma 9.3 (Permanence Principles). *Let E and F be normed spaces and let $T : E \to F$ be a bounded linear mapping.*

a) *If $G \subseteq F$ is a closed linear subspace of F and $Tf \in G$ for all $f \in C$ for some fundamental set C in E, then* $\operatorname{ran}(T) \subseteq G$.

b) *If* $S : E \to F$ *is another bounded linear mapping and* $Sf = Tf$ *for all* $f \in C$ *for some fundamental set* C *in* E*, then* $S = T$.

Proof. a) Since T is linear and continuous, $T^{-1}(G)$ is a closed subspace of E (Lemma 4.7.(v)). As $C \subseteq T^{-1}(G)$ by hypothesis, $\mathrm{span}(C) \subseteq T^{-1}(G)$ and hence $E = \overline{\mathrm{span}}(C) \subseteq T^{-1}(G)$. b) is a consequence of a) by passing to $T - S$ and the closed linear subspace $G = \{0\}$. □

We shall employ Lemma 9.3 in the proof of the Riemann–Lebesgue lemma in Section 9.5 below. One of the most important permanence principles is the "strong convergence lemma", to be treated in Section 9.6.

9.1. Lebesgue's Proof of Weierstrass' Theorem

Weierstrass' Theorem 3.22 has a long and interesting history, beginning with Weierstrass' original work [**Wei85**] and extending until the present. From the many known proofs we chose Lebesgue's to be presented here, since it is intuitive in its stepwise approach. Bernstein's proof in Appendix C may be less intuitive in comparison, but it is short, elegant, and moreover, *effective.* Namely, it is based on an explicit formula for the approximating polynomials. For many other proofs and generalizations of Weierstrass' theorem we recommend the nice article [**Pin00**] by Pinkus.

The idea behind Lebesgue's proof is to approximate a given $f \in \mathrm{C}[a, b]$ first by a piecewise linear function g, and then to show that g can be approximated by polynomials. Here $g : [a, b] \to \mathbb{C}$ is called **piecewise linear** if there is a partition

$$a = t_0 < t_1 < \cdots < t_n = b$$

such that g is continuous on $[a, b]$ and a linear polynomial on each $[t_{j-1}, t_j]$. Let us denote the space of piecewise linear functions by $\mathrm{PL}[a, b]$. It is obviously a linear subspace of $\mathrm{C}[a, b]$.

Lemma 9.4. *The space* $\mathrm{PL}[a, b]$ *of piecewise linear functions is* $\|\cdot\|_\infty$*-norm dense in* $\mathrm{C}[a, b]$.

Proof. Let $f \in \mathrm{C}[a, b]$ and $\epsilon > 0$. Since f is uniformly continuous (Exercise 4.24), there is $\delta > 0$ such that $|f(s) - f(t)| \le \epsilon$ whenever $|s - t| \le \delta$. Choose $n \in \mathbb{N}$ so large that ${}^1\!/_n(b-a) < \delta$ and define $t_j := a + {}^j\!/_n(b-a)$ for $j = 0, \dots, n$. Hence

$$s \in [t_{j-1}, t_j] \quad \Longrightarrow \quad |f(s) - f(t_{j-1})| \le \epsilon.$$

Now let $g \in \mathrm{PL}[a, b]$ be the unique piecewise linear function satisfying $g(t_j) = f(t_j)$ for all $j = 0, \dots n$, i.e.,

$$g(s) = f(t_{j-1}) + (s - t_{j-1}) \frac{f(t_j) - f(t_{j-1})}{t_j - t_{j-1}} \qquad (t_{j-1} \le s \le t_j).$$

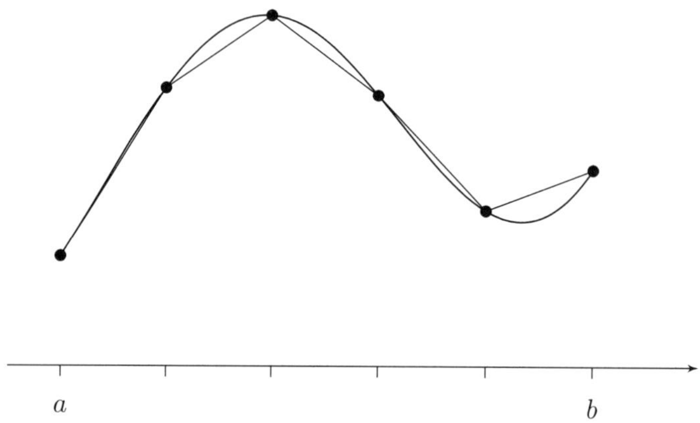

Figure 14. A piecewise linear approximation of a continuous function.

We claim that $\|f-g\|_\infty \le 2\epsilon$. Indeed, if $s \in [t_{j-1}, t_j]$, then

$$|g(s) - f(t_{j-1})| = (s - t_{j-1})\frac{|f(t_j) - f(t_{j-1})|}{t_j - t_{j-1}} \le |f(t_j) - f(t_{j-1})| \le \epsilon.$$

Since $f(t_{j-1}) = g(t_{j-1})$ it follows that

$$|f(s) - g(s)| \le |f(s) - f(t_{j-1})| + |g(s) - f(t_{j-1})| \le 2\epsilon.$$

This establishes our claim and concludes the proof. □

In the next step we observe that the space $\mathrm{PL}[a,b]$ is linearly spanned by a set of very simple functions. Let

$$h_0(s) := \frac{s+|s|}{2} = \begin{cases} 0 & \text{if} \quad s \le 0, \\ s & \text{if} \quad s \ge 0, \end{cases}$$

and $h_t(s) := h_0(s-t)$ for $t \in \mathbb{R}$. Clearly, each h_t is piecewise linear; see Figure 15.

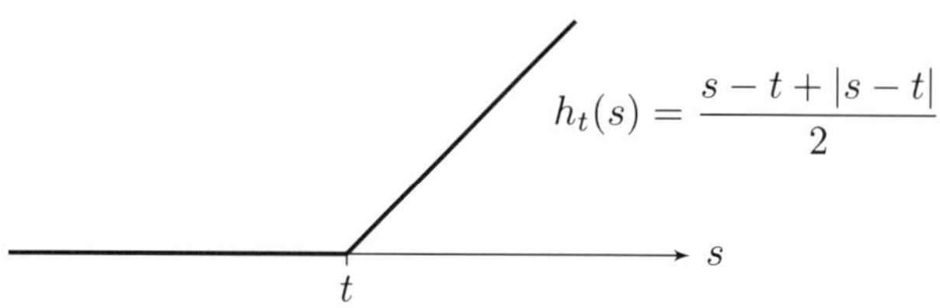

Figure 15. The function h_t.

Lemma 9.5. *The space* PL$[a,b]$ *is spanned by* $\mathbf{1}$ *and all the functions* h_t.

Proof. Let $g \in \mathrm{PL}[a,b]$ and suppose that $a = t_0 < \cdots < t_n = b$ is a partition such that g is a linear polynomial on each $[t_{j-1}, t_j]$ with slope

$$c_j := \frac{g(t_j) - g(t_{j-1})}{t_j - t_{j-1}}.$$

The function $g_1 := g(t_0)\mathbf{1} + c_1 h_{t_0}$ coincides with g on $[a, t_1]$. Then the function

$$g_2 := g(t_0)\mathbf{1} + c_1 h_{t_0} + (c_2 - c_1) h_{t_1}$$

coincides with g on $[a, t_2]$. Continuing in this way eventually yields a representation of g as a linear combination of $\mathbf{1}$ and functions h_t. □

Up to now we have shown that the set $\{\mathbf{1}\} \cup \{h_t \mid t \in [a,b]\}$ is fundamental in C$[a,b]$. Hence by Theorem 9.2.a) all that remains is to show that each h_t can be uniformly approximated by polynomials.

Lemma 9.6. *For* $t \in [a,b]$ *the function* h_t *is approximable by polynomials uniformly on* $[a,b]$.

Proof. Shifting a polynomial yields again a polynomial, hence we may suppose that $t = 0 \in [a,b]$. Since $h_0(s) = \frac{1}{2}(s + |s|)$ it suffices to approximate the modulus function $|s|$. Enlarging the interval we may suppose that $[a,b]$ is symmetric around 0, and then scaling the variable we may suppose that $[a,b] = [-1,1]$. Now we write Ex.9.4

$$|s| = \sqrt{s^2} = \sqrt{1 - (1 - s^2)} = \sum_{n=0}^{\infty} \binom{1/2}{n} (1 - s^2)^n$$

and note that the convergence of the power series is uniform for $0 \le 1 - s^2 \le 1$, i.e., for $|s| \le \sqrt{2}$. (See Exercise 9.14 for a detailed analysis or Exercise 9.15 for an alternative.) Hence the partial sums of the power series are polynomials that approximate $|s|$ uniformly on $[-1,1]$ as desired. □

9.2. Truncation

Truncation is one of the most important approximation techniques. In its simplest form it is used to approximate functions on infinite domains by functions that "live" only on a compact subset. We shall illustrate this for the domain $\mathbb{R}$, but similar results hold for any unbounded interval.

In the following we shall derive some approximation results for L^p-spaces. Later we shall use results only for the cases $p \in \{1, 2, \infty\}$ and you can restrict without harm the exponent p to these cases.

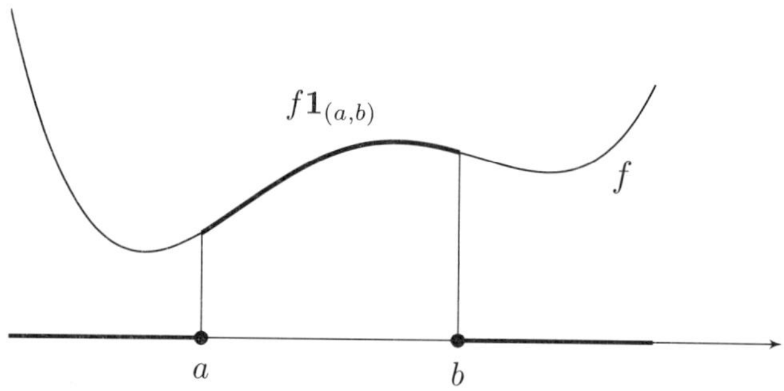

Figure 16. The function f is truncated outside (a, b).

A function $f : \mathbb{R} \to \mathbb{C}$ is said to have **compact support** if f vanishes outside some finite interval (a, b). Equivalently, $f = f\mathbf{1}_{(a,b)}$ for some finite interval $[a, b]$. For $1 \le p \le \infty$ we write

$$\mathrm{L}^p_c(\mathbb{R}) := \{f \in \mathrm{L}^p(\mathbb{R}) \mid f \text{ has compact support}\}$$

and note that this is a linear subspace of $\mathrm{L}^p(\mathbb{R})$. In working with these spaces it is convenient to regard $\mathrm{L}^p(a, b)$ as a subspace of $\mathrm{L}^p(\mathbb{R})$ by identifying a function $f \in \mathrm{L}^p(a, b)$ with its extension

$$\tilde{f}(s) := \begin{cases} f(s) & \text{if } s \in (a, b), \\ 0 & \text{if } s \notin (a, b) \end{cases}$$

to all of $\mathbb{R}$ by 0 outside of $[a, b]$. Then the embedding $\mathrm{L}^p(a, b) \subseteq \mathrm{L}^p(\mathbb{R})$ is $\|\cdot\|_p$-isometric and we may write

$$\mathrm{L}^p_c(\mathbb{R}) = \bigcup\nolimits_{a<b} \mathrm{L}^p(a, b).$$

Clearly, multiplying a function $f \in \mathrm{L}^p(\mathbb{R})$ by a characteristic function $\mathbf{1}_{(a,b)}$ yields the function

$$f\mathbf{1}_{(a,b)} = \begin{cases} f & \text{on } (a, b), \\ 0 & \text{outside of } (a, b), \end{cases}$$

which is called a **truncation** of f for obvious reasons; see Figure 16.

The characteristic function $\mathbf{1}_{(a,b)}$ plays the role of a **cutoff** function — multiplying by it "cuts off" those parts of f that are outside (a, b).

Lemma 9.7. *For $1 \le p < \infty$ the space $\mathrm{L}^p_c(\mathbb{R})$ is $\|\cdot\|_p$-dense in $\mathrm{L}^p(\mathbb{R})$. More precisely, if $f \in \mathrm{L}^p(\mathbb{R})$, then $\left\|f - f\mathbf{1}_{(-n,n)}\right\|_p \to 0$ as $n \to \infty$.*

Proof. We treat the case $p = 2$. Let $f \in \mathrm{L}^2(\mathbb{R})$ and define $f_n := f \cdot \mathbf{1}_{(-n,n)}$ as an approximation. Since $|f_n| \le |f|$ it is clear that $f_n \in \mathrm{L}^2(\mathbb{R})$ for each $n \in \mathbb{N}$, and $|f - f_n|^2 \to 0$ pointwise. Moreover,

$$|f - f_n|^2 \le |f|^2 \in \mathrm{L}^1(\mathbb{R})$$

and hence the dominated convergence theorem (Theorem 7.16) yields

$$\|f - f_n\|_2^2 = \int_{\mathbb{R}} |f - f_n|^2 \to 0 \quad \text{as } n \to \infty.$$

The case $p = 1$ is left as an exercise; see also Exercise 7.18. □

Lemma 9.7 becomes false for $p = \infty$: take $f = \mathbf{1}$ and note that one has $\|\mathbf{1} - g\|_{\mathrm{L}^\infty} \ge 1$ whenever g has compact support.

The truncation method described above is rough in the sense that the cutoff at the boundary is abrupt. That is to say, $f\mathbf{1}_{(a,b)}$ is in general not continuous even if f is. Let us write

$$\mathrm{C_c}(\mathbb{R}) := \{f \in \mathrm{C}(\mathbb{R}) \mid f \text{ has compact support}\}$$

for the space of continuous functions with compact support. Then we have

$$\mathrm{C_c}(\mathbb{R}) \subseteq \mathrm{L}^1(\mathbb{R}) \cap \mathrm{C}_b(\mathbb{R}) \subseteq \mathrm{L}^p(\mathbb{R})$$

(see Exercise 7.29). The canonical extension $\tilde{f}$ by 0 to $\mathbb{R}$ of any $f \in \mathrm{C}[a, b]$ is continuous precisely if $f(a) = f(b) = 0$, i.e., $f \in \mathrm{C}_0[a, b]$. Therefore, we can regard $\mathrm{C}_0[a, b] \subseteq \mathrm{C_c}(\mathbb{R})$ canonically, and hence can write

$$\mathrm{C_c}(\mathbb{R}) = \bigcup\nolimits_{a<b} \mathrm{C}_0[a, b].$$

In order to preserve continuity after truncation, one has to use *continuous* cutoff functions. The simplest are piecewise linear, namely

$$\varphi_n(t) := \begin{cases} n(t-a), & t \in [a, a + 1/n], \\ 1, & t \in [a + 1/n, b - 1/n], \\ -n(t-b), & t \in [b - 1/n, b]; \end{cases}$$

see Figure 17. Then $\varphi_n \in \mathrm{C}_0[a, b]$ and

$$0 \le \varphi_n \le \mathbf{1} \quad \text{and} \quad \varphi_n = \mathbf{1} \quad \text{on } [a + 1/n, b - 1/n]. \tag{9.1}$$

Using these functions we arrive at the following result.

Lemma 9.8. *If $f \in \mathrm{C}[a, b]$, then $f\varphi_n \in \mathrm{C}_0[a, b]$ and $\|f - f\varphi_n\|_p \to 0$ for $1 \le p < \infty$.*

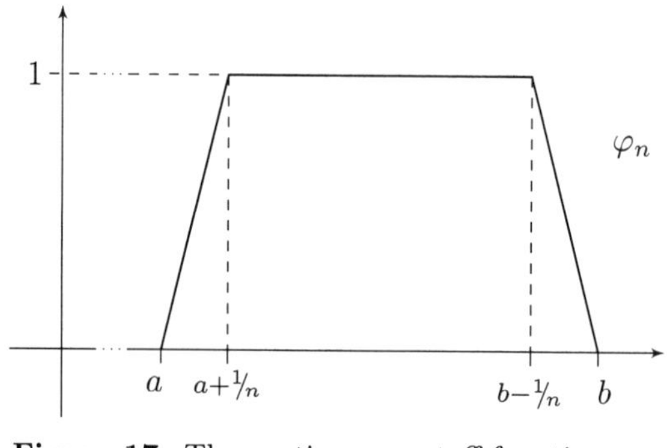

Figure 17. The continuous cutoff function φ_n.

Proof. By (9.1) we have

$$\|f - f\varphi_n\|_p^p = \int_a^{a+1/n} |f - f\varphi_n|^p + \int_{b-1/n}^b |f - f\varphi_n|^p$$
$$\leq \int_a^{a+1/n} |f|^p + \int_{b-1/n}^b |f|^p \leq \frac{2}{n} \|f\|_\infty^p \to 0 \quad \text{as } n \to \infty. \qquad \square$$

Using step-by-step approximation as in the proof of Weierstrass' theorem, we hence can formulate the following central result.

Theorem 9.9. *Let $1 \leq p < \infty$. Then the following assertions hold.*

a) *The space $\mathrm{C}_0^1[a,b]$ is $\|\cdot\|_p$-dense in $\mathrm{L}^p(a,b)$.*

b) *The set $D := \bigcup_{a<b} \mathrm{C}_0^1[a,b]$ is $\|\cdot\|_p$-dense in $\mathrm{L}^p(\mathbb{R})$.*

Proof. a) By Example 4.15, the space $\mathrm{C}_0^1[a,b]$ is $\|\cdot\|_\infty$-dense in $\mathrm{C}_0[a,b]$, hence a fortiori also $\|\cdot\|_p$-dense. By Lemma 9.8, $\mathrm{C}_0[a,b]$ is $\|\cdot\|_p$-dense in $\mathrm{C}[a,b]$, which in turn is $\|\cdot\|_p$-dense in $\mathrm{L}^p(a,b)$ by Theorem 7.24.

b) By a), D is $\|\cdot\|_p$-dense in $\bigcup_{a<b} \mathrm{L}^p(a,b) = \mathrm{L}_c^p(\mathbb{R})$, which is $\|\cdot\|_p$-dense in $\mathrm{L}^p(\mathbb{R})$ by Lemma 9.7. $\square$

We note that the set $D = \bigcup_{a<b} \mathrm{C}_0^1[a,b]$ considered in Theorem 9.9 is merely a *set* and *not* a vector space of functions. This is because if $f \in \mathrm{C}_0^1[a,b]$, then its canonical extension is in general not differentiable neither at a nor at b, only the one-sided derivatives exist. The linear span of D consists of all continuous *piecewise* C^1-functions with compact support.

For completeness we note the following corollary, weaker than Theorem 9.9 but sufficient in many situations.

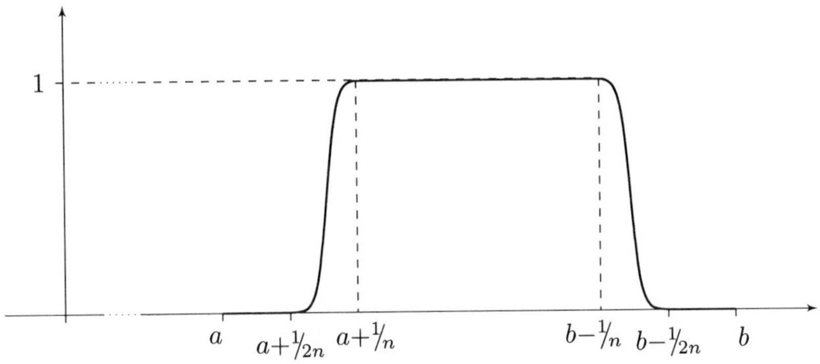

Figure 18. The functions ρ_n.

Corollary 9.10. *The space* $\mathrm{C_c}(\mathbb{R})$ *is* $\|\cdot\|_p$*-dense in* $\mathrm{L}^p(\mathbb{R})$ *for* $1 \le p < \infty$. Ex.9.5

***Approximation by Smooth Functions.** Theorem 9.9 is all we need in the remainder of this chapter, but in many cases it is still too weak. When dealing with boundary value problems one wants to control not only the values of a function at the boundary, but also the values of its derivatives. For this using a cutoff function in $\mathrm{C}_0^1[a,b]$ is not enough, one needs *smooth* cutoff functions as in the following theorem; cf. Figure 18.

Theorem 9.11. *Given any finite interval* $[a,b] \subseteq \mathbb{R}$ *there are functions* $\rho_n \in \mathrm{C}^\infty[a,b]$ *satisfying*

1) $0 \le \rho_n \le 1$;
2) $\rho_n = 1$ *on* $[a+1/n, b-1/n]$;
3) $\rho_n = 0$ *on* $[a, a+1/2n]$ *and on* $[b-1/2n, b]$.

Proof. The proof belongs to real analysis and is given in Appendix D. □

The functions ρ_n from above are contained in the space

$$\mathrm{C}_\mathrm{c}^\infty(a,b) := \{f \in \mathrm{C}^\infty[a,b] \mid f \equiv 0 \text{ in a neighborhood of } a \text{ and } b\}$$

of smooth functions of compact support *within* (a,b). Extending a function $f \in \mathrm{C}_\mathrm{c}^\infty(a,b)$ by 0 to all of $\mathbb{R}$ yields an element of

$$\mathrm{C}_\mathrm{c}^\infty(\mathbb{R}) := \mathrm{C}_\mathrm{c}(\mathbb{R}) \cap \mathrm{C}^\infty(\mathbb{R}),$$

the space of **smooth functions of compact support**. Hence we can write

$$\mathrm{C}_\mathrm{c}^\infty(\mathbb{R}) = \bigcup\nolimits_{a<b} \mathrm{C}_\mathrm{c}^\infty(a,b).$$

We arrive at the following strengthening of Theorem 9.9.

Theorem 9.12. *For $1 \le p < \infty$ the following assertions hold.*

a) *The space $\mathrm{C}_c^\infty(a,b)$ is $\|\cdot\|_\infty$-dense in $\mathrm{C}_0[a,b]$.*

b) *The space $\mathrm{C}_c^\infty(a,b)$ is $\|\cdot\|_p$-dense in $\mathrm{L}^p(a,b)$.*

c) *The space $\mathrm{C}_c^\infty(\mathbb{R})$ is $\|\cdot\|_p$-dense in $\mathrm{L}^p(\mathbb{R})$.*

Proof. a) Let $f \in \mathrm{C}^\infty[a,b] \cap \mathrm{C}_0[a,b]$. Then $\rho_n f \in \mathrm{C}_c^\infty(a,b)$ and $f\rho_n - f$ vanishes on $[a + 1/n, b - 1/n]$. Hence

$$\|f\rho_n - f\|_\infty \le 2 \sup\{|f(x)| \mid \max(|x-a|, |x-b|) \le 1/n\}.$$

Since $f(a) = f(b) = 0$ and f is continuous, it follows that $\|f\rho_n - f\|_\infty \to 0$ as $n \to \infty$. By Example 4.15, $\mathrm{C}^\infty[a,b] \cap \mathrm{C}_0[a,b]$ is $\|\cdot\|_\infty$-dense in $\mathrm{C}_0[a,b]$, hence a) is established.
b) and c) follow from a) as in Theorem 9.9. □

9.3. Classical Fourier Series

Recall from Chapter 1 that the *trigonometric system* consists of the functions

$$e_n(t) = \mathrm{e}^{2\mathrm{i}\pi n t} \qquad (n \in \mathbb{Z}, t \in [0,1]).$$

Every finite linear combination of these functions e_n is called a **trigonometric polynomial**. It was proved in Section 1.3 that $(e_n)_{n\in\mathbb{Z}}$ is an orthonormal system in $\mathrm{L}^2(0,1)$. This system does not come with a canonical enumeration by natural numbers, but rather as a two-sided sequence. In principle one can reduce this case to the one treated by means of any bijection $\mathbb{N} \to \mathbb{Z}$ (see Appendix F), but the notation becomes a little clumsy. So we rather consider

$$\sum_{n=-\infty}^{\infty} \langle f, e_n \rangle e_n$$

instead. Clearly, there is a choice to make here how to interpret this series. The usual interpretation would be as a *double series*

$$\sum_{n=-\infty}^{\infty} \langle f, e_n \rangle e_n = \lim_{N,M\to\infty} \sum_{n=-M}^{N} \langle f, e_n \rangle e_n;$$

see Exercises 3.17 and 5.22. It is then routine to check that the results of Section 5.4 on series in Banach and Hilbert spaces as well as the results of the previous section on abstract Fourier expansions all hold *mutatis mutandis* for such double series; see Exercises 8.11, 8.12 and 8.13.

Alternatively, one may favour the interpretation using *symmetric* partial sums

$$\sum_{n=-\infty}^{\infty} \langle f, e_n \rangle e_n = \lim_{N\to\infty} \sum_{n=-N}^{N} \langle f, e_n \rangle e_n.$$

That is what we will do in the following, for the mere sake of simplicity. It leads to slightly weaker results, but no essential features are lost.

In the context of classical Fourier series the abstract Fourier coefficients of $f \in \mathrm{L}^2(0,1)$ with respect to the trigonometric system are denoted by

$$\widehat{f}(k) := \langle f, e_k \rangle_{\mathrm{L}^2} = \int_0^1 f(s)\mathrm{e}^{-2\pi \mathrm{i} k\, s}\, \mathrm{d}s \qquad (k \in \mathbb{Z}).$$

They are simply called the **Fourier coefficients** of f. Bessel's inequality then reads

$$\sum_{k=-N}^{M} |\widehat{f}(k)|^2 \le \sum_{k=-\infty}^{\infty} |\widehat{f}(k)|^2 \le \int_0^1 |f(s)|^2\, \mathrm{d}s \qquad (f \in \mathrm{L}^2(0,1)).$$

The question whether and in which sense an arbitrary function f on $[0,1]$ is given by its Fourier series

$$f(t) \stackrel{?}{=} \sum_{k=-\infty}^{\infty} \widehat{f}(k)\mathrm{e}^{2\pi \mathrm{i} k\, t} \qquad (t \in (0,1))$$

goes back to Fourier's classical text on the theory of heat, and was one of the main stimuli for the development of analysis in the 19th century. Before the advent of functional analysis, the concept of a norm was unknown and hence the focus laid on pointwise convergence. Among the wealth of results from this period we mention only a single one, with a modern proof by Chernoff[1] [**Che80**]. To formulate it we introduce the space

$$\mathrm{C}_{\mathrm{per}}[0,1] := \{f \in \mathrm{C}[0,1] \mid f(0) = f(1)\}$$

of **1-periodic** continuous functions. This is a $\|\cdot\|_\infty$-closed subspace of $\mathrm{C}[0,1]$ and hence a Banach space; see Exercise 4.11. Its name stems from the fact that each function $f \in \mathrm{C}_{\mathrm{per}}[0,1]$ can be extended uniquely to a continuous function $\tilde{f} \in \mathrm{C}(\mathbb{R})$ on the whole real line satisfying $\tilde{f}(s+1) = \tilde{f}(s)$ for all $s \in \mathbb{R}$.

Theorem 9.13. *Let $f \in \mathrm{C}_{\mathrm{per}}[0,1] \cap \mathrm{C}^1[0,1]$. Then $\sum_{k=-\infty}^{\infty} |\widehat{f}(k)| < \infty$ and*

$$f(t) = \sum_{k=-\infty}^{\infty} \widehat{f}(k) e_k(t)$$

uniformly in $t \in [0,1]$.

Proof. We compute the k-th Fourier coefficient for $k \neq 0$, integrating by parts:

$$\widehat{f}(k) = \int_0^1 f(s)\mathrm{e}^{-2\pi \mathrm{i} k\, s}\, \mathrm{d}s = \frac{1}{2\pi \mathrm{i} k} \int_0^1 f'(s)\mathrm{e}^{-2\pi \mathrm{i} k\, s}\, \mathrm{d}s = \frac{1}{2\pi \mathrm{i} k} \langle f', e_k \rangle .$$

---

[1] Paul Chernoff (1942–), American mathematician.

(The boundary terms vanish because of the periodicity.) This yields

$$\sum_{k\neq 0} |\widehat{f}(k)| = \sum_{k\neq 0} \frac{|\langle f', e_k\rangle|}{2\pi k} \le \left(\sum_{k\neq 0} \frac{1}{4\pi^2 k^2}\right)^{1/2} \left(\sum_{k\neq 0} |\langle f', e_k\rangle|^2\right)^{1/2}$$
$$\le \left(\sum_{k=1}^{\infty} \frac{2}{4\pi^2 k^2}\right)^{1/2} \|f'\|_2 < \infty$$

by Cauchy–Schwarz in $\ell^2(\mathbb{Z})$ and Bessel's inequality in $\mathrm{L}^2(0,1)$. Since $\|e_k\|_\infty = 1$ for all $k \in \mathbb{Z}$, we obtain that the double series

$$g := \sum_{k=-\infty}^{\infty} \widehat{f}(k) e_k$$

is absolutely convergent in the Banach space $\mathrm{C}_{\mathrm{per}}[0,1]$. Consequently, the series converges uniformly to g, and it remains to show that $g = f$.

To this end, fix $t \in [0,1]$ and compute

$$f(t) - \sum_{k=-m}^{n} \widehat{f}(k) e_k(t) = \int_0^1 (f(t) - f(s)) \sum_{k=-m}^{n} \mathrm{e}^{2\pi \mathrm{i} k(t-s)}\, \mathrm{d}s$$
$$= \int_0^1 \frac{f(t)-f(s)}{1-\mathrm{e}^{2\pi\mathrm{i}\,(t-s)}} (1-\mathrm{e}^{2\pi\mathrm{i}\,(t-s)}) \sum_{k=-m}^{n} \mathrm{e}^{2\pi\mathrm{i} k(t-s)}\, \mathrm{d}s$$
$$= \int_0^1 \frac{f(t)-f(s)}{1-\mathrm{e}^{2\pi\mathrm{i}\,(t-s)}} \sum_{k=-m}^{n} \left(\mathrm{e}^{2\pi\mathrm{i}\, k(t-s)} - \mathrm{e}^{2\pi\mathrm{i}\,(k+1)(t-s)}\right) \mathrm{d}s.$$

The sum is telescoping and equals

$$\mathrm{e}^{-2\pi\mathrm{i}\, m(t-s)} - \mathrm{e}^{2\pi\mathrm{i}\,(n+1)(t-s)} = e_m(s) e_{-m}(t) - e_{-(n+1)}(s) e_{n+1}(t).$$

Define the function $h_t(s) := \frac{f(t)-f(s)}{1-\mathrm{e}^{2\pi\mathrm{i}\,(t-s)}}$. Then it is routine to check that
Ex.9.6 $h_t \in \mathrm{C}[0,1]$; see Exercise 9.6. Hence we can write

$$\left|f(t) - \sum_{k=-m}^{n} \widehat{f}(k) e_k(t)\right| = |\langle h_t, e_{-m}\rangle\, e_{-m}(t) - \langle h_t, e_{n+1}\rangle\, e_{n+1}(t)|$$
$$\le |\langle h_t, e_{-m}\rangle| + |\langle h_t, e_{n+1}\rangle| .$$

By Bessel's inequality one has

$$\sum_m |\langle h_t, e_{-m}\rangle|^2 + \sum_n |\langle h_t, e_{n+1}\rangle|^2 < \infty$$

and hence $\lim_{m\to\infty} |\langle h_t, e_{-m}\rangle| = 0 = \lim_{n\to\infty} \left|\langle h_t, e_{n+1}\rangle\right|$. It follows that

$$\lim_{n\to\infty} \left|f(t) - \sum_{k=-n}^{n} \widehat{f}(k) e_k(t)\right| = 0$$

and hence $f(t) = g(t)$. Since $t \in [0,1]$ was arbitrary, it follows that $f = g$, and this concludes the proof. □

Theorem 9.13 can be improved towards functions that are *piecewise* C^1 on $[0,1]$. By relaxing the differentiability assumption, one may lose the uniform convergence but still has pointwise convergence, the classical result being the *Dirichlet-Dini criterion*[2]; see [**Kat04**, 2.5] or [**SS03b**, Theorem 2.1].

It came as a surprise to most mathematicians when in 1873 Paul du Bois-Reymond[3] presented an example of a *continuous* one-periodic function whose Fourier series does not converge pointwise to f. Although the precise construction of such a function (see e.g. [**Kör89**, Chapter 18] or [**SS03b**, Section 3.2.2]) has a certain interest, the mere *existence* of it can nowadays be understood from general functional analytic principles. This will be the subject of Section 15.3 below.

The impact of du Bois-Reymond's result is not to be underrated: it shows that even such "nice" functions as continuous ones may behave badly when it comes to Fourier approximation. So, what should become of Fourier's belief that "every" function is representable by its Fourier series? The answer came in the form of Hilbert space theory.

Theorem 9.14. *The trigonometric system* $(e_n)_{n\in\mathbb{Z}}$ *is a maximal orthonormal system, i.e., an orthonormal basis in* $\mathrm{L}^2(0,1)$.

Proof. We use the characterization (i) from Corollary 8.16 and define

$$F := \overline{\operatorname{span}}\{e_k \mid k \in \mathbb{Z}\},$$

where the closure is in the L^2-norm. We have to show that $F = \mathrm{L}^2(0,1)$. Since F is a subspace and contains each e_k, it contains all trigonometric polynomials. If $f \in \mathrm{C}^1_0[0,1]$, then by Theorem 9.13 its Fourier series converges *uniformly* to f. Since the L^2-norm is weaker, we have

$$f = \lim_{n\to\infty} \sum_{k=-n}^{n} \widehat{f}(k) e_k$$

in the L^2-norm. In particular, we conclude that $f \in F$, whence we obtain $\mathrm{C}^1_0[0,1] \subseteq F$. But $\mathrm{C}^1_0[0,1]$ is $\|\cdot\|_2$-dense in $\mathrm{L}^2(0,1)$ by Theorem 9.9, and this concludes the proof. □

[2]Gustav Lejeune Dirichlet (1805–1859), German mathematician.

[3]Paul Gustave Du Bois-Reymond (1831–1889), German mathematican.

We can now apply Corollary 8.16 to the trigonometric system.

Corollary 9.15. *For every $f \in L^2(0,1)$:*

a) $f(t) = \sum_{k=-\infty}^{\infty} \widehat{f}(k) e^{2k\pi i\, t}$ *as a limit in* $L^2(0,1)$, *and*

b) (Parseval's Identity) $\sum_{k=-\infty}^{\infty} |\widehat{f}(k)|^2 = \int_0^1 |f(s)|^2 \, ds.$

It is very important to note that the convergence in the expression $f(t) = \sum_{k\in\mathbb{Z}} \widehat{f}(k) e^{2\pi i\, t}$ is *not* to be understood pointwise. Corollary 9.15 just states a convergence in square mean, i.e., in the L^2-norm. It is true, however, that the convergence is *pointwise almost everywhere*, but this is a very deep result by Carleson[4] from 1966 [**Car66**].

Finally, we note another consequence of Theorem 9.13.

Theorem 9.16 (Weierstrass). *The space of trigonometric polynomials is dense in* $C_{per}[0,1]$ *with the supremum norm.* $\|\cdot\|_\infty$*-dense in* $C_{per}[0,1]$.

Proof. As above, let $e_n(s) = e^{2\pi i n s}$ and let $F := \overline{\operatorname{span}}\{e_n \mid n \in \mathbb{Z}\}$, the $\|\cdot\|_\infty$-closure of the space of trigonometric polynomials. Then $F \subseteq C_{per}[0,1]$ and

$$C_0^1[0,1] \subseteq C_{per}[0,1] \cap C^1[0,1] \subseteq F,$$

by Theorem 9.13. By Example 4.15 it follows that $C_0[0,1] \subseteq F$. Hence, if $f \in C_{per}[0,1]$ is arbitrary, then

$$f = (f - f(0)\mathbf{1}) + f(0)\mathbf{1} \in F + F \subseteq F,$$

since $f - f(0)\mathbf{1} \in C_0[0,1]$. □

Both theorems of Weierstrass, the polynomial and the trigonometric approximation theorem, are special cases of the so-called Stone–Weierstrass theorem. This beautiful and powerful result is (unfortunately) beyond the scope of this book; see [**Lan93**] or [**Con90**].

[4]Lennart Carleson (1928–), Swedish mathematician, Abel prize 2006.

9.4. Fourier Coefficients of L^1-Functions

We now leave the L^2-Fourier theory for a while and turn to the L^1-theory. For a function $f \in \mathrm{L}^1(0,1)$ we define its **Fourier coefficients** by

$$\widehat{f}(k) := \int_0^1 f(s)\mathrm{e}^{-2\pi \mathrm{i} k\, s}\,\mathrm{d}s \qquad (k \in \mathbb{Z}).$$

The aim is to show that f is completely determined up to a.e.-equality by the two-sided sequence $(\widehat{f}(k))_{k\in\mathbb{Z}}$. Note that for $f \in \mathrm{L}^2(0,1)$ this follows from Corollary 9.15.

We start with an auxiliary result, interesting in its own right. Its proof is again based on an approximation argument.

Lemma 9.17. *Let $f \in \mathrm{L}^1(a,b)$. Then*

$$\|f\|_1 = \sup\left\{\left|\int_a^b f(s)g(s)\,\mathrm{d}s\right| \mid g \in \mathrm{C}[a,b],\ \|g\|_\infty \le 1\right\}. \tag{9.2}$$

In particular, $f = 0$ a.e. if and only if $\int_a^b f(s)g(s)\,\mathrm{d}s = 0$ for all $g \in \mathrm{C}[a,b]$.

Proof. Clearly the second assertions follows from the first. Denote by $p(f)$ the right-hand side of (9.2). Then $p(f) \le \|f\|_1$ by Hölder's inequality. In addition, $p : \mathrm{L}^1(a,b) \to \mathbb{R}_+$ satisfies the triangle inequality and is positively homogeneous. Hence it satisfies the second triangle inequality, i.e.,

$$|p(f) - p(h)| \le p(f-h) \le \|f-h\|_1\,.$$

It follows that p is continuous on $\mathrm{L}^1(a,b)$. But $p(f) = \|f\|_1$ for each $f \in \mathrm{C}[a,b]$, by Example 2.27. Since $\mathrm{C}[a,b]$ is $\|\cdot\|_1$-dense in $\mathrm{L}^1[a,b]$, the claim follows. □

With the help of the lemma we obtain the uniqueness of Fourier coefficients for L^1-functions.

Theorem 9.18 (Uniqueness Theorem for Fourier Series). *If $f, g \in \mathrm{L}^1(0,1)$ satisfy $\widehat{f}(k) = \widehat{g}(k)$ for all $k \in \mathbb{Z}$, then $f = g$ (a.e.).*

Proof. By linearity we may suppose that $g = 0$. Then it follows from Theorem 9.16 and approximation that

$$\int_a^b f(s)g(s)\,\mathrm{d}s = 0 \qquad \text{for every } g \in \mathrm{C_{per}}[0,1].$$

Every $g \in \mathrm{C}[a,b]$ can be written as $g = h + \lambda k$ where h is periodic, $\lambda \in \mathbb{C}$ and $k(s) = s$. By Lemma 9.17 it hence suffices to show that

$$\int_0^1 f(s)s\,\mathrm{d}s = 0.$$

For $0 < n < 1$ define $k_n \in \mathrm{C_{per}}[0,1]$ by

$$k_n(s) := \begin{cases} 1 + s - ns, & 0 \le s \le \frac{1}{n}, \\ s, & \frac{1}{n} \le s \le 1. \end{cases}$$

Then, since $\int_0^1 f(s)k_n(s)\,\mathrm{d}s = 0$,

$$\left|\int_0^1 f(s)s\mathrm{d}s\right| = \left|\int_0^{1/n} f(s)(1-ns)\,\mathrm{d}s\right| \le \int_0^{1/n} |f(s)|\,\mathrm{d}s \to 0$$

as $n \to \infty$, by the dominated convergence theorem. □

Next, we turn to the asymptotics of the sequence of Fourier coefficients.

Theorem 9.19 (Riemann–Lebesgue). *If $f \in \mathrm{L}^1(0,1)$, then* $\lim_{k\to\pm\infty} \widehat{f}(k) = 0$.

Proof. We consider the limit as $k \to \infty$, the other case is treated analogously. Note that by Hölder's inequality,

$$\left|\widehat{f}(k)\right| \le \|f\|_1 \quad \text{for all } k \in \mathbb{Z}.$$

Hence we can define a linear and contractive operator $T : \mathrm{L}^1(0,1) \to \ell^\infty$ by

$$Tf := (\widehat{f}(k))_{k\in\mathbb{N}} \qquad (f \in \mathrm{L}^1(0,1)).$$

If $f \in \mathrm{C}[0,1]$, then by Bessel's inequality we have $Tf \in \ell^2 \subseteq \mathrm{c}_0$. Since c_0 is a closed subspace of ℓ^∞ and $\mathrm{C}[0,1]$ is $\|\cdot\|_1$-dense in $\mathrm{L}^1(0,1)$, it follows by approximation (Lemma 9.3.a)) that $Tf \in \mathrm{c}_0$ for all $f \in \mathrm{L}^1(0,1)$, and that had to be shown. □

The L^1-theory of Fourier series is quite intricate. Theorem 9.19 is the best we know in general about the Fourier coefficients of L^1-functions, but it yields only a sufficient condition: there are in fact sequences $(a_k)_{k\in\mathbb{Z}}$ that are not the Fourier coefficients of any L^1-function but nevertheless satisfy $\lim_{|k|\to\infty} a_k = 0$; see [**Kat04**, 4.2].

9.5. The Riemann–Lebesgue Lemma

For a function $f \in \mathrm{L}^1(\mathbb{R})$ its **Fourier transform** $\mathcal{F}f$ is defined by

$$(\mathcal{F}f)(t) = \int_{\mathbb{R}} f(s)\,\mathrm{e}^{-\mathrm{i}st}\,\mathrm{d}s \qquad (t \in \mathbb{R}). \tag{9.3}$$

The integral is well-defined since

$$\int_{\mathbb{R}} \left|f(s)\,\mathrm{e}^{-\mathrm{i}st}\right| \mathrm{d}s = \int_{\mathbb{R}} |f(s)|\,\mathrm{d}s = \|f\|_1 < \infty.$$

Moreover, by the triangle inequality for integrals it follows that $|(\mathcal{F}f)(t)| \le \|f\|_1$ and taking the supremum over $t \in \mathbb{R}$ we arrive at

$$\|\mathcal{F}f\|_\infty \le \|f\|_1 \qquad (f \in \mathrm{L}^1(\mathbb{R})). \tag{9.4}$$

This shows that the Fourier transform is a bounded linear operator

$$\mathcal{F} : (\mathrm{L}^1(\mathbb{R}), \|\cdot\|_1) \longrightarrow (\mathcal{B}(\mathbb{R}), \|\cdot\|_\infty).$$

Applying the dominated convergence theorem one can show that $\mathcal{F}f$ is a continuous function for every $f \in \mathrm{L}^1(\mathbb{R})$; see Exercise 7.21. Regarding the asympotic behaviour of $\mathcal{F}f(t)$ for large values of $|t|$ we have the following analogue of Theorem 9.19.

Theorem 9.20 (Riemann–Lebesgue[5]). *If $f \in \mathrm{L}^1(\mathbb{R})$, then $\mathcal{F}f \in \mathrm{C}(\mathbb{R})$ and*

$$\lim_{|t|\to\infty} (\mathcal{F}f)(t) = 0.$$

For the proof we proceed as Theorem 9.19. Define

$$\mathrm{C}_0(\mathbb{R}) := \{g \in \mathrm{C}(\mathbb{R}) \mid \lim_{|t|\to\infty} g(t) = 0\}.$$

Then we have

$$\mathrm{C}_c(\mathbb{R}) \subseteq \mathrm{C}_0(\mathbb{R}) \subseteq \mathrm{C}_b(\mathbb{R})$$

since continuous functions are bounded on compact sets. We have to show that $\operatorname{ran}(\mathcal{F}) \subseteq \mathrm{C}_0(\mathbb{R})$. Let us note the following result.

Lemma 9.21. *The space $\mathrm{C}_0(\mathbb{R})$ is $\|\cdot\|_\infty$-closed in $\mathcal{B}(\mathbb{R})$, and hence a Banach space with respect to $\|\cdot\|_\infty$.*

Proof. The proof is a simple 3ϵ-argument and is left as Exercise 9.7. □ Ex.9.7

In the next step we try to find a dense subset $D \subseteq \mathrm{L}^1(\mathbb{R})$ with $\mathcal{F}(D) \subseteq \mathrm{C}_0(\mathbb{R})$. Then Theorem 9.20 follows by approximation (Lemma 9.3.a).
To this aim, fix any interval $[a, b]$, take $f \in \mathrm{C}_0^1[a, b]$ and regard it as a function in $\mathrm{L}^1(\mathbb{R})$ by extending it to all of $\mathbb{R}$ by 0 outside of $[a, b]$. For such a function f it is quite elementary to show that $\mathcal{F}f \in \mathrm{C}(\mathbb{R})$; see Exercise 9.8. Moreover, for $t \neq 0$, integration by parts yields Ex.9.8

$$(\mathcal{F}f)(t) = \int_a^b f(s)\mathrm{e}^{-\mathrm{i}st}\,\mathrm{d}s = \frac{1}{\mathrm{i}t}\int_a^b f'(s)\mathrm{e}^{-\mathrm{i}st}\,\mathrm{d}s.$$

Hence $|(\mathcal{F}f)(t)| \le {}^1\!/\!_{|t|}\, \|f'\|_{\mathrm{L}^1(a,b)} \to 0$ as $|t| \to \infty$.

---

[5]In the literature this is usually called the "Riemann–Lebesgue *lemma*".

It follows that $\mathcal{F}f \in \mathrm{C}_0(\mathbb{R})$ for each f from the set $D = \bigcup_{a<b} \mathrm{C}_0^1[a,b]$, dense in $\mathrm{L}^1(\mathbb{R})$ by Theorem 9.9. As already said, Theorem 9.20 follows by virtue of Lemma 9.3.a). □

9.6. *The Strong Convergence Lemma and Fejér's Theorem

In this (optional) section we prove an important permanence principle and apply it to an approximation result in Fourier analysis.

Let E, F be normed spaces. A sequence $(T_n)_{n\in\mathbb{N}}$ of linear operators $T_n : E \to F$, is called **uniformly bounded** if $\sup_{n\in\mathbb{N}} \|T_n\| < \infty$. (Equivalently, the set $\{T_n \mid n \in \mathbb{N}\}$ is bounded in $\mathcal{L}(E;F)$ with respect to the operator norm.)

Lemma 9.22. *Let E, F be normed spaces, let $T \in \mathcal{L}(E;F)$, and let $(T_n)_{n\in\mathbb{N}}$ be a uniformly bounded sequence of linear operators $T_n : E \to F$. Then*

$$G := \{f \in E \mid \lim_{n\to\infty} T_n f = Tf\}$$

is a closed linear subspace of E.

Proof. The set G is a linear subspace of E by Theorem 4.13. Let $f \in E$ and suppose that $f_n \to f$ with $f_n \in G$ for all $n \in \mathbb{N}$. Denote $c := \sup_{n\in\mathbb{N}} \|T_n\|$, which is a finite number by hypothesis. Then

$$\begin{aligned}\|T_n f - Tf\| &\le \|T_n(f - f_m)\| + \|T_n f_m - Tf_m\| + \|T(f_m - f)\| \\ &\le c\,\|f - f_m\| + \|T\|\,\|f - f_m\| + \|T_n f_m - Tf_m\|\end{aligned}$$

for all $n, m \in \mathbb{N}$. Given $\epsilon > 0$ we can pick m such that each of the first two summands is less than $\epsilon/3$. Since $f_m \in G$ we can then find N such that the third summand is $\le \epsilon/3$ for all $n \ge N$. Hence $\|T_n f - Tf\| \le \epsilon$ for $n \ge N$, and since $\epsilon > 0$ was arbitrary, we conclude that $T_n f \to Tf$, i.e., $f \in G$. □

We say that a sequence $(T_n)_{n\in\mathbb{N}}$ of operators $T_n : E \to F$ **converges strongly on** a set $C \subseteq E$ to an operator $T : E \to F$ if $\lim_{n\to\infty} T_n f = Tf$ for all $f \in C$.

Corollary 9.23 (Strong Convergence Lemma)**.** *Let E, F be normed spaces, let $T \in \mathcal{L}(E;F)$, and let $(T_n)_{n\in\mathbb{N}}$ be a uniformly bounded sequence of linear operators $T_n : E \to F$. If $\lim_{n\to\infty} T_n f = Tf$ for all $f \in C$ for some fundamental subset C of E, then $\lim_{n\to\infty} T_n f = Tf$ for all $f \in E$.*

So in the terminology from above, under the hypothesis of uniform boundedness, if $T_n \to T$ strongly on some fundamental set C, then $T_n \to T$ strongly on all of E. See Exercise 9.17 for a slight strengthening of the strong convergence lemma. We shall encounter strong convergence again in Section 11.3 below. Ex.9.9 Ex.9.10

It turns out that the hypothesis of uniform boundedness of the sequence $(T_n)_{n\in\mathbb{N}}$ is necessary in the strong convergence lemma if E is a Banach space. This follows from the Banach–Steinhaus theorem (Theorem 15.6).

In this section we apply the strong convergence lemma to establish another important result from Fourier analysis, Fejér's theorem.

Convolution Operators and Fejér's Theorem. Every function $f \in \mathrm{C}[0,1]$ with $f(0) = f(1)$ can be uniquely extended to a 1-periodic continuous function on the whole of $\mathbb{R}$. For this reason we usually do not distinguish between f and this extension. In particular, we may integrate f over each finite interval. Then, by the periodicity, we have

$$\int_a^{a+1} f(s)\,\mathrm{d}s = \int_0^1 f(s)\,\mathrm{d}s \qquad (f \in \mathrm{C}_{\mathrm{per}}[0,1],\ a \in \mathbb{R}). \tag{9.5}$$

For a given $k \in \mathrm{C}_{\mathrm{per}}[0,1]$ and $f \in \mathrm{L}^1(0,1)$ we consider their **convolution product**

$$(k * f)(t) := \int_0^1 k(t-s)f(s)\,\mathrm{d}s \qquad (t \in \mathbb{R}).$$

Actually, the resulting function $k * f$ is continuous:

Lemma 9.24. *If $k \in \mathrm{C}_{\mathrm{per}}[0,1]$ and $f \in \mathrm{L}^1(0,1)$, then $k * f \in \mathrm{C}_{\mathrm{per}}[0,1]$.*

Proof. By the 1-periodicity of k it is clear that $k * f$ is 1-periodic, too. For the continuity, let $t, t' \in \mathbb{R}$ and estimate

$$\left|(k * f)(t) - (k * f)(t')\right| \le \int_0^1 \left|k(t-s) - k(t'-s)\right| \left|f(s)\right| \,\mathrm{d}s.$$

The function k is continuous, hence *uniformly continuous* on $[-1,1]$. Thus, for given $\epsilon > 0$ one can find $\delta > 0$ such that $x, x' \in [-1,1]$ with $|x - x'| \le \delta$ implies that $|k(x) - k(x')| \le \epsilon$. Consequently, if $t, t' \in [0,1]$ with $|t - t'| \le \delta$, then $t - s, t' - s \in [-1,1]$ for all $s \in [0,1]$. Moreover, $|(t-s) - (t'-s)| \le \delta$, and hence the estimation from above can be continued with

$$\le \epsilon \int_0^1 |f(s)| \,\mathrm{d}s = \epsilon \left\| f \right\|_1 .$$

It follows that $k * f$ is continuous. □

By Lemma 9.24 we can consider the **convolution operator**

$$T_k : \mathrm{L}^1(0,1) \longrightarrow \mathrm{C}_{\mathrm{per}}[0,1] \subseteq \mathrm{L}^1(0,1), \qquad T_k f := k * f.$$

The following lemma summarizes the properties of this linear(!) operator.

Lemma 9.25. *If $f \in \mathrm{L}^1(0,1)$ and $k \in \mathrm{C}_{\mathrm{per}}[0,1]$ the following assertions hold.*

a) $\|T_k f\|_\infty = \|k * f\|_\infty \le \|k\|_\infty \|f\|_1$.

b) $\|T_k f\|_\infty = \|k * f\|_\infty \le \|k\|_1 \|f\|_\infty$ *if f is bounded.*

c) $\|T_k f\|_1 = \|k * f\|_1 \le \|k\|_1 \|f\|_1$.

d) $\widehat{k * f}(m) = \widehat{k}(m) \cdot \widehat{f}(m)$ *for all $m \in \mathbb{Z}$.*

Proof. We leave a) and b) as an exercise. The proof of c) is simple if f is continuous, because then we can interchange the order of integration and obtain

$$\begin{aligned}\|k * f\|_1 &\le \int_0^1 \int_0^1 |k(t-s)|\,|f(s)|\,\mathrm{d}s\,\mathrm{d}t = \int_0^1 \int_0^1 |k(t-s)|\,\mathrm{d}t\,|f(s)|\,\mathrm{d}s \\ &= \int_0^1 \int_0^1 |k(t)|\,\mathrm{d}t\,|f(s)|\,\mathrm{d}s = \|k\|_1 \|f\|_1 .\end{aligned}$$

This argument still works if $f \in \mathrm{L}^1(0,1)$ but needs *Fubini's theorem* from integration theory (see Section 11.1). However, one can do without this tool and instead employ an approximation argument. Note that from a) we obtain the estimate $\|T_k f\|_1 \le \|T_k f\|_\infty \le \|k\|_\infty \|f\|_1$ for all $f \in \mathrm{L}^1(0,1)$. That is, the convolution operator $T_k : \mathrm{L}^1 \to \mathrm{L}^1$ is bounded. It follows that the subset $A := \{f \in \mathrm{L}^1(0,1) \mid \|T_k f\|_1 \le \|k\|_1 \|f\|_1\}$ is closed in $\mathrm{L}^1(0,1)$. But we have just proved that $\mathrm{C}[0,1] \subseteq A$, and hence $A = \mathrm{L}^1(0,1)$ by approximation.
d) If $f \in \mathrm{C}[0,1]$, then

$$\begin{aligned}\widehat{k * f}(m) &= \int_0^1 (k * f)(t) \mathrm{e}^{2\pi \mathrm{i} m t}\,\mathrm{d}t = \int_0^1 \int_0^1 k(t-s) f(s) \mathrm{e}^{2\pi \mathrm{i} m t}\,\mathrm{d}s\,\mathrm{d}t \\ &= \int_0^1 \int_0^1 k(t-s) \mathrm{e}^{2\pi \mathrm{i} m t}\,\mathrm{d}t\, f(s) \mathrm{d}s \\ &= \int_0^1 \int_0^1 k(t) \mathrm{e}^{2\pi \mathrm{i} m (t+s)}\,\mathrm{d}t\, f(s) \mathrm{d}s \\ &= \int_0^1 \int_0^1 k(t) \mathrm{e}^{2\pi \mathrm{i} m t}\,\mathrm{d}t\, f(s) \mathrm{e}^{2\pi \mathrm{i} m s}\,\mathrm{d}s = \widehat{k}(m) \widehat{f}(m).\end{aligned}$$

To pass from $f \in \mathrm{C}[0,1]$ to a general $f \in \mathrm{L}^1(0,1)$ one can again employ an approximation argument (or Fubini's theorem). We leave this as an
Ex.9.11 exercise. □

We shall now focus on a very special family of convolution operators. For $n \in \mathbb{N}$ let

$$K_n := \frac{1}{n}\sum_{j=0}^{n-1}\sum_{k=-j}^{j} e_k = \frac{1}{n}\sum_{k=-(n-1)}^{n-1}(n-|k|)e_k,$$

where $e_k(s) = \mathrm{e}^{2\pi i k s}$ as before. Note that each function K_n is a trigonometric polynomial. The following lemma lists some of its properties.

Lemma 9.26. *Let K_n be defined as above. Then the following assertions hold.*

a) $K_n(s) = \frac{1}{n}\left|\sum_{k=0}^{n-1} e_k(s)\right|^2 = \frac{1}{n}\left|\frac{\sin \pi n s}{\sin \pi s}\right|^2 \ge 0.$

b) $\|K_n\|_1 = \int_0^1 K_n(s)\,\mathrm{d}s = 1.$

c) $K_n * e_j = e_j$ *for each* $j \in \mathbb{Z}$ *with* $|j| \le n$.

Proof. a) We note that $e_j e_k = e_{j+k}$ and $\overline{e_k} = e_{-k}$ for all $j, k \in \mathbb{Z}$. Therefore

$$\left|\sum_{k=0}^{N} e_k\right|^2 = \left(\sum_{k=0}^{N} e_k\right)\left(\sum_{j=0}^{N} \overline{e_j}\right) = \sum_{j,k=0}^{N} e_{k-j}$$
$$= \sum_{n=-N}^{N} (N+1-|n|)e_k = (N+1)\,K_{N+1}.$$

Putting $N = n-1$ yields the first part of assertion a). The remaining part is left as Exercise 9.12. Ex.9.12

b) Since $K_n \ge 0$ we have $\|K_n\|_1 = \int_0^1 K_n(s)\,\mathrm{d}s$ and the rest is simple.

c) This follows from $e_k * e_j = \langle e_j, e_k\rangle_{\mathrm{L}^2} e_k$ for all $k, j \in \mathbb{Z}$. □

The sequence $(K_n)_{n\in\mathbb{N}}$ is called the **Fejér**[6] **kernel**. (The use of the word "kernel" here is historical, and has nothing to do with a kernel of a linear mapping.) It is important for the following reason.

Theorem 9.27 (Fejér). *If $f \in \mathrm{C}_{\mathrm{per}}[0,1]$, then $K_n * f \to f$ uniformly on $[0,1]$. If $f \in \mathrm{L}^1(0,1)$, then $K_n * f \to f$ in $\|\cdot\|_1$.*

In most books on Fourier analysis, the first assertion in Fejér's theorem is proved directly; see e.g., [**SS03b**, Theorem 5.2]. Here we derive it from our previous results by means of the strong convergence lemma.

Proof. Let us define $T_n f := K_n * f$ and consider these operators on $\mathrm{C}_{\mathrm{per}}[0,1]$ with the supremum norm. By Lemma 9.25.a), $\|T_n\| \le \|K_n\|_1 = 1$ for all $n \in \mathbb{N}$, whence the operators T_n are uniformly bounded. By Lemma 9.26, $T_n f = K_n * f = f$ for all large $n \in \mathbb{N}$, if f is one of the functions e_j. But by

[6]Lipót Fejér (1880–1959), formerly Leopold Weiss, Hungarian mathematician.

the trigonometric Weierstrass theorem (Theorem 9.16) these functions form a fundamental set in $\mathrm{C}_{\mathrm{per}}[0,1]$, hence it follows from the strong convergence lemma that $T_n \to \mathrm{I}$ strongly on $\mathrm{C}_{\mathrm{per}}[0,1]$.

The L^1-part of Fejér's theorem is proved the same way, with the only difference that the operators T_n are now considered on the space $\mathrm{L}^1(0,1)$. Note
Ex.9.13 that the e_j also form a fundamental set in the space $\mathrm{L}^1(0,1)$ (why?). □

9.7. *Extension of a Bounded Linear Mapping

Let $f : \Omega \to \Omega'$ be a continuous function between metric spaces, and $D \subseteq \Omega$ dense in Ω. Then Lemma 9.3.b) says that f is uniquely determined by its values on the points $x \in D$. That is, to know f it suffices to know $f(x)$ for all $x \in D$, and $f(y)$ can be reconstructed (by means of an approximation) from these values.

In many situations, however, one starts with a continuous mapping

$$f_0 : D \longrightarrow \Omega'$$

defined only on D. Is it possible to extend this mapping continuously to the whole of Ω? That is, does there exist a continuous mapping $f : \Omega \to \Omega'$ such that $f(x) = f_0(x)$ for all $x \in D$? At least, the result above tells that there is *at most one* such function f.

The answer to this question is: no, not in general. Just consider

$$f_0 : (0,1] \longrightarrow \mathbb{R}, \qquad f_0(x) = \frac{1}{x}$$

with $\Omega = [0,1]$. What goes wrong here is readily understood when one realizes that $[0,1]$ is compact. Hence every continuous extension f of f_0 to $[0,1]$ must be uniformly continuous, and therefore f_0 must be uniformly continuous as well. This yields, at least in this example, a necessary condition for f_0.

In the context of Banach spaces, we have the following result.

Theorem 9.28 (Extension Theorem). *Let E be a normed space and let $E_0 \subseteq E$ be a dense subspace. Furthermore, let $T_0 : E_0 \to F$ be a bounded linear operator into a Banach space F. Then T_0 extends uniquely to a bounded linear operator $T : E \to F$. Moreover, T has the same norm as T_0, i.e.,*

$$\|T\|_{E\to F} = \|T_0\|_{D\to F}\,.$$

Proof. The uniqueness is clear from Lemma 9.3.b) since bounded operators are continuous. For the existence, let $f \in E$ be an arbitrary element. Since E_0 is dense in E we find elements $f_n \in E_0$ with $f_n \to f$. Then we would like to define $Tf := \lim_{n\to\infty} T_0 f_n$.

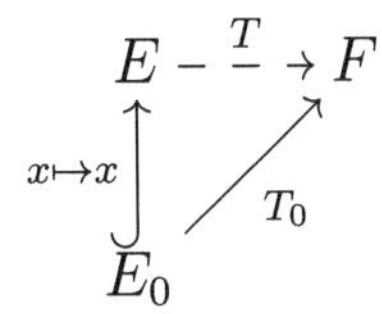

Figure 19. Extension of a bounded linear mapping

In order to do this we have to ensure (1) that the limit exists, and (2) that it depends only on f and not on the chosen sequence $(f_n)_{n\in\mathbb{N}}$. Now, $(f_n)_{n\in\mathbb{N}}$ is Cauchy, whence

$$\|T_0 f_n - T_0 f_m\| = \|T_0(f_n - f_m)\| \le \|T_0\| \, \|f_n - f_m\| \to 0 \qquad (n, m \to \infty).$$

So $(T_0 f_n)_{n\in\mathbb{N}}$ is Cauchy as well, and since F is complete, the limit $Tf := \lim_{n\to\infty} T_0 f_n$ exists in F.

If also $g_n \to f$, then $g_n - f_n \to 0$, and by the boundedness of T_0 we conclude that

$$T_0 f_n - T_0 g_n = T_0(f_n - g_n) \to 0$$

as well. It follows that $\lim_{n\to\infty} T_0 f_n = \lim_{n\to\infty} T_0 g_n$.

Hence our mapping $T : E \to F$ is well-defined. Since for $f \in E_0$ we can choose the constant sequence $f_n := f$, it follows that T coincides with T_0 on E_0. To show linearity, we pick $f, g \in E$ and $\lambda \in \mathbb{K}$, and approximating sequences $f_n \to f$ and $g_n \to g$ with $f_n, g_n \in E_0$. Then $f_n + \lambda g_n \to f + \lambda g$, and by definition of T and the linearity of T_0,

$$\begin{aligned} T(f + \lambda g) &= \lim_n T_0(f_n + \lambda g_n) = \lim_n T_0 f_n + \lambda T_0 g_n \\ &= \lim_n T_0 f_n + \lambda \lim_n T_0 g_n = Tf + \lambda Tg. \end{aligned}$$

Finally, by taking the limit in the estimate $\|T_0 f_n\| \le \|T_0\| \, \|f_n\|$, we obtain

$$\|Tf\| \le \|T_0\| \, \|f\| \qquad \text{for every } f \in E.$$

This implies $\|T\| \le \|T_0\|$, and since the converse inequality is trivial, the proof is complete. □

We note that the theorem is trivially false if the target space F is not complete.

Example 9.29. Suppose that $\varphi : \mathrm{C}[a,b] \to \mathbb{C}$ is a linear functional, bounded for the L^2-norm: $|\varphi(f)| \le c \, \|f\|_2$. Since $\mathrm{C}[a,b]$ is dense in $\mathrm{L}^2(a,b)$, Theorem 9.28 yields that φ can be uniquely extended to a bounded linear functional on $\mathrm{L}^2(a,b)$ — denoted by φ again — with $\|\varphi\| \le c$. Then by the Riesz–Fréchet theorem there is $g \in \mathrm{L}^2(a,b)$ with $\varphi(f) = \int_a^b fg$ for all $f \in \mathrm{L}^2(a,b)$.

In this way, the extension leads to a concrete *representation* of the functional. More results about concrete representations of functionals can be found in Section 16.3 below.

In the following we shall study a very important application of the extension theorem.

Plancherel's Theorem. Recall that the Fourier transform is defined on $L^1(\mathbb{R})$ by means of the integral

$$(\mathcal{F}f)(t) = \int_{\mathbb{R}} f(s)e^{-its}\,ds \quad (t \in \mathbb{R}, f \in L^1(\mathbb{R})).$$

We have seen in Section 9.5 that this constitutes a bounded linear operator $\mathcal{F} : L^1(\mathbb{R}) \to C_0(\mathbb{R})$. Plancherel's theorem[7] states that the Fourier transform acts (almost) isometrically on $L^2(\mathbb{R})$. The basis is the following result from elementary Fourier analysis. A proof can be found in Appendix E.1.

Theorem 9.30 (Plancherel's Identity). *Let $f \in C_c(\mathbb{R})$. Then $\mathcal{F}f \in C_0(\mathbb{R}) \cap L^2(\mathbb{R})$ and*

$$\|\mathcal{F}f\|_2 = \sqrt{2\pi}\,\|f\|_2 .$$

Plancherel's identity yields in particular that the Fourier transform is a bounded operator

$$\mathcal{F} : (C_c(\mathbb{R}), \|\cdot\|_2) \longrightarrow (L^2(\mathbb{R}), \|\cdot\|_2).$$

Since $C_c(\mathbb{R})$ is $\|\cdot\|_2$-dense in $L^2(\mathbb{R})$ by Corollary 9.10, by the extension theorem (Theorem 9.28) the Fourier transform extends uniquely to a bounded linear operator

$$\tilde{\mathcal{F}} : L^2(\mathbb{R}) \longrightarrow L^2(\mathbb{R})$$

also called the **Fourier transform** (on L^2). For $f \in L^1(\mathbb{R}) \cap L^2(\mathbb{R})$ we thus have now "two" Fourier transforms, namely $\mathcal{F}f$ and $\tilde{\mathcal{F}}f$. The following result shows that this distinction is unneccessary.

Lemma 9.31. *If $f \in L^1(\mathbb{R}) \cap L^2(\mathbb{R})$, then $\tilde{\mathcal{F}}f = \mathcal{F}f$. In particular, $\mathcal{F}f \in L^2(\mathbb{R})$ with $\|\mathcal{F}f\|_2 = \sqrt{2\pi}\,\|f\|_2$.*

For the proof we need, once more, an approximation result.

Lemma 9.32. *The space $C_c(\mathbb{R})$ is simultaneously dense in $L^1(\mathbb{R})$ and $L^2(\mathbb{R})$. That means, given $f \in L^1(\mathbb{R}) \cap L^2(\mathbb{R})$ there is a sequence $f_n \in C_c(\mathbb{R})$ such that*

$$\|f_n - f\|_1 + \|f_n - f\|_2 \to 0 \qquad \text{as } n \to \infty.$$

[7] Michel Plancherel (1885–1967), Swiss mathematician.

Proof. We have $f\mathbf{1}_{(-n,n)} \to f$ in both norms by Lemma 9.7. Hence one can find $b > 0$ and $g \in \mathrm{L}^2(-b,b)$ which is arbitrarily close to f in both norms. This g can in turn be $\|\cdot\|_2$-approximated from within $\mathrm{C}_0[-b,b]$, by Theorem 9.9. But on a finite interval the L^1-norm is weaker than the L^2-norm, and hence the chosen approximation will be arbitrarily close to g in both norms simultaneously; cf. Theorem 9.2.c). □

Proof of Lemma 9.31. Let $f \in \mathrm{L}^1(\mathbb{R}) \cap \mathrm{L}^2(\mathbb{R})$ and define $g := \mathcal{F}f$ and $h := \tilde{\mathcal{F}}f$ its Fourier transforms in the L^1-sense and the L^2-sense, respectively. By Lemma 9.32 pick a sequence $f_n \in \mathrm{C}_c(\mathbb{R})$ such that $f_n \to f$ in both L^p-norms. Then $\mathcal{F}f_n \to g$ uniformly on $\mathbb{R}$ and $\mathcal{F}f_n \to h$ in L^2-norm. Fix any interval $[a,b]$. Then

$$\Big(\int_a^b |g-h|^2\Big)^{1/2} \le \Big(\int_a^b |g-\mathcal{F}f_n|^2\Big)^{1/2} + \Big(\int_a^b |\mathcal{F}f_n - h|^2\Big)^{1/2}$$
$$\le \sqrt{b-a}\,\|g-\mathcal{F}f_n\|_\infty + \|\mathcal{F}f_n - h\|_2 \to 0.$$

Hence $g = h$ almost everywhere on $[a,b]$, and since the interval was arbitrary, $g = h$ almost everywhere on $\mathbb{R}$. □

Because of Lemma 9.31 there is no harm in dropping the symbol $\tilde{\mathcal{F}}$ and writing $\mathcal{F}f$ also for the Fourier transform on L^2.

Finally, and again as a consequence of an approximation result, we state and prove Plancherel's theorem.

Theorem 9.33 (Plancherel). *The operator*

$$(1/\sqrt{2\pi})\mathcal{F} : \mathrm{L}^2(\mathbb{R}) \longrightarrow \mathrm{L}^2(\mathbb{R})$$

is an isometric isomorphism. Its inverse operator is given by

$$\Big((1/\sqrt{2\pi})\mathcal{F}\Big)^{-1} = (1/\sqrt{2\pi})S\mathcal{F},$$

where S is the reflection map defined by $Sf(t) = f(-t)$, $t \in \mathbb{R}$.

Proof. By approximation from within $\mathrm{C}_c(\mathbb{R})$ and Plancherel's identity, we have $\|\mathcal{F}f\|_2 = \sqrt{2\pi}\,\|f\|_2$ for all $f \in \mathrm{L}^2(\mathbb{R})$. Hence the operator $T := (1/\sqrt{2\pi})\mathcal{F}$ is an isometry on $\mathrm{L}^2(\mathbb{R})$. In particular, T is injective, and since $\mathrm{L}^2(\mathbb{R})$ is complete, T has closed range $\mathrm{ran}(T)$, see Exercise 5.15. By Corollary E.7 $\mathrm{ran}(T)$ contains the set $D = \bigcup_{a<b} \mathrm{C}^1_0[a,b]$. This is dense in $\mathrm{L}^2(\mathbb{R})$ by Theorem 9.9, so T is surjective.

By Theorems E.5 and E.4, the Fourier inversion formula

$$f(s) = \frac{1}{2\pi}\int_{\mathbb{R}} (\mathcal{F}f)(t)\,\mathrm{e}^{\mathrm{i}ts}\,\mathrm{d}t = (STTf)(s)$$

holds for each $f \in \mathrm{C}_0^1[a,b]$. Since the operators I and $(ST)T$ are bounded, it follows by approximation and Theorem 9.9 that they coincide everywhere, i.e., $(ST)T = \mathrm{I}$; cf. Lemma 9.3.b). Hence $T^{-1} = ST$, as claimed. □

Exercises 9A

Exercise 9.1. Prove the following statements:

a) Let (Ω, d) be a metric space and $A, B \subseteq \Omega$ such that $A \subseteq \overline{B}$. If A is dense in Ω, then so is B.

b) Let $(\Omega, d), (\Omega', d')$ be metric spaces and $f : \Omega \to \Omega'$ be a continuous mapping. Prove that if $A \subseteq \Omega$ is dense in Ω, then $f(A)$ is dense in $f(\Omega)$.

c) Let d_w, d_s be two metrics on a set Ω, such that d_w is weaker than d_s in the sense that
$$d_s(x_n, x_\infty) \to 0 \quad \Longrightarrow \quad d_w(x_n, x_\infty) \to 0$$
for every sequence $(x_n)_{n\in\mathbb{N}\cup\{\infty\}}$ in Ω. If $A \subseteq \Omega$ is d_s-dense in Ω, it is also d_w-dense in Ω.

Exercise 9.2. We consider the Banach space $\mathrm{C}[a,b]$ with the supremum norm. Let the polynomials q_a, q_b be defined by
$$q_a(s) = \frac{b-s}{b-a}, \qquad q_b(s) := \frac{b-s}{b-a} \qquad (s \in [a,b]).$$
Then consider the linear(!) mapping
$$T : \mathrm{C}[a,b] \longrightarrow \mathrm{C}[a,b], \qquad Tf = f - f(a)q_a - f(b)q_b.$$
Show that T is bounded and $\operatorname{ran}(T) = \mathrm{C}_0[a,b]$. Combine the Weierstrass polynomial approximation Theorem 3.22 with b) of Theorem 9.2 to conclude that $\mathrm{P}[a,b] \cap \mathrm{C}_0[a,b]$ is $\|\cdot\|_\infty$-dense in $\mathrm{C}_0[a,b]$. Then observe that this is essentially the proof given in Example 4.15.

Exercise 9.3. Let (Ω, d) and (Ω', d') be two metric spaces, and let $f, g : \Omega \to \Omega'$ be continuous mappings. Furthermore, let $D \subseteq \Omega$ and $C \subseteq \Omega'$. Give a proof for the following assertions.

a) If C is closed and $f(x) \in C$ for all $x \in D$, then this holds for all $x \in \overline{D}$.

b) If $f(x) = g(x)$ for all $x \in D$, then this holds for all $x \in \overline{D}$.

Exercise 9.4. Given $r > 0$ and a sequence of polynomials $(p_n)_{n\in\mathbb{N}}$ such that $p_n(s) \to |s|$ uniformly on $[-1,1]$, show that $rp_n(s/r) \to |s|$ uniformly on $[-r,r]$.

Exercise 9.5. Give a proof of Corollary 9.10 that avoids the Weierstrass approximation theorem.

Exercise 9.6. Let $f \in \mathrm{C}^1[0,1]$ and $t \in [0,1]$. Show that the function
$$h_t(s) := \frac{f(t) - f(s)}{1 - \mathrm{e}^{2\pi \mathrm{i}(t-s)}} \qquad (s \in [0,1])$$
is continuous on $[0,1]$. (Of course, for $s = t$ one has to define $h_t(s)$ appropriately.)

Exercise 9.7. Show that the space

$$\mathrm{C}_0(\mathbb{R}) = \{f \in \mathrm{C}(\mathbb{R}) \mid \lim_{|t|\to\infty} f(t) = 0\}$$

is a $\|\cdot\|_\infty$-closed subspace of $\mathrm{C}_\mathrm{b}(\mathbb{R})$.

Exercise 9.8. Show by elementary arguments, i.e., without the use of the dominated convergence theorem, that for $f \in \mathrm{L}^1(a,b)$

$$\mathcal{F}f(t) = \int_a^b f(s)\mathrm{e}^{-\mathrm{i}ts}\,\mathrm{d}s$$

is a continuous function of $t \in \mathbb{R}$.

Exercise 9.9. Let E, F be a normed spaces and let $(T_n)_{n\in\mathbb{N}}$ be a uniformly bounded sequence in $\mathcal{L}(E;F)$. Show that the set

$$F := \{f \in E \mid (T_n f)_{n\in\mathbb{N}} \text{ is a Cauchy sequence}\}$$

is a closed subspace of E.

Exercise 9.10. Let E be a normed space and F a Banach space. Let $(T_n)_{n\in\mathbb{N}}$ be a uniformly bounded sequence in $\mathcal{L}(E;F)$. Suppose that $A \subseteq E$ is such that $\operatorname{span}(A)$ is a dense subspace of E, and that the limit

$$\lim_{n\to\infty} T_n f \tag{9.6}$$

exists for each $f \in A$. Show that the limit (9.6) exists for each $f \in E$, and a bounded linear operator $T \in \mathcal{L}(E;F)$ is defined by $TF := \lim_{n\to\infty} T_n f$, $f \in E$.

Exercise 9.11. Complete the proof of Lemma 9.25 by means of an approximation argument.

Exercise 9.12. Show that $K_n(s) = \frac{1}{n}\left|\frac{\sin \pi ns}{\sin \pi s}\right|^2$. [Hint: Note that $e_k = (e_1)^k$ and use the formula for a geometric sum.]

Exercise 9.13. Employ Fejér's theorem to give an alternative proof of Theorem 9.18 about the uniqueness of Fourier coefficients.

Exercises 9B

Exercise 9.14 (Binomial Series). Let $0 < \alpha < 1$ and consider the power series

$$f(x) := \sum\nolimits_{n=0}^\infty \binom{\alpha}{n} (-1)^n x^n$$

where $\binom{\alpha}{n} = \frac{\alpha(\alpha-1)\cdots(\alpha-n+1)}{n!}$ for $n \ge 1$ and $\binom{\alpha}{0} := 1$.

a) Show that the series f has radius of convergence equal to 1.

b) Show that $(1-x)f'(x) = -\alpha f(x)$ for $x \in (-1,1)$.

c) Show, e.g., by differentiating $f(x)(1-x)^{-\alpha}$, that $f(x) = (1-x)^\alpha$ for $x \in (-1,1)$.

d) Show that $(-1)^n\binom{\alpha}{n} \le 0$ for all $n \ge 1$, and conclude that $\sum_{n=1}^\infty \left|\binom{\alpha}{n}\right| = 1$.

Finally, prove that the power series $f(x)$ converges to $(1-x)^\alpha$ uniformly on $[-1,1]$.

Exercise 9.15. Consider in $\mathrm{C}[0,1]$ the sequence of polynomials defined recursively by

$$p_0 := 0 \quad \text{and} \quad p_{n+1}(s) := p_n + \frac{1}{2}(s - p_n(s)^2) \qquad (n \geq 0).$$

Show that $0 \leq p_n(s) \leq \sqrt{s}$ for each $s \in [0,1]$. Conclude that $p_n \leq p_{n+1}$. Then apply Dini's theorem from Exercise 4.33 to prove that $p_n(s) \to \sqrt{s}$ uniformly in $s \in [0,1]$. (Remark: One can replace Dini's theorem in the argument by the estimate

$$(\sqrt{s} - p_n(s))(2 + n\sqrt{s}) \leq 2\sqrt{s} \qquad (n \in \mathbb{N})$$

which implies that $|\sqrt{s} - p_n(s)| \leq 2/n$.)

Exercise 9.16. Formulate and prove a result analogous to Theorem 9.9.b) for the spaces $\mathrm{L}^p(0,\infty)$, $p \in \{1,2\}$.

Exercise 9.17. Let E, F be normed spaces, let $(T_n)_{n\in\mathbb{N}}$ be a uniformly bounded sequence in $\mathcal{L}(E;F)$ such that

$$T_n f \to Tf \quad \text{as } n \to \infty \quad \text{for each } f \in E.$$

Show that then the convergence $T_n f \to Tf$ is *uniform for f from compact subsets of E*. By this we mean that given a compact subset $K \subseteq E$, then

$$\lim_{n\to\infty} \sup_{f\in K} \|T_n f - Tf\| = 0.$$

Exercise 9.18. Show that $\mathrm{C_c}(\mathbb{R})$ is $\|\cdot\|_\infty$-dense in $\mathrm{C}_0(\mathbb{R})$.

Exercise 9.19. Thinking in analogies, how would you define the function spaces $\mathrm{C}_0[0,\infty)$ and $\mathrm{C}_0(0,\infty)$?

Exercise 9.20. A function f on $\mathbb{R}$ is called **locally integrable** if $\mathbf{1}_{(a,b)} f \in \mathrm{L}^1(a,b)$ for every interval $[a,b]$. Show that if f is locally integrable and $\int_{\mathbb{R}} f\varphi = 0$ for every $\varphi \in \mathrm{C_c}(\mathbb{R})$, then $f = 0$.

Exercise 9.21 (Moment Sequence). For $f \in \mathrm{L}^1(0,1)$ its **moment sequence** is

$$a_n(f) := \int_0^1 f(t)t^n \, \mathrm{d}t \quad (n \geq 0).$$

Show that f is uniquely determined by its moment sequence. Then show that $(a_n(f))_{n\in\mathbb{N}} \in \mathrm{c}_0$:

1) by virtue of the dominated convergence theorem;
2) by an approximation argument.

Exercises 9C

Exercise 9.22 (Asymptotic Fourier Inversion Formula on L^2). Show that for $f \in \mathrm{L}^2(\mathbb{R})$ one has

$$f(t) = \lim_{n\to\infty} \frac{1}{2\pi} \int_{-n}^{n} (\mathcal{F}f)(s)\mathrm{e}^{\mathrm{i}st} \, \mathrm{d}s$$

(not pointwise but) as a limit in the L^2-norm. (Note that by Plancherel's theorem $\mathcal{F}f \in \mathrm{L}^2(\mathbb{R})$ and hence $\mathbf{1}_{[-n,n]}(\mathcal{F}f) \in \mathrm{L}^1[-n,n]$ for each $n \in \mathbb{N}$.)

Exercise 9.23 (Riemann–Stieltjes Type Integral of Regulated Functions). Let E be a Banach space and let $[a,b] \subseteq \mathbb{R}$ be a compact interval. Recall from Exercises 5.27 and 2.34 the definitions of the spaces $\mathrm{Reg}([a,b];E)$ of E-valued *regulated* functions and $\mathrm{BV}[a,b]$ of scalar functions of *bounded variation*. The purpose of this exercise is to construct an *integral*

$$\int_a^b f(t)\,\mathrm{d}g(t)$$

for $f \in \mathrm{Reg}([a,b];E)$ in such a way that this coincides with the Riemann integral if E is finite-dimensional and $g(t) = t$, $t \in [a,b]$.

a) Let $f \in \mathrm{St}([a,b];E)$ be a step function. Then there are a partition $a = t_0 < t_1 < \cdots < t_n = b$ and vectors $x_j \in E$ such that

$$f(t) = x_j \quad \text{if} \quad t_{j-1} < t < t_j \qquad (j = 1,\dots,n).$$

Show that the elementary integral

$$\int_a^b f(t)\,\mathrm{d}t := \sum\nolimits_{j=1}^{n} x_j(g(t_j) - g(t_{j-1}))$$

is independent of the chosen partition.

b) Show that the so-constructed mapping

$$\mathrm{St}([a,b];E) \to E, \qquad f \mapsto \int_a^b f(t)\,\mathrm{d}g(t)$$

is linear and satisfies

$$\left\| \int_a^b f(t)\,\mathrm{d}g(t) \right\| \le \|g\|_v \, \|f\|_\infty \, . \tag{9.7}$$

c) Employ Theorem 9.28 to extend the elementary integral from b) to a bounded linear mapping $\mathrm{Reg}([a,b];E) \to E$. We still use the notation $f \mapsto \int_a^b f(t)\,\mathrm{d}g(t)$ for this mapping.

d) Let F be another Banach space, and let $T : E \to F$ be a bounded linear mapping. Show that if $f \in \mathrm{Reg}([a,b];E)$, then $T \circ f \in \mathrm{Reg}([a,b];F)$ and

$$T\Big(\int_a^b f(t)\,\mathrm{d}g(t)\Big) = \int_a^b T(f(t))\,\mathrm{d}g(t).$$

e) Show that each $\mathbb{K}$-valued regulated function is Riemann integrable and that the "new" integral coincides, for the choice $g(t) = t$, with the Riemann integral.

[A word of caution: This integral deviates from the classical Riemann–Stieltjes integral as developed, e.g., in [**Lan93**, Chap. X]. One can show that $f \in \mathrm{Reg}([a,b];E)$ is RS-integrable with respect to $g \in \mathrm{BV}[a,b]$ if all points of discontinuity of f are points of continuity of g; in this case the two integrals coincide.]

Exercise 9.24 (Young's Inequality). Let $1 < p < \infty$, and let $k \in \mathrm{C}_{\mathrm{per}}[0,1]$. Prove **Young's inequality**

$$\|k * f\|_p \le \|k\|_1 \, \|f\|_p \qquad (f \in \mathrm{L}^p(0,1))$$

by the following steps (cf. the proof of Lemma 9.25.c)):

a) First, take $f \in \mathrm{C}[0,1]$; perform a (q/p)-Hölder estimate to the pair

$$|k(t-s)f(s)| = |k(t-s)|^{1/q} \cdot |k(t-s)|^{1/p}\, |f(s)|$$

(integral with respect to s), then raise to the p-th power and integrate with respect to t.

b) Approximate a general $f \in \mathrm{L}^p(0,1)$ with continuous functions.

Exercise 9.25. Let $(K_n)_{n\in\mathbb{N}}$ be the Fejér kernel as in Section 9.6. Let $1 < p < \infty$. Show that $K_n * f \to f$ in $\|\cdot\|_p$ for each $f \in \mathrm{L}^p(0,1)$. [Hint: Imitate the proof of Fejér's theorem employing Young's inequality from Exercise 9.24.]

Exercise 9.26. Formulate and prove an extension theorem for *bilinear* mappings (see Exercise 4.40) analogous to Theorem 9.28.

Exercise 9.27 (Convolution on $\mathbb{R}$). For $f, g \in \mathrm{C_c}(\mathbb{R})$ we define their **convolution** by

$$(f * g)(t) := \int_{\mathbb{R}} f(t-s)g(s)\,\mathrm{d}s \quad (t \in \mathbb{R}). \tag{9.8}$$

Show that $f * g \in \mathrm{C_c}(\mathbb{R})$ and

$$\|f * g\|_1 \le \|f\|_1 \, \|g\|_1 .$$

Clearly, convolution is a bilinear mapping: $* : \mathrm{C_c}(\mathbb{R}) \times \mathrm{C_c}(\mathbb{R}) \to \mathrm{C_c}(\mathbb{R})$. Show that there is a unique extension to a continuous bilinear mapping $* : \mathrm{L}^1(\mathbb{R}) \times \mathrm{L}^1(\mathbb{R}) \to \mathrm{L}^1(\mathbb{R})$, still called **convolution**. [Hint: Exercise 9.26.]

Finally, show that convolution is associative, i.e.,

$$(f * g) * h = f * (g * h) \qquad (f, g, h \in \mathrm{L}^1(\mathbb{R})).$$

Remark: using *Fubini's theorem* one can show that the formula (9.8) is meaningful for $f, g \in \mathrm{L}^1(\mathbb{R})$ in the "almost everywhere"-sense.

Chapter 10

Sobolev Spaces and the Poisson Problem

In this chapter we shall apply Hilbert space methods to differential equations, most prominently the *Poisson problem*[1]

$$u'' = -f \quad \text{on} \quad (a, b), \qquad u(a) = u(b) = 0.$$

In order to do so, we have to pass from the "classical" $\mathrm{C}[a, b]$-setting to the $\mathrm{L}^2(a, b)$-setting; in particular, we need the notion of a derivative of an $\mathrm{L}^2(a, b)$-"function". Since L^2-elements are not really functions, we cannot do this by using the elementary definition via differential quotients.

10.1. Weak Derivatives

Let $[a, b] \subseteq \mathbb{R}$ be a finite interval. Each function $\psi \in \mathrm{C}^1[a, b]$ satisfying $\psi(a) = \psi(b) = 0$ is called a **test function**. The space of all test functions is

$$\mathrm{C}^1_0[a, b] = \{\psi \in \mathrm{C}^1[a, b] \mid \psi(a) = \psi(b) = 0\}.$$

Note that if ψ is a test function, then so is $\overline{\psi}$, with $(\overline{\psi})' = \overline{\psi'}$.

Lemma 10.1. *The space of test functions $\mathrm{C}^1_0[a, b]$ is dense in $\mathrm{L}^2(a, b)$. If $g, h \in \mathrm{L}^2(a, b)$ are such that*

$$\int_a^b g(s)\psi(s)\,\mathrm{d}s = \int_a^b h(s)\psi(s)\,\mathrm{d}s,$$

for all test functions $\psi \in \mathrm{C}^1_0[a, b]$, then $g = h$ (almost everywhere).

[1] Siméon Denis Poisson (1781–1840), French mathematician and physicist.

Proof. The first assertion is just Theorem 9.9. The second follows from the first by approximation (how?). □

Given a function $f \in \mathrm{L}^2(a,b)$ we want to define its derivative f' without using a differential quotient but still in such a way that for $f \in \mathrm{C}^1[a,b]$ the symbol f' retains its classical meaning. Now, *if* $f \in \mathrm{C}^1[a,b]$, then integration by parts yields

$$(10.1) \qquad \int_a^b f'(s)\psi(s)\,\mathrm{d}s = -\int_a^b f(s)\psi'(s)\,\mathrm{d}s \qquad \text{for all } \psi \in \mathrm{C}_0^1[a,b].$$

(The boundary terms vanish since $\psi(a) = 0 = \psi(b)$.) This shows that the classical derivative f' is a "weak derivative" in the following sense.

Definition 10.2. Let $f \in \mathrm{L}^2(a,b)$. A function $g \in \mathrm{L}^2(a,b)$ is called a **weak derivative** of f if

$$(10.2) \qquad \int_a^b g(s)\psi(s)\,\mathrm{d}s = -\int_a^b f(s)\psi'(s)\,\mathrm{d}s \qquad \text{for all } \psi \in \mathrm{C}_0^1[a,b].$$

If f has a weak derivative, we call f **weakly differentiable**. The space

$$\mathrm{H}^1(a,b) := \{f \in \mathrm{L}^2(a,b) \mid f \text{ has a weak derivative}\}$$

is called the (first order) **Sobolev space**.[2]

Note that the condition for a weak derivative can be equivalently written as

$$\langle g, \psi \rangle = -\left\langle f, \psi' \right\rangle \qquad (\psi \in \mathrm{C}_0^1[a,b]).$$

A function $f \in \mathrm{L}^2(a,b)$ can have *at most one* weak derivative. Indeed, if g_1, g_2 are weak derivatives, then

$$\int_a^b g_1(s)\psi(s)\,\mathrm{d}s = -\int_a^b f(s)\psi'(s)\,\mathrm{d}s = \int_a^b g_2(s)\psi(s)\,\mathrm{d}s$$

for all $\psi \in \mathrm{C}_0^1[a,b]$, and hence $g_1 = g_2$ (almost everywhere) by Lemma 10.1.

Since classical derivatives are weak derivatives (as observed above), it is therefore unambiguous to write f' for a (=the) weak derivative of f,
Ex.10.1 *whenever* such a weak derivative exists.
Ex.10.2

The next example shows that there exist weakly differentiable functions that are not (classically) differentiable.

[2] Sergei Lvovich Sobolev (1908–1989), Russian mathematician.

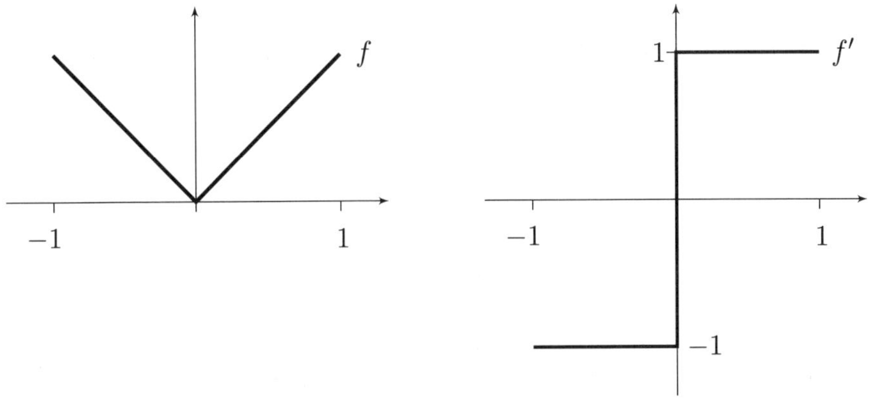

Figure 20. The function $f(t) = |t|$ and its weak derivative on $[-1, 1]$.

Example 10.3. Let $[a, b] = [-1, 1]$ and $f(t) := |t|$, $t \in [-1, 1]$. Then f has weak derivative $f' = \mathbf{1}_{(0,1)} - \mathbf{1}_{(-1,0)}$; indeed,

$$\begin{aligned}
-\int_{-1}^{1} f(t)\psi'(t)\,\mathrm{d}t &= -\int_{-1}^{0} (-t)\psi'(t)\,\mathrm{d}t - \int_{0}^{1} t\psi'(t)\,\mathrm{d}t \\
&= t\psi(t)\Big|_{-1}^{0} - \int_{-1}^{0} \psi(t)\,\mathrm{d}t - t\psi(t)\Big|_{0}^{1} + \int_{0}^{1} \psi(t)\,\mathrm{d}t \\
&= 0 - \int_{-1}^{0} \psi(t)\,\mathrm{d}t - 0 + \int_{0}^{1} \psi(t)\,\mathrm{d}t \\
&= \int_{-1}^{1} \big(\mathbf{1}_{(0,1)} - \mathbf{1}_{(-1,0)}\big)(t)\psi(t)\,\mathrm{d}t \qquad \text{for all } \psi \in \mathrm{C}_0^1[-1, 1].
\end{aligned}$$

Note f' is determined only almost everywhere, and hence it does not matter which value it takes at the single point 0. See Figure 20 and Exercise 10.3. Ex.10.3

It follows from Corollary 10.7 below that there exist functions $f \in \mathrm{L}^2(a, b)$ that are not weakly differentiable.

10.2. The Fundamental Theorem of Calculus

In this section we shall establish basic properties of weak derivatives. It turns out that important results for classical derivatives remain valid in the weak setting. Our main tool is the **integration operator**

$$(Jf)(t) := \int_a^t f(x)\,\mathrm{d}x = \langle f, \mathbf{1}_{(a,t)} \rangle \qquad (t \in [a, b],\, f \in \mathrm{L}^2(a, b));$$

cf. Example 7.17.

Note: This is in general *not* a Riemann integral, even if it looks like one.

Classically, J is a *right inverse* to differentiation, i.e.,

$$(10.3) \qquad (Jf)'(t) = \frac{d}{dt}\int_a^t f(s)\,\mathrm{d}s = f(t) \qquad (t \in [a,b], f \in \mathrm{C}[a,b]).$$

The next result shows that this formula remains true for weak differentiation.

Lemma 10.4. *The operator $J : \mathrm{L}^2(a,b) \to \mathrm{C}[a,b]$ is linear and bounded. Moreover, $(Jf)' = f$ in the weak sense for all $f \in \mathrm{L}^2(a,b)$.*

Ex.10.4 **Proof.** Since $\mathrm{L}^2(a,b) \subseteq \mathrm{L}^1(a,b)$ and by Example 7.17, we have $Jf \in \mathrm{C}[a,b]$ and

$$\|Jf\|_\infty \le \|f\|_1 \le \sqrt{b-a}\,\|f\|_2 \,.$$

Hence, J is bounded as an operator from $(\mathrm{L}^2(a,b), \|\cdot\|_2)$ to $(\mathrm{C}[a,b], \|\cdot\|_\infty)$. The claim $(Jf)' = f$ is, by definition, equivalent to

$$(10.4) \qquad \langle Jf, \psi' \rangle = -\langle f, \psi \rangle$$

for all test functions ψ. We employ a density argument to establish this. Fix a test function ψ and consider the linear functional

$$T : \mathrm{L}^2(a,b) \to \mathbb{C}, \quad Tf := \langle Jf, \psi' \rangle + \langle f, \psi \rangle \,.$$

Then T is bounded (why?), and $Tf = 0$ for all $f \in \mathrm{C}^1[a,b]$, by classical integration by parts. Since $\mathrm{C}^1[a,b]$ is dense in $\mathrm{L}^2(a,b)$, it follows that $Tf = 0$ for all $f \in \mathrm{L}^2(a,b)$ (cf. Lemma 9.3), which is (10.4). □

Classically, i.e., for $f \in \mathrm{C}^1[a,b]$, the function $J(f') - f$ is constant. This comes from the fundamental theorem of calculus, and it implies that if the (classical) derivative of a function is zero, then the function is constant. We shall see that these results remain valid in the setting of weak derivatives.

As in earlier chapters, we write $\mathbf{1}$ for the function being constant to one on $[a,b]$. Then a constant function has the form $f = c\mathbf{1}$ for some number $c \in \mathbb{K}$, and hence $\mathbb{C}\mathbf{1}$ is the one-dimensional subspace of constant functions. A spanning unit vector for this space is $e := (b-a)^{-1/2}\mathbf{1}$, and so

$$P_{\mathbb{C}\mathbf{1}} f = \langle f, e \rangle\, e = \frac{\langle f, \mathbf{1} \rangle}{b-a}\mathbf{1} \qquad (f \in \mathrm{L}^2(a,b))$$

is the orthogonal projection onto $\mathbb{C}\mathbf{1}$. In particular, if $f = c\mathbf{1}$ is constant (almost everywhere) then c is given by

$$c = \frac{\langle f, \mathbf{1} \rangle}{b-a}.$$

We now determine the orthogonal complement $\mathbf{1}^\perp$ of $\mathbb{C}\mathbf{1}$.

Lemma 10.5. *The space* $\mathrm{L}^2(a,b)$ *decomposes orthogonally into*

$$\mathrm{L}^2(a,b) = \mathbb{C}\mathbf{1} \oplus \overline{\{\psi' \mid \psi \in \mathrm{C}_0^1[a,b]\}},$$

with $\|\cdot\|_2$*-closure on the right-hand side.*

Proof. We let $G := \mathbf{1}^\perp$ and $F := \{\psi' \mid \psi \in \mathrm{C}_0^1[a,b]\}$. Then it remains to show that $G = \overline{F}$. Now $\mathbf{1} \perp F$, since

$$\int_a^b \mathbf{1} \cdot \psi'(s)\,\mathrm{d}s = \psi(b) - \psi(a) = 0 - 0 = 0 \qquad \text{for each } \psi \in \mathrm{C}_0^1[a,b].$$

This yields $F \subseteq \mathbf{1}^\perp = G$, and hence $\overline{F} \subseteq G$, because G is closed.

For the converse inclusion, note that the complementary projection is

$$P_G f = (\mathrm{I} - P_{\mathbb{C}\mathbf{1}})f = f - \frac{\langle f, \mathbf{1}\rangle}{b-a}\mathbf{1} \qquad (f \in \mathrm{L}^2(a,b)).$$

Since $\operatorname{ran}(P_G) = G$, we have to show that $\operatorname{ran}(P_G) \subseteq \overline{F}$. To this aim, fix $f \in \mathrm{C}[a,b]$. Then obviously $P_G f \in \mathrm{C}[a,b]$ as well, hence

$$JP_G f = Jf - \frac{\langle f, \mathbf{1}\rangle}{b-a} J\mathbf{1} \in \mathrm{C}^1[a,b].$$

Moreover, $JP_G f(a) = 0$ (by definition of J) and

$$\begin{aligned}(JP_G f)(b) &= (Jf)(b) - \frac{\langle f, \mathbf{1}\rangle}{b-a}(J\mathbf{1})(b) = \int_a^b f(s)\,\mathrm{d}s - \frac{\langle f, \mathbf{1}\rangle}{b-a}\int_a^b 1\,\mathrm{d}s \\ &= \langle f, \mathbf{1}\rangle - \langle f, \mathbf{1}\rangle = 0.\end{aligned}$$

But this means that $JP_G f \in \mathrm{C}_0^1[a,b]$, whence $P_G f = (JP_G f)' \in F$.

For general $f \in \mathrm{L}^2(a,b)$ we find a sequence $(f_n)_{n\in\mathbb{N}}$ in $\mathrm{C}[a,b]$ with $f_n \to f$ in $\mathrm{L}^2(a,b)$. Then $P_G f_n \to P_G f$ since any orthogonal projection is bounded. It follows that $P_G f \in \overline{F}$; cf. Lemma 9.3. □

Corollary 10.6. *Let* $f \in \mathrm{L}^2(a,b)$ *such that* f *has weak derivative* $f' = 0$. *Then* f *is a constant (a.e.).*

Proof. Let $f \in \mathrm{L}^2(a,b)$ and suppose that $f' = 0$ in the weak sense. By definition, this means

$$\int_a^b f(s)\psi'(s)\,\mathrm{d}s = 0 \qquad (\psi \in \mathrm{C}_0^1[a,b]),$$

and this can be rephrased as $f \perp F$, with F being as in the previous proof. Hence $f \perp \overline{F} = (\mathbb{C}\mathbf{1})^\perp$, and thus $f \in \mathbb{C}\mathbf{1}$, by Corollary 8.11. □

Let us formulate an interesting corollary.

Corollary 10.7. *One has* $\mathrm{H}^1(a,b) \subseteq \mathrm{C}[a,b]$. *More precisely,* $f \in \mathrm{H}^1(a,b)$ *if and only if* f *has a representation*

$$f = Jg + c\mathbf{1}$$

with $g \in \mathrm{L}^2(a,b)$ *and* $c \in \mathbb{K}$. *Such a representation is unique, namely*

$$g = f' \quad \textit{and} \quad c = \frac{\langle f - Jf', \mathbf{1}\rangle}{b-a}.$$

Moreover, the **fundamental theorem of calculus** *holds, i.e.,*

$$\int_c^d f'(s)\,\mathrm{d}s = f(d) - f(c) \quad \textit{for every interval } [c,d] \subseteq [a,b].$$

Proof. If $f = Jg + c\mathbf{1}$ is represented as described, then $f' = (Jg)' + 0 = g$ and $f \in \mathrm{H}^1(a,b)$. Conversely let $f \in \mathrm{H}^1(a,b)$ and set $g := f'$ and $h := Jg$. Then $(f-h)' = f' - (Jg)' = f' - g = 0$, and by Corollary 10.6 there exists $c \in \mathbb{K}$ such that $f - Jg = c\mathbf{1}$. The remaining statement follows from

$$\int_c^d f'(s)\,\mathrm{d}s = (Jg)(d) - (Jg)(c) = f(d) - f(c). \qquad \square$$

In particular, a function which does not coincide almost everywhere with a continuous function cannot have a weak derivative. For example, within $(-1,1)$ the characteristic function $\mathbf{1}_{(0,1)}$ does not have a weak derivative; see Exercise 10.5.

On the other hand, *not every* continuous function is weakly differentiable. For instance, the function $f(s) = \sqrt{s}$ on $[0,1]$ is continuous, but is not contained in $\mathrm{H}^1(0,1)$ (why?).

10.3. Sobolev Spaces

On $\mathrm{H}^1(a,b)$ we define the inner product(!)

$$\langle f, g\rangle_{\mathrm{H}^1} := \langle f, g\rangle_{\mathrm{L}^2} + \langle f', g'\rangle_{\mathrm{L}^2} = \int_a^b f(s)\overline{g(s)}\,\mathrm{d}s + \int_a^b f'(s)\overline{g'(s)}\,\mathrm{d}s$$

for $f, g \in \mathrm{H}^1(a,b)$. Writing out the norm of $f \in \mathrm{H}^1(a,b)$ we obtain

$$\|f\|_{\mathrm{H}^1} = \left(\|f\|_2^2 + \|f'\|_2^2\right)^{1/2}.$$

Hence, *convergence of a sequence* $(f_n)_{n\in\mathbb{N}}$ *in* $\mathrm{H}^1(a,b)$ *is the same as the simultaneous* L^2*-convergence of* $(f_n)_{n\in\mathbb{N}}$ *and of* $(f_n')_{n\in\mathbb{N}}$.

Theorem 10.8. *The first Sobolev space* $\mathrm{H}^1(a,b)$ *is a Hilbert space, and the mapping*

$$\mathrm{H}^1(a,b) \longrightarrow \mathrm{L}^2(a,b), \qquad f \longmapsto f'$$

is linear and bounded. Moreover, the inclusion $\mathrm{H}^1(a,b) \subseteq \mathrm{C}[a,b]$ *is continuous, i.e., there is a constant* $c > 0$ *such that*

$$\|f\|_\infty \le c\,\|f\|_{\mathrm{H}^1} \qquad (f \in \mathrm{H}^1(a,b)).$$

Proof. By Exercise 10.2, $\mathrm{H}^1(a,b)$ is a vector space, and the derivative is a linear mapping. By definition of the norm

$$\|f'\|_2^2 \le \|f\|_2^2 + \|f'\|_2^2 = \|f\|_{\mathrm{H}^1}^2\,,$$

and so $(f \mapsto f')$ is a bounded mapping.
Take a Cauchy sequence $(f_n)_{n\in\mathbb{N}} \subseteq \mathrm{H}^1(a,b)$. This means that

$$\|f_n - f_m\|_2^2 + \|f_n' - f_m'\|_2^2 = \|f_n - f_m\|_{\mathrm{H}^1}^2 \to 0 \qquad \text{as } n,m \to \infty.$$

Hence both sequences $(f_n)_{n\in\mathbb{N}}$ and $(f_n')_{n\in\mathbb{N}}$ are $\|\cdot\|_2$-Cauchy sequences. By the completeness of $\mathrm{L}^2(a,b)$, there are functions $f, g \in \mathrm{L}^2(a,b)$ such that

$$\|f_n - f\|_2 \to 0, \quad \text{and} \quad \|f_n' - g\|_2 \to 0.$$

It remains to show that g is a weak derivative of f. In order to establish this, let $\psi \in \mathrm{C}_0^1[a,b]$. Then

$$\langle g, \psi\rangle_{\mathrm{L}^2} = \lim_n \langle f_n', \psi\rangle_{\mathrm{L}^2} = \lim_n -\langle f_n, \psi'\rangle = -\langle f, \psi'\rangle\,,$$

where we used that f_n' is a weak derivative of f_n. The remaining statement is left as an exercise. □ Ex.10.6

Corollary 10.9. *The space* $\mathrm{C}^1[a,b]$ *is dense in* $\mathrm{H}^1(a,b)$.

Proof. Let $f \in \mathrm{H}^1(a,b)$ and $g := f'$. Then $f = Jg + c\mathbf{1}$ for some $c \in \mathbb{K}$, by Corollary 10.7. Since $\mathrm{C}[a,b]$ is dense in $\mathrm{L}^2(a,b)$, there is a sequence $(g_n)_{n\in\mathbb{N}} \subseteq \mathrm{C}[a,b]$ such that $\|g_n - g\|_2 \to 0$. By Exercise 10.8, $J : \mathrm{L}^2(a,b) \to \mathrm{H}^1(a,b)$ is bounded, and hence $Jg_n \to Jg$ in $\mathrm{H}^1(a,b)$. But then $f_n := Jg_n + c\mathbf{1} \to Jg + c\mathbf{1} = f$ in $\mathrm{H}^1(a,b)$ as well. Evidently, $f_n \in \mathrm{C}^1[a,b]$ and the proof is complete. □

When one has digested the definition of H^1 and the fact that some classically nondifferentiable functions are weakly differentiable, it is then only a small step towards higher (weak) derivatives. One defines, recursively,

$$\mathrm{H}^n(a,b) := \{f \in \mathrm{H}^1(a,b) \mid f' \in \mathrm{H}^{n-1}(a,b)\} \qquad (n \in \mathbb{N},\ n \ge 2)$$

with norm

$$\|f\|_{\mathrm{H}^n}^2 := \|f\|_2^2 + \|f'\|_2^2 + \cdots + \|f^{(n)}\|_2^2 \qquad (f \in \mathrm{H}^n(a,b)).$$

It can be shown by induction on $n \in \mathbb{N}$ that $\mathrm{H}^n(a,b)$ is a Hilbert space with respect to the inner product Ex.10.7

$$\langle f, g\rangle_{\mathrm{H}^n} := \sum\nolimits_{k=0}^n \left\langle f^{(k)}, g^{(k)}\right\rangle_{\mathrm{L}^2} \qquad (f, g \in \mathrm{H}^n(a,b)).$$

The space $\mathrm{H}^n(a,b)$ is called (n-th order) **Sobolev space**.

10.4. The Variational Method for the Poisson Problem

We shall give a typical application in the field of partial differential equations. Actually, no *partial* derivatives will appear, since we are treating the toy case of dimension one. The higher-dimensional case is sketched in Section 10.5 below.

Consider for given $f \in \mathrm{L}^2(a,b)$ the boundary value problem ("Poisson problem")

$$u'' = -f, \qquad u \in \mathrm{H}^2(a,b),\ u(a) = u(b) = 0. \tag{10.5}$$

Of course, there is a straightfoward way to solve this problem. Integrating twice we obtain

$$u(t) = -[J^2 f](t) + c(t-a) + d \qquad (t \in [a,b])$$

for some scalars $c, d \in \mathbb{K}$. The boundary conditions $u(a) = u(b) = 0$ are then equivalent to

$$d = 0 \quad \text{and} \quad c = \frac{1}{b-a}[J^2 f](b).$$

Hence we obtain

$$u(t) = -[J^2 f](t) + \frac{[J^2 f](b)}{b-a}(t-a).$$

Employing the formula

$$(J^2 f)(t) = \int_a^t (s-a) f(s)\,\mathrm{d}s \qquad (a \le t \le b)$$

(see Lemma 11.3 below) we then conclude

$$\begin{aligned} u(t) &= -\int_a^t (t-s) f(s)\,\mathrm{d}s + \frac{t-a}{b-a}\int_a^b (b-s) f(s)\,\mathrm{d}s \\ &= \int_a^b \left(-\mathbf{1}_{[a,t]}(s)(t-s) + \frac{(t-a)(b-s)}{b-a} \right) f(s)\,\mathrm{d}s \\ &= \int_a^b g(t,s) f(s)\,\mathrm{d}s, \end{aligned}$$

where

$$g(t,s) = \frac{1}{b-a}\begin{cases} (b-t)(s-a) & \text{if } s \le t, \\ (b-s)(t-a) & \text{if } t \le s. \end{cases}$$

The function g is the so-called **Green's function**[3] for the Poisson problem.

[3] George Green (1793–1841), English mathematician and physicist.

This explicit solution method is, unfortunately, bound to the one-dimensional situation. In higher dimensions (cf. Section 10.5 below) there is no analogue of the integration operator and one has to employ other solution methods. The aim of the remaining part of this section is to develop such an alternative approach, called the **variational method**. For simplicity, we stay in a one-dimensional setting, but shall sketch the higher-dimensional case in Section 10.5.

Since the differential equation $-u'' = f$ is to be understood in the weak sense, we can equivalently write

$$\langle \psi', u' \rangle_{\mathrm{L}^2} = \langle \psi, f \rangle_{\mathrm{L}^2} \qquad (\psi \in \mathrm{C}_0^1[a,b]), \tag{10.6}$$

where u is supposed to be contained in the space

$$\mathrm{H}_0^1(a,b) := \mathrm{H}^1(a,b) \cap \mathrm{C}_0[a,b] = \{u \in \mathrm{H}^1(a,b) \mid u(a) = u(b) = 0\}.$$

(Note that (10.6) implies that $u \in \mathrm{H}^2(0,1)$.) On $\mathrm{H}_0^1(a,b)$ we define the **energy norm**

$$\|u\|_{\mathrm{H}_0^1} := \|u'\|_{\mathrm{L}^2} \quad \text{for } u \in \mathrm{H}_0^1(a,b),$$

which comes from the sesquilinear form

$$\langle u, v \rangle_{\mathrm{H}_0^1} := \langle u', v' \rangle_{\mathrm{L}^2} = \int_a^b u'(s)\overline{v'(s)}\,\mathrm{d}s \tag{10.7}$$

appearing in the Poisson problem (10.6). The term "norm" here is justified, by the next result.

Lemma 10.10 (Poincaré Inequality[4]). *There is a constant $C \geq 0$ depending on $b - a$ such that*

$$\|u\|_{\mathrm{L}^2} \leq C\|u'\|_{\mathrm{L}^2} \tag{10.8}$$

for all $u \in \mathrm{H}_0^1(a,b)$. In particular, (10.7) *is an inner product and $\|\cdot\|_{\mathrm{H}_0^1}$ is a norm on $\mathrm{H}_0^1(a,b)$.*

Proof. Let $u \in \mathrm{H}_0^1(a,b)$. We claim that $u = Ju'$. Indeed, if $(Ju')' = u'$ and by Corollary 10.6 $Ju' - u = c$ is a constant. But $Ju' - u$ vanishes at a and hence $c = 0$. Using $Ju' = u$ we finally obtain

$$\|u\|_{\mathrm{L}^2} = \|Ju'\|_{\mathrm{L}^2} \leq C\|u'\|_{\mathrm{L}^2}$$

for some constant C, since by Lemma 10.4 the integration operator J is bounded on $\mathrm{L}^2(a,b)$. The remaining statements follow readily. □

[4]Henri Poincaré (1854–1912), French physicist, mathematician and philosopher of science.

Remarks 10.11. 1) The Poincaré inequality (10.8) as it stands cannot hold for $\mathrm{H}^1(a,b)$, since for a constant function f one has $f' = 0$. Consequently, the "energy norm" is not a norm on $\mathrm{H}^1(a,b)$.

2) It requires some work to determine the *optimal* constant C in the Poincaré inequality; see Section 14.5 below.

It follows from the Poincaré inequality that on $\mathrm{H}^1_0(a,b)$ *the energy norm is equivalent to the* H^1*-norm.* Indeed, we have $\|u\|_{\mathrm{H}^1_0} \le \|u\|_{\mathrm{H}^1}$ trivially, and

$$\|u\|_{\mathrm{H}^1}^2 = \|u\|_2^2 + \|u'\|_2^2 \le (C^2+1)\|u'\|_2^2 = (C^2+1)\,\|u\|_{\mathrm{H}^1_0}^2$$

by the Poincaré inequality, hence $\|u\|_{\mathrm{H}^1} \le \sqrt{1+C^2}\,\|u\|_{\mathrm{H}^1_0}$.

Theorem 10.12. *The space* $\mathrm{H}^1_0(a,b)$ *is* $\|\cdot\|_{\mathrm{H}^1}$*-closed in* $\mathrm{H}^1(a,b)$ *and a Hilbert space with respect to the inner product* (10.7).

Proof. Since $(\mathrm{H}^1(a,b), \|\cdot\|_{\mathrm{H}^1}) \subseteq (\mathrm{C}[a,b], \|\cdot\|_\infty)$ continuously (Theorem 10.8), the point evaluations are bounded on $\mathrm{H}^1(a,b)$. This proves the first assertion. It follows that $\mathrm{H}^1_0(a,b)$ is complete with respect to the H^1-norm and — by equivalence of the two norms — also with respect to the energy norm. □

After all these preparatory remarks, let us come back to the Poisson problem. In terms of the notions introduced above, we can reformulate it as follows: for given $f \in \mathrm{L}^2(a,b)$ find $u \in \mathrm{H}^1_0(a,b)$ such that

$$\langle \psi, u\rangle_{\mathrm{H}^1_0} = \langle \psi, f\rangle_{\mathrm{L}^2} \quad \text{for all } \psi \in \mathrm{C}^1_0[a,b].$$

The right-hand side can be written as $\varphi(\psi)$ for the linear functional

$$\varphi : \mathrm{H}^1_0(a,b) \longrightarrow \mathbb{C}, \qquad \varphi(v) := \langle v, f\rangle_{\mathrm{L}^2}\,.$$

The functional φ is bounded for the energy norm, since

$$|\varphi(v)| \le \|v\|_2\,\|f\|_2 \le C\,\|f\|_2\,\|v\|_{\mathrm{H}^1_0}$$

by the Poincaré inequality. The Riesz–Fréchet theorem 8.12 yields a unique $u \in \mathrm{H}^1_0(a,b)$ such that

$$\langle v', u'\rangle_{\mathrm{L}^2} = \langle v, u\rangle_{\mathrm{H}^1_0} = \varphi(v) = \langle v, f\rangle_{\mathrm{L}^2}$$

for all $v \in \mathrm{H}^1_0(a,b)$. In particular, this is true for all $v \in \mathrm{C}^1_0[a,b]$, whence u
Ex.10.9 is a solution of our original problem.

This method of obtaining solutions of boundary value problems is called *variational* because it is based on the Riesz–Fréchet theorem, which can be rephrased as a quadratic minimization result; cf. Exercise 8.21. For the special case of the Poisson problem, this is known as the **Dirichlet principle**; see Exercise 10.17 and cf. also Exercises 10.15 and 10.18 on the Dirichlet problem.

10.5. *Poisson's Problem in Higher Dimensions

The Poisson problem and its treatment here has an analogue in higher dimensions. We shall only give a rough sketch without any proofs. One starts with a *bounded, open* set $\Omega \subseteq \mathbb{R}^d$. On Ω one considers the d-dimensional Lebesgue measure and the Hilbert space $\mathrm{L}^2(\Omega)$. Then the *Poisson problem* is

$$\Delta u = -f, \quad u\big|_{\partial\Omega} = 0,$$

were Δ is the *Laplace operator* defined by

$$\Delta u = \sum_{j=1}^{d} \frac{\partial^2 u}{\partial x_j^2}.$$

In the *classical* case, $f \in \mathrm{C}(\Omega)$, and one wants a solution $u \in \mathrm{C}^2(\Omega) \cap \mathrm{C}(\overline{\Omega})$ satisfying literally the PDE above. The functional analytic way to treat this is a two-step procedure: first find a solution within $\mathrm{L}^2(\Omega)$, where the derivatives are interpreted in a "weak" manner, then try to find conditions on f such that this solution is a classical solution. The second step belongs properly to the realm of PDE, but the first step can be done by our abstract functional analytic methods.

As the space of *test functions* one takes

$$\mathrm{C}_0^1(\Omega) := \{\psi \in \mathrm{C}^1(\overline{\Omega}) \mid \psi\big|_{\partial\Omega} = 0\}.$$

A *weak gradient* of a function $f \in \mathrm{L}^2(\Omega)$ is a d-tuple $g = (g_1, \dots, g_d) \in \mathrm{L}^2(\Omega)^d$ such that

$$\int_\Omega f(x) \frac{\partial \psi}{\partial x_j}(x)\,\mathrm{d}x = -\int_\Omega g_j(x)\psi(x)\,\mathrm{d}x \qquad (j = 1, \dots, d)$$

for all $\psi \in \mathrm{C}_0^1(\Omega)$. One can prove that such a weak gradient is unique, and writes $\nabla f := g$ and ${}^{\partial}\!/_{\partial x_j} f := g_j$. Then one defines

$$\mathrm{H}^1(\Omega) := \{u \in \mathrm{L}^2(\Omega) \mid u \text{ has a weak gradient}\},$$

which is a Hilbert space for the scalar product

$$\begin{aligned}\langle u, v\rangle_{\mathrm{H}^1} &= \langle u, v\rangle_{\mathrm{L}^2} + \int_\Omega \langle \nabla u(x), \nabla v(x)\rangle_{\mathbb{C}^d}\,\mathrm{d}x \\ &= \int_\Omega u(x)v(x)\,\mathrm{d}x + \sum_{j=1}^d \int_\Omega \frac{\partial u}{\partial x_j}(x)\frac{\partial v}{\partial x_j}(x)\,\mathrm{d}x.\end{aligned}$$

The boundary condition is incorporated into a closed subspace $\mathrm{H}^1_0(\Omega)$ of $\mathrm{H}^1(\Omega)$:

$$\mathrm{H}^1_0(\Omega) = \overline{\mathrm{C}^1_0(\Omega)}$$

(closure within $\mathrm{H}^1(\Omega)$). One then has to establish a Poincaré inequality:

$$\int_\Omega |u|^2\,\mathrm{d}x \le c \int_\Omega |\nabla u|^2\,\mathrm{d}x \qquad (u \in \mathrm{H}^1_0(\Omega)) \tag{10.9}$$

for some constant c depending on Ω. In the end, Riesz–Fréchet is applied to obtain a unique $u \in \mathrm{H}^1_0(\Omega)$ such that

$$\int_\Omega \langle \nabla u(x), \nabla \psi(x)\rangle\,\mathrm{d}x = \int_\Omega f(x)\psi(x)\,\mathrm{d}x \qquad (\psi \in \mathrm{C}^1_0(\Omega)),$$

that is, $\Delta u = -f$ in the weak sense.

Of course one would like to have $u \in \mathrm{H}^2(\Omega)$, but this will be true only if Ω is sufficiently regular, e.g., if $\partial\Omega$ is smooth. See [**Eva98**, Chapter 6] for further information.

See also Exercise 10.18 on the Dirichlet problem in higher dimensions.

Exercises 10A

Exercise 10.1. Let $f \in \mathrm{H}^1(a, b)$ with weak derivative $g \in \mathrm{L}^2(a, b)$. Show that $\overline{g}$ is a weak derivative of $\overline{f}$.

Exercise 10.2. Show that $\mathrm{H}^1(a, b)$ is a vector space and that

$$\mathrm{H}^1(a, b) \longrightarrow \mathrm{L}^2(a, b), \qquad f \longmapsto f'$$

is a linear mapping.

Exercise 10.3. Consider $[a, b] = [-1, 1]$ and $f(t) = |t|^q$ for some $1/2 < q$. Show that $f \in \mathrm{H}^1(-1, 1)$ and compute its weak derivative. Is $q = 1/2$ also possible?

Exercise 10.4. Show that for $f \in \mathrm{L}^2(a, b)$ one has

$$|Jf(t) - Jf(s)| \le \sqrt{t - s}\,\|f\|_2 \qquad (s, t \in [a, b]).$$

I.e., Jf is *Hölder* continuous with exponent $1/2$. (This improves on the mere continuity of Jf established in Example 7.17.)

Exercise 10.5. Show that there is no $f \in \mathrm{C}[-1,1]$ such that $f = \mathbf{1}_{(0,1)}$ almost everywhere. (This is somehow similar to the proof of Theorem 5.8.) Conclude that $\mathbf{1}_{(0,1)}$ does not have a weak derivative in $\mathrm{L}^2(-1,1)$.

Exercise 10.6. Find a constant $c \geq 0$ such that

$$\|f\|_\infty \leq c\,\|f\|_{\mathrm{H}^1} \qquad (f \in \mathrm{H}^1(a,b)).$$

[Hint: Corollary 10.7.]

Exercise 10.7. Show that, for every $n \in \mathbb{N}$, $\mathrm{H}^n(a,b)$ is a Hilbert space with respect to the inner product

$$\langle f, g\rangle_{\mathrm{H}^n} := \sum\nolimits_{k=0}^{n} \left\langle f^{(k)}, g^{(k)}\right\rangle_{\mathrm{L}^2} \qquad (f, g \in \mathrm{H}^n(a,b)).$$

Exercise 10.8. Consider the integration operator J on the interval $[a,b]$. In the proof of Lemma 10.4 it was shown that

$$\|Jf\|_\infty \leq \sqrt{b-a}\,\|f\|_2 \qquad (f \in \mathrm{L}^2(a,b)).$$

Show that

$$\|Jf\|_2 \leq \frac{b-a}{\sqrt{2}}\,\|f\|_2 \qquad (f \in \mathrm{L}^2(a,b)).$$

Then determine $c \geq 0$ such that

$$\|Jf\|_{\mathrm{H}^1} \leq c\,\|f\|_2 \qquad (f \in \mathrm{L}^2(a,b)).$$

Exercise 10.9. Show that $\mathrm{C}_0^1[a,b]$ is dense in $\mathrm{H}_0^1(a,b)$. Conclude that the solution in $\mathrm{H}_0^1(a,b)$ of Poisson's equation is *unique*.

Exercises 10B

Exercise 10.10. Show that on $\mathrm{H}^2(a,b)$,

$$|\!|\!| f |\!|\!| := \left(\|f\|_2 + \|f''\|_2\right)^{1/2}$$

is an equivalent Hilbert space norm, i.e., is a norm which comes from an inner product, and this norm is equivalent to the norm given in the main text.

Exercise 10.11 (Product Rule)**.** Show (e.g., by virtue of the definition of a weak derivative) that if $u \in \mathrm{H}^1(a,b)$ and $v \in \mathrm{C}^1[a,b]$, then $uv \in \mathrm{H}^1(a,b)$ with

$$(uv)' = u'v + uv'. \tag{10.10}$$

Then show by approximation (Corollary 10.9) that these assertions are true for all $u,\, v \in \mathrm{H}^1(a,b)$.

Exercise 10.12 (Integration by Parts)**.** Show that if $f, g \in \mathrm{H}^1(a,b)$, then

$$\int_a^b f(s)g'(s)\,\mathrm{d}s = f(b)g(b) - f(a)g(a) - \int_a^b f'(s)g(s)\,\mathrm{d}s.$$

[Hint: Exercise 10.11.] Explain why all the ingredients in this formula are meaningful.

Exercise 10.13 (The General Poincaré Inequality). a) Determine a constant c such that

$$\|Jf - \langle Jf, \mathbf{1}\rangle_{L^2} \mathbf{1}\|_2 \le c\,\|f\|_2$$

for all $f \in L^2(0,1)$.

b) Use a) to establish the general Poincaré inequality

$$\|u - \langle u, \mathbf{1}\rangle_{L^2} \mathbf{1}\|_2 \le c\,\|u'\|_2 \qquad (u \in H^1(0,1)). \tag{10.11}$$

[Hint: Corollary 10.7.]

c) How would (10.11) have to be modified if the interval $(0,1)$ is replaced by a general interval (a,b)?

Exercise 10.14 (Neumann Boundary Conditions[5]). Fix a real number $\lambda > 0$.

a) Show that

$$\|u\|_\lambda := \left(\lambda\,\|u\|_2^2 + \|u'\|_2^2\right)^{1/2} \qquad (u \in H^1(a,b))$$

is a norm on $H^1(a,b)$ which is induced by an inner product $\langle\cdot,\cdot\rangle_\lambda$. Then show that this norm is equivalent to the standard norm $\|\cdot\|_{H^1}$ on $H^1(a,b)$.

b) Show that for $f \in L^2(a,b)$ and $u \in H^1(a,b)$ the following statements are equivalent:

(i) $u \in H^2(a,b)$, $\lambda u - u'' = f$ and $u'(a) = 0 = u'(b)$.

(ii) $\lambda \int_a^b u\varphi + \int_a^b u'\varphi' = \int_a^b f\varphi$ for all $\varphi \in C^1[a,b]$.

(iii) $\lambda \int_a^b uv + \int_a^b u'v' = \int_a^b fv$ for all $\varphi \in H^1(a,b)$.

c) Show that given $f \in L^2(a,b)$ there is a unique $u \in H^1(a,b)$ satisfying (i)–(iii). (Hint: Exercise 10.12.]

Exercise 10.15 (Dirichlet Problem in Dimension One). The **Dirichlet problem** on the interval $[a,b]$ is

$$u'' = 0, \quad u(a) = g(a), \quad u(b) = g(b)$$

where $g : \{a,b\} \to \mathbb{C}$ is a given function, and u is supposed to be contained in $H^1(a,b)$. Clearly the linear function

$$u_0(s) = g(a) + \frac{g(b)}{b-a}(s-a)$$

is a solution. Show that for $u \in H^1(a,b)$ the following assertions are equivalent:

(i) $u = u_0$.

(ii) u is the unique element within the affine subspace

$$G = \{v \in H^1(a,b) \mid v = g \text{ on } \{a,b\}\}$$

that has minimal energy "norm" $\|v'\|_2$.

[5]Carl Neumann (1832–1925), German mathematician.

Exercise 10.16. Show that $C^2[a,b]$ is $\|\cdot\|_{H^2}$-dense in $H^2(a,b)$.

Exercises 10C

Exercise 10.17 (Dirichlet Principle). Show that the solution u of the Poisson problem

$$-u'' = f, \quad u \in H_0^1(a,b) \cap H^2(a,b)$$

is the unique minimizer in $H_0^1(a,b)$ of the quadratic functional

$$\Phi(v) = \int_a^b \tfrac{1}{2} |v'(t)|^2 - \operatorname{Re} v(t)\overline{f(t)}\, dt \qquad (v \in H_0^1(a,b)).$$

More generally, in the framework of Section 10.5, the solution $u \in H_0^1(\Omega)$ to the Poisson problem

$$-\Delta u = f, \qquad u\big|_{\partial\Omega} = 0$$

is the unique minimizer in $H_0^1(\Omega)$ of the quadratic functional

$$\Phi(v) = \int_\Omega \tfrac{1}{2} |\nabla v|^2 - \operatorname{Re} v\overline{f}\, dx.$$

[Hint: Exercise 8.21.]

Exercise 10.18 (Dirichlet Problem in Higher Dimensions). Let $\Omega \subseteq \mathbb{R}^d$ be a bounded open set. We use the notation and definitions from Section 10.5. The **Dirichlet problem** consists in finding a function u on Ω satisfying

$$\Delta u = 0, \quad u|_{\partial\Omega} = f$$

where f is a given function on the topological boundary of Ω. In order to employ Hilbert space methods, one reads Laplace's equation $\Delta u = 0$ in the weak sense, i.e., one requires $u \in H^1(\Omega)$ and

$$\int_\Omega \langle \nabla u(x), \nabla\psi(x)\rangle_{\mathbb{C}^d}\, dx = 0 \qquad \text{for all } \psi \in C_0^1(\Omega). \tag{10.12}$$

But then a diffculty arises: in general, a function $u \in H^1(\Omega)$ need not extend continuously to the boundary (as it does if $d = 1$). That means, the "boundary values" $u|_{\partial\Omega}$ need to be defined.

One way to do this is to require $f \in H^1(\Omega)$ (and not just defined on the boundary) and read $u|_{\partial\Omega} = f$ as

$$u - f \in H_0^1(\Omega).$$

In this formulation the problem can be solved. Define

$$\langle u, v\rangle_e := \int_\Omega \langle \nabla u(x), \nabla v(x)\rangle\, dx$$

and the *energy norm*

$$\|u\|_e := \sqrt{\langle u, u\rangle_e} = \int_\Omega |\nabla u(x)|^2\, dx$$

for $u, v \in \mathrm{H}^1(\Omega)$. Then $\langle \cdot, \cdot \rangle_e$ is a positive semi-definite symmetric sesquilinear form on $\mathrm{H}^1(\Omega)$. We write $u \perp_e v$ for orthogonality with respect to $\langle \cdot, \cdot \rangle_e$. Define

$$G := \{v \in \mathrm{H}^1(\Omega) \mid v - f \in \mathrm{H}^1_0(\Omega)\}.$$

(This is an affine subspace of $\mathrm{H}^1(\Omega)$.) Prove that for $u \in G$ the following assertions are equivalent:

(i) $\Delta u = 0$ weakly, i.e., (10.12).

(ii) $u \perp_e \mathrm{H}^1_0(\Omega)$.

(iii) u minimizes the energy norm on G.

[Hint: For (ii) $\Longleftrightarrow$ (iii) adapt the proof of Lemma 8.6; observe that the definiteness of the inner product was not needed there.]

Adapt the proof of Theorem 8.5 to show that there is exactly one $u \in G$ satisfying (i)–(iii). (As in Section 10.5 one needs that $\mathrm{H}^1_0(\Omega)$ is a Hilbert space with respect to $\langle \cdot, \cdot \rangle_e$.)

Chapter 11

Operator Theory I

In this chapter we shall have a closer look at bounded linear operators, in particular at those that arise in solving linear differential and integral equations.

11.1. Integral Operators and Fubini's Theorem

Let us begin with a simple example. On an interval $[a, b]$ consider the differential equation

$$u' = f, \quad u(a) = 0$$

where f is a given function in $L^2(a, b)$. The derivative here is to be understood in the weak sense, i.e., we look for $u \in H^1(a, b)$ that solves the equation. Since $H^1(a, b) \subseteq C[a, b]$, the boundary condition is meaningful.

Of course we know the solution: $u = Jf$, where J is the integration operator

$$Jf(t) = \int_a^t f(s)\, ds = \int_a^b \mathbf{1}_{[a,t]}(s) f(s)\, ds \qquad (t \in [a, b]).$$

Using the notation

$$k(t, s) := \mathbf{1}_{[a,t]}(s) = \begin{cases} 1 & \text{if} \quad a \le s \le t \le b, \\ 0 & \text{if} \quad a \le t < s \le b, \end{cases}$$

we can write

$$Jf(t) = \int_a^b k(t, s) f(s)\, ds$$

for every function $f \in L^2(a, b)$. Hence J is an integral operator in the following sense.

Definition 11.1. Let X, Y be intervals of $\mathbb{R}$. An operator A is called an **integral operator** if there is a function

$$k : X \times Y \longrightarrow \mathbb{K}$$

such that A is given by

$$(Af)(t) = (A_{[k]}f)(t) := \int_Y k(t,s)f(s)\,\mathrm{d}s \qquad (t \in X) \tag{11.1}$$

for functions f where this is meaningful. The function k is called the **integral kernel** or the **kernel function** of A.

Attention: don't mix up the "integral kernel" k of the integral operator A with the "kernel" (= null space) $\ker(A)$ of A.

To work with integral operators properly, one has to know something about the theory of product measure spaces and, in particular, Fubini's theorem. Roughly speaking, Lebesgue outer measure on $\mathbb{R}^2$ is defined as in Definition 7.1, where intervals are replaced by *rectangles*, i.e., Cartesian products of intervals. Then Theorem 7.2 holds if $\mathbb{R}$ is replaced by $\mathbb{R}^2$ and one obtains the **two-dimensional Lebesgue measure** λ^2. The notion of measurable functions (Definition 7.4) carries over, with the interval X being replaced by the rectangle $X \times Y$. (A function $f \in \mathcal{M}(X \times Y)$ is then called **product measurable**.) Then the whole theory of null sets, $\mathcal{L}^p$ and L^p-spaces carries over to the 2-dimensional setting. In particular, we can form the Banach spaces

$$\mathrm{L}^1(X \times Y) \quad \text{and} \quad \mathrm{L}^2(X \times Y)$$

and one has a dominated convergence and a completeness theorem.

The integral of an integrable function $f \in \mathrm{L}^1(X \times Y)$ with respect to two-dimensional Lebesgue measure is computed via iterated integration in either order:

$$\int_{X \times Y} f(\cdot,\cdot)\,\mathrm{d}\lambda^2 = \int_X \int_Y f(x,y)\,\mathrm{d}y\,\mathrm{d}x.$$

This is called **Fubini's theorem**[1] and it includes the statement that if one integrates out just one variable, the function

$$x \longmapsto \int_Y f(x,y)\,\mathrm{d}y$$

is again measurable.

[1]Guido Fubini (1879-1943), Italian mathematician.

Actually, it is not that simple, and there are quite some measure-theoretical subtleties here. However, we shall boldly ignore them and refer to a book on measure theory instead.

For measurable functions $f \in \mathcal{M}(X)$ and $g \in \mathcal{M}(Y)$ we define the function

$$f \otimes g : X \times Y \longrightarrow \mathbb{K}, \qquad (f \otimes g)(x, y) := f(x)g(y). \tag{11.2}$$

From general measure theory we take the result for granted that $f \otimes g$ is again measurable.

Theorem 11.2. *Let $1 \le p < \infty$. If $f \in \mathrm{L}^p(X)$ and $g \in \mathrm{L}^p(Y)$, then $f \otimes g \in \mathrm{L}^p(X \times Y)$ with $\|f \otimes g\|_{\mathrm{L}^p(X\times Y)} = \|f\|_{\mathrm{L}^p(X)} \|g\|_{\mathrm{L}^p(Y)}$. Moreover, the space*

$$\operatorname{span}\{f \otimes g \mid f \in \mathrm{L}^p(X),\, g \in \mathrm{L}^p(Y)\}$$

is dense in $\mathrm{L}^p(X \times Y)$.

Proof. Since $f \otimes g$ is measurable we only have to show $\int_{X\times Y} |f \otimes g|^p < \infty$. We compute

$$\int_{X\times Y} |f \otimes g|^p \, \mathrm{d}\lambda^2 = \int_X \int_Y |f(x)g(y)|^p \, \mathrm{d}y \, \mathrm{d}x = \int_X |f(x)|^p \int_Y |g(y)|^p \, \mathrm{d}y \, \mathrm{d}x$$

$$= \int_X |f(x)|^p \, \|g\|^p_{\mathrm{L}^p(Y)} \, \mathrm{d}x = \|f\|^p_{\mathrm{L}^p(X)} \, \|g\|^p_{\mathrm{L}^p(Y)} < \infty.$$

Hence $f \otimes g \in \mathrm{L}^p(X \times Y)$ and taking p-th roots concludes the proof of the norm identity. The remaining statement of the theorem must be taken for granted, since a proof would require to go deeper into measure theory. □ Ex.11.1

It is interesting that the approximation part of Theorem 11.2 is hard to find in the literature. However, the proof is not difficult if one knows a certain basic approximation principle from measure theory; see [**Bau01**, Thm. 5.7] or [**Lan93**, Ch. VI,§4]. In the case $p = \infty$ the assertion is false; see Exercise 12.20.

Let us return from general theory to the study of integral operators. For future reference we note the following simple fact.

Lemma 11.3. *Let $f \in \mathrm{L}^1(a, b)$ and $n \in \mathbb{N}$. Then*

$$(J^n f)(t) = \frac{1}{(n-1)!} \int_a^t (t-s)^{n-1} f(s) \, \mathrm{d}s \quad \textit{for all } t \in [a, b].$$

In particular, J^n is again an integral operator, with kernel function

$$k_n(t,s) = \frac{1}{(n-1)!}\mathbf{1}_{[a,t]}(s)(t-s)^{n-1} \qquad (s,t \in [a,b]).$$

Ex.11.2 **Proof.** This is proved by induction and Fubini's theorem. □

11.2. The Dirichlet Laplacian and Hilbert–Schmidt Operators

In Chapter 10 we proved that for each $f \in \mathrm{L}^2(a,b)$ the Poisson problem

$$u'' = -f, \quad u(a) = u(b) = 0 \tag{11.3}$$

has a unique solution $u \in \mathrm{H}^2(a,b)$. Let us rephrase this result in an operator-theoretic way. The **Dirichlet Laplacian** on (a,b) is the operator

$$\Delta_D : \mathrm{H}^2(a,b) \cap \mathrm{H}^1_0(a,b) \longrightarrow \mathrm{L}^2(a,b), \qquad \Delta_D u := u''$$

whose **domain** is abbreviated by

$$\operatorname{dom}(\Delta_D) := \mathrm{H}^2(a,b) \cap \mathrm{H}^1_0(a,b).$$

This is a closed subspace of $\mathrm{H}^2(a,b)$ with respect to the H^2-norm (why?), and Δ_D is bounded. The existence and uniqueness result for the Poisson problem is hence equivalent to saying that Δ_D is bijective.

Now, an equation is called **well-posed** if it has a unique solution for each input parameter *and* the solution *depends continuously* on that input. Existence and uniqueness of solutions account for the existence of a **solution operator**, which maps the input onto the unique solution for that input; well-posedness means in addition that this solution operator is continuous. For the Poisson problem from above, the solution operator is the inverse of $-\Delta_D$, and well-posedness would mean that Δ_D^{-1} is continuous. That is, Δ_D is *invertible* in the following sense (see also Exercise 5.16).

Definition 11.4. A bounded linear operator $T : E \to F$ between normed spaces E, F is called **invertible** if T is bijective and T^{-1} is bounded.

Now what is the inverse of Δ_D? Recall that $(\Delta_D)^{-1}f = -u$, where u is the unique solution of the Poisson problem. Hence, by our computation at the beginning of Section 10.4 above,

$$(\Delta_D)^{-1}f(t) = -\int_a^b g(t,s)f(s)\,\mathrm{d}s,$$

where

$$g(t,s) = \frac{1}{b-a}\begin{cases}(b-t)(s-a) & \text{if } s \le t,\\ (b-s)(t-a) & \text{if } t \le s\end{cases}$$

Ex.11.3 is the Green's function for the Poisson problem.

The proper abstract definition of a "Green's function" needs the framework of distributions, developed by Laurent Schwartz[2] in the 1950's. Here we make an *ad hoc* use of the term for kernel functions of integral operators that arise in the context of boundary value problems.

The Green's function for the Poisson problem satisfies $|g(t,s)| \le b-a$ for all s,t, and hence $\int_a^b \int_a^b |g(t,s)|^2 \, \mathrm{d}s\mathrm{d}t < \infty$. This property has a special name.

Definition 11.5. Let $X, Y \subseteq \mathbb{R}$ be intervals and let $k : X \times Y \to \mathbb{K}$ be product measurable. If $k \in \mathrm{L}^2(X \times Y)$, i.e., if

$$\int_X \int_Y |k(x,y)|^2 \, \mathrm{d}y \, \mathrm{d}x < \infty,$$

then k is called a **Hilbert–Schmidt kernel function**.

The next result shows that a Hilbert–Schmidt kernel function induces indeed a bounded operator on the respective L^2-spaces.

Theorem 11.6. *Let $k \in \mathrm{L}^2(X \times Y)$ be a Hilbert–Schmidt kernel function. Then the associated Hilbert–Schmidt integral operator $A_{[k]}$ given by (11.1) satisfies*

$$\|A_{[k]} f\|_{\mathrm{L}^2(X)} \le \|k\|_{\mathrm{L}^2(X\times Y)} \|f\|_{\mathrm{L}^2(Y)}$$

for all $f \in \mathrm{L}^2(Y)$, hence $A_{[k]} : \mathrm{L}^2(Y) \to \mathrm{L}^2(X)$ is bounded. Moreover, the kernel function k is uniquely determined (λ^2-a.e.) by $A_{[k]}$.

Proof. If $f \in \mathrm{L}^2(Y)$, then by Cauchy–Schwarz

$$\left| \int_Y k(x,y) f(y) \, \mathrm{d}y \right| \le \int_Y |k(x,y) f(y)| \, \mathrm{d}y \le \left[\int_Y |k(x,y)|^2 \, \mathrm{d}y \right]^{1/2} \|f\|_{\mathrm{L}^2(Y)}$$

for all $x \in [a,b]$. Hence

$$\begin{aligned} \|A_{[k]} f\|_{\mathrm{L}^2(X)}^2 &= \int_X \left| \int_Y k(x,y) f(y) \, \mathrm{d}y \right|^2 \mathrm{d}x \\ &\le \left(\int_X \int_Y |k(x,y)|^2 \, \mathrm{d}y \, \mathrm{d}x \right) \|f\|_{\mathrm{L}^2(Y)}^2 . \end{aligned}$$

Taking square-roots we arrive at

$$\|A_{[k]} f\|_{\mathrm{L}^2(X)} \le \left(\int_X \int_Y |k(x,y)|^2 \, \mathrm{d}y \, \mathrm{d}x \right)^{1/2} \|f\|_{\mathrm{L}^2(Y)}$$

[2]Laurent-Moïse Schwartz (1915–2002), French mathematician.

as claimed. That k is determined by $A_{[k]}$ amounts to saying that if $A_{[k]} = 0$, then $k = 0$ in $\mathrm{L}^2(X \times Y)$ (why?). This can be proved by using Theorem
Ex.11.4 11.2; see Exercise 11.4. □

The integral operator $A_{[k]} : \mathrm{L}^2(Y) \to \mathrm{L}^2(X)$ associated with a Hilbert–Schmidt kernel function $k \in \mathrm{L}^2(X \times Y)$ is called a **Hilbert–Schmidt integral operator**. By Theorem 11.6, the inverse Δ_D^{-1} of the Dirichlet Laplacian is a Hilbert–Schmidt integral operator on $\mathrm{L}^2(a,b)$. It is then an exercise to show that

$$\Delta_D^{-1} : \mathrm{L}^2(a,b) \longrightarrow \mathrm{H}^2(a,b)$$

Ex.11.5 is bounded, and hence that *the Poisson problem* (11.3) *is well-posed.*

The next example shows that there are natural examples of integral operators that are *not* of Hilbert–Schmidt type.

Example 11.7 (The Laplace Transform). Consider the case $X = Y = \mathbb{R}_+$ and the kernel function

$$k(x,y) := \mathrm{e}^{-xy} \qquad (x, y > 0).$$

The associated integral operator is the **Laplace transform**

$$(\mathcal{L}f)(x) = \int_0^\infty \mathrm{e}^{-xy} f(y)\, \mathrm{d}y \qquad (x > 0).$$

In Exercise 7.19 it is shown that $\mathcal{L}$ is a bounded operator from $\mathrm{L}^1(\mathbb{R}_+)$ to $\mathrm{C_b}(\mathbb{R}_+)$. Here we are interested in its behaviour on $\mathrm{L}^2(\mathbb{R}_+)$. Neglecting (as usual) the measurability questions, we estimate

$$|\mathcal{L}f(x)| \le \int_0^\infty \mathrm{e}^{-xy/2}\, y^{-1/4} \cdot \mathrm{e}^{-xy/2}\, y^{1/4}\, |f(y)|\, \mathrm{d}y$$

and hence by Cauchy–Schwarz

$$|\mathcal{L}f(x)|^2 \le \int_0^\infty \mathrm{e}^{-xy} y^{-1/2}\, \mathrm{d}y \cdot \int_0^\infty \mathrm{e}^{-xy}\, y^{1/2}\, |f(y)|^2\, \mathrm{d}y.$$

We evaluate the first integral by change of variables ($y \mapsto y/x$, $y \mapsto y^2$),

$$\begin{aligned}\int_0^\infty \mathrm{e}^{-xy}\, y^{-1/2}\, \mathrm{d}y &= x^{-1/2} \int_0^\infty \mathrm{e}^{-y}\, y^{-1/2}\, \mathrm{d}y = x^{-1/2}\, 2 \int_0^\infty \mathrm{e}^{-y^2} \mathrm{d}y \\ &= \sqrt{\pi} x^{-1/2}.\end{aligned}$$

If we use this information and integrate, we obtain

$$\begin{aligned}\|\mathcal{L}f\|_2^2 &= \int_0^\infty |\mathcal{L}f(x)|^2\, \mathrm{d}x \le \sqrt{\pi} \int_0^\infty \int_0^\infty x^{-1/2}\, \mathrm{e}^{-xy}\, y^{1/2}\, |f(y)|^2\, \mathrm{d}y\, \mathrm{d}x \\ &= \sqrt{\pi} \int_0^\infty \int_0^\infty x^{-1/2}\, \mathrm{e}^{-xy}\, \mathrm{d}x\, y^{1/2}\, |f(y)|^2\, \mathrm{d}y = \pi\, \|f\|_2^2 .\end{aligned}$$

Hence $\mathcal{L} : \mathrm{L}^2(\mathbb{R}_+) \to \mathrm{L}^2(\mathbb{R}_+)$ is a bounded operator, with $\|\mathcal{L}f\|_2 \le \sqrt{\pi}\,\|f\|_2$ for each $f \in \mathrm{L}^2(\mathbb{R}_+)$. However, *the Laplace transform is not a Hilbert–Schmidt operator.* Ex.11.6

11.3. Approximation of Operators

In the case of the Poisson problem the solution operator $-\Delta_D^{-1}$ can be determined by direct computation, and there is an explicit formula for it. In more general situations, however, this may be impossible and one has to employ abstract methods in order to establish properties of the solution operator, or even its very existence. One possible approach here, the variational method, has already been treated in Section 10.4.

A different technique is to approximate A, the solution operator of the original "difficult" problem, by operators A_n that are solution operators to "simpler" problems. (Typical instances of this situation are the discretization techniques in numerical analysis, e.g., the finite element or the Galerkin method; cf. [**Che01**, Chap. 4.4] or [**Zei95a**, Sec. 2.7.2].)

Given our discussion in Section 3.3 of different notions of convergence for sequences of functions, it should not come as a surprise that a similar variety can be found when dealing with operators.

Operator Norm Convergence. Let E, F be normed spaces. Recall that a linear operator $T : E \to F$ is continuous if and only if it is bounded (Theorem 4.16), if and only if its (operator) norm

$$\|T\| = \sup\nolimits_{f\in E, \|f\|\le 1} \|Tf\|$$

is a finite number, in which case we have the important formula

$$\|Tf\| \le \|T\| \cdot \|f\| \qquad (f \in E) \tag{11.4}$$

(Lemma 2.12). In Section 2.3 we have encountered a list of examples, including some where the operator norm is not attained. Moreover, the operator norm turns the space $\mathcal{L}(E; F)$ of bounded linear operators from E to F into a normed vector space (Theorem 2.13).

Theorem 11.8. *If F is a Banach space, then $\mathcal{L}(E; F)$ is also a Banach space.*

Proof. Let $(T_n)_{n\in\mathbb{N}}$ be a Cauchy sequence in $\mathcal{L}(E; F)$. For each $f \in E$ by (11.4) we have

$$\|T_n f - T_m f\| = \|(T_n - T_m)f\| \le \|T_n - T_m\|\,\|f\| \tag{11.5}$$

for all $n, m \in \mathbb{N}$. This shows that $(T_n f)_{n\in\mathbb{N}}$ is a Cauchy sequence in F, and since F is complete, the limit

$$Tf := \lim_{n\to\infty} T_n f$$

exists. The linearity of T follows by letting $n \to \infty$ in the equation

$$T_n(\alpha f + \beta g) = \alpha T_n f + \beta T_n g$$

(where $\alpha, \beta \in \mathbb{K}$ and $f, g \in E$). Since every Cauchy sequence is bounded (Lemma 5.2), there is $M \geq 0$ such that $\|T_n\| \leq M$ for all $n \in \mathbb{N}$. If we let $n \to \infty$ in the inequality

$$\|T_n f\| \leq M \|f\| \qquad (f \in E)$$

we see that T is also bounded, with $\|T\| \leq M$.

Now fix $\epsilon > 0$ and find $N \in \mathbb{N}$ so that $\|T_n - T_m\| \leq \epsilon$ for all $n, m \geq N$. This means that

$$\|T_n f - T_m f\| \leq \epsilon$$

for all f with $\|f\| \leq 1$, and $n, m \geq N$. Letting $m \to \infty$ yields

$$\|T_n f - Tf\| \leq \epsilon$$

for all $n \geq N$ and unit vectors f, and taking the supremum over the f we arrive at

$$\|T_n - T\| \leq \epsilon$$

for $n \geq N$. Since $\epsilon > 0$ was arbitrary, we find that $T_n \to T$ in $\mathcal{L}(E; F)$. □

One immediate consequence of (11.4) is the inequality

$$\|ST\| \leq \|S\| \, \|T\|$$

for $T \in \mathcal{L}(E; F)$ and $S \in \mathcal{L}(F; G)$, with G a third normed space, (Lemma 2.14). In the case $E = F = G$ and $S = T$ this yields

$$\|T^n\| \leq \|T\|^n \qquad (n \in \mathbb{N}).$$

However, it may happen that $\|T^n\| \neq \|T\|^n$ for every $n \geq 2$.

Example 11.9 (Integration Operator). The n-th power of the integration operator J on $E = \mathrm{C}[a, b]$ is induced by the integral kernel

$$k_n(t, s) = \mathbf{1}_{\{s \leq t\}}(t, s) \frac{(t-s)^{n-1}}{(n-1)!}.$$

Ex.11.7 From this it follows that $\|J^n\|_{\mathcal{L}(E)} = {}^1\!/_{n!} \neq 1^n = \|J\|^n$. (See Exercise 11.7.)

The following result completes the list of formal properties of the operator norm.

Lemma 11.10. *Let E, F, G be normed space, $T, T_n \in \mathcal{L}(E;F)$, $S, S_n \in \mathcal{L}(F;G)$ and $f, f_n \in E$ for $n \in \mathbb{N}$. Then*

$$T_n \to T,\ S_n \to S \quad \Longrightarrow \quad S_nT_n \to ST \qquad \textit{and}$$
$$T_n \to T,\ f_n \to f \quad \Longrightarrow \quad T_nf \to Tf.$$

Proof. This is proved analogously to Theorem 4.13. Write

$$S_nT_n - ST = (S_n - S)(T_n - T) + S(T_n - T) + (S_n - S)T,$$

then take norms and estimate

$$\begin{aligned}\|S_nT_n - ST\| &\le \|(S_n - S)(T_n - T)\| + \|S(T_n - T)\| + \|(S_n - S)T\| \\ &\le \|S_n - S\|\,\|T_n - T\| + \|S\|\,\|T_n - T\| + \|S_n - S\|\,\|T\| \to 0.\end{aligned}$$

The proof of the second assertion is analogous; see also Exercise 11.25. □

Strong Convergence. A sequence of operators $(T_n)_{n\in\mathbb{N}} \subseteq \mathcal{L}(E;F)$ is said to converge **strongly** to $T \in \mathcal{L}(E;F)$ if for each $f \in E$,

$$T_nf \to Tf \quad \text{in the norm of } F.$$

(Compare with the terminology introduced before the strong convergence lemma, Corollary 9.23). Since $\|T_nf - Tf\|_F \le \|T_n - T\|\,\|f\|$, operator norm convergence implies strong convergence. So in spite of the name, "strong convergence" of operators is *weaker* than norm convergence. (The reason for the name is that there exists an even weaker notion of convergence; cf. Section 16.2 below.) The following example shows that strong convergence is in fact *strictly* weaker.

Example 11.11. Suppose that $(e_j)_{j\in\mathbb{N}}$ is an orthonormal basis of the Hilbert space H. The orthogonal projection onto $F_n := \operatorname{span}\{e_1, \dots, e_n\}$ is

$$P_n := \sum\nolimits_{j=1}^{n} \langle \cdot, e_j\rangle\, e_j$$

and we know from Chapter 8 that $P_nf \to f$ for each $f \in H$. So $(P_n)_{n\in\mathbb{N}}$ approximates the identity operator I strongly. However, $\mathrm{I} - P_n$ is the projection onto $F_n^\perp$ and so $\|\mathrm{I} - P_n\| = 1$. In particular, $P_n \not\to \mathrm{I}$ in operator norm.

The Hilbert–Schmidt Norm. Let $X, Y \subseteq \mathbb{R}$ be intervals and let $k \in \mathrm{L}^2(X \times Y)$ be a Hilbert–Schmidt kernel. This determines a Hilbert–Schmidt operator $A_{[k]}$ by (11.1). We call

$$\|A_{[k]}\|_{\mathrm{HS}} := \|k\|_{\mathrm{L}^2(X\times Y)} = \left(\int_X \int_Y |k(x,y)|^2 \,\mathrm{d}y\,\mathrm{d}x\right)^{1/2}$$

the **Hilbert–Schmidt norm** of the operator $A_{[k]}$. This is a valid definition since by Theorem 11.6, k is uniquely determined by $A_{[k]}$. Moreover, by Ex.11.8

Theorem 11.6 we obtain

$$\|A_{[k]}\|_{\mathcal{L}} \le \|A_{[k]}\|_{\mathrm{HS}}$$

i.e., on the class of Hilbert–Schmidt operators the Hilbert–Schmidt norm is *stronger* than the operator norm. In fact, it is strictly stronger, i.e., the
Ex.11.9 norms are *not equivalent*, see Exercise 11.9.
Ex.11.10

11.4. The Neumann Series

Approximation methods are particularly useful to study perturbed equations. As an example, take any bounded linear operator $T : \mathrm{H}^2(a,b) \to \mathrm{L}^2(a,b)$ and consider the problem

$$u'' - Tu = -f \qquad u \in \mathrm{H}^2(a,b),\ u(a) = u(b) = 0 \tag{11.8}$$

for given $f \in \mathrm{L}^2(a,b)$. If $T = 0$, this is just the Poisson problem, which we know is well-posed. For $T \neq 0$ we consider (11.8) as a **perturbation** of the case $T = 0$, i.e., of the Poisson problem. We shall see in the following that if T is "small enough" then the problem (11.8) is also well-posed.

As a first step, recall that $\Delta_D^{-1} : \mathrm{L}^2(a,b) \to \mathrm{H}^2(a,b)$ is the solution operator of the Poisson problem. Hence we may rewrite (11.8) as

$$-f = u'' - Tu = u'' - T\Delta_D^{-1}u'' = (\mathrm{I} - T\Delta_D^{-1})\Delta_D u$$

for $u \in \mathrm{H}^2(a,b) \cap \mathrm{H}^1(a,b)$. Therefore, since Δ_D is bijective and bounded, the well-posedness of (11.8) is equivalent to the invertibility of the operator

$$\mathrm{I} - T\Delta_D^{-1} : \mathrm{L}^2(a,b) \to \mathrm{L}^2(a,b).$$

Note that $T\Delta_D^{-1}$ indeed maps $\mathrm{L}^2(a,b)$ boundedly into itself since Δ_D^{-1} is bounded from $\mathrm{L}^2(a,b)$ to $\mathrm{H}^2(a,b)$.

The problem is best studied in an abstract setting. Let E be a Banach space, and let $A \in \mathcal{L}(E)$. Then we look for conditions such that the equation

$$(\mathrm{I} - A)u = f \tag{11.9}$$

for given $f \in E$ has a (unique) solution $u \in E$. As a first step we rewrite (11.9) as a fixed point problem

$$u = f + Au, \qquad u \in E.$$

Then, as in Chapter 6, we set up an iteration

$$u_{n+1} := f + Au_n, \qquad \text{where } u_0 = 0, \text{ say.}$$

If $u = \lim_{n\to\infty} u_n$ exists, then by the boundedness of A one has

$$u = \lim_n u_{n+1} = \lim_n f + Au_n = f + \lim_n Au_n = f + Au,$$

i.e., u solves (11.9). The first iterations here are

$$u_0 = 0, \quad u_1 = f, \quad u_2 = f + Af, \quad u_3 = f + Af + A^2 f, \quad \ldots$$

and by an easy induction we obtain

$$u_n = \sum_{j=0}^{n-1} A^j f \qquad (n \in \mathbb{N}).$$

In effect, we have proved the following lemma.

Lemma 11.12. *Let E be a normed space and $A \in \mathcal{L}(E)$. If $f \in E$ is such that the series $u := \sum_{n=0}^{\infty} A^n f$ converges in E, then $u - Au = f$.*

Based on this lemma, we can formulate a very useful criterion.

Theorem 11.13. *Let E be a Banach space and let $A \in \mathcal{L}(E)$ be such that*

$$\sum_{n=0}^{\infty} \|A^n\| < \infty.$$

Then the operator $\mathrm{I} - A$ is invertible and its inverse is given by the so-called **Neumann series**[3]

$$(\mathrm{I} - A)^{-1} = \sum_{n=0}^{\infty} A^n.$$

Proof. If $(\mathrm{I} - A)f = 0$, then $Af = f$ and hence

$$\sum_{n=0}^{\infty} \|f\| = \sum_{n=0}^{\infty} \|A^n f\| \le \sum_{n=0}^{\infty} \|A^n\| \, \|f\| < \infty,$$

which is only possible if $f = 0$. Hence $\mathrm{I} - A$ is injective.
If E is a Banach space, so is $\mathcal{L}(E)$ (Theorem 11.8). Hence by Theorem 5.16 the hypothesis implies that

$$B := \sum_{n=0}^{\infty} A^n$$

converges in $\mathcal{L}(E)$ with respect to the operator norm. Since operator norm convergence implies strong convergence, we obtain for every $f \in E$

$$Bf = \Big(\sum_{n=0}^{\infty} A^n\Big) f = \sum_{n=0}^{\infty} A^n f$$

as a convergent series in E. By Lemma 11.12 it follows that $(\mathrm{I} - A)Bf = f$, whence $\mathrm{I} - A$ is bijective and B is its inverse. $\square$

The Neumann series is absolutely convergent for instance in the case that A is a **strict contraction**, i.e., $\|A\| < 1$; indeed, in this case we have

$$\sum_{n=0}^{\infty} \|A^n\| \le \sum_{n=0}^{\infty} \|A\|^n < \infty.$$

However, strict contractivity is by no means a necessary condition, see Example 11.9. Ex.11.11

[3]Named after Carl Neumann; cf. page 190.

Returning to our introductory example, we see that the perturbed problem (11.8) is well-posed if $\|T\Delta_D^{-1}\| < 1$, i.e., if there is $0 \le q < 1$ such that

$$\|Tu\|_{\mathrm{L}^2} \le q \|u''\|_{\mathrm{L}^2} \quad \text{for all} \quad u \in \mathrm{H}^2(a,b) \cap \mathrm{H}^1_0(a,b).$$

This is what was meant by "small enough" in our claim from above. In terms of norms, a sufficient condition is

$$\|T\| < 1/\|\Delta_D^{-1}\|,$$

where T is considered as an operator of H^2 to L^2 and $\|\Delta_D^{-1}\|$ is the operator norm of Δ_D^{-1} as an operator from L^2 to H^2.

Volterra Integral Equations. We turn to another important application of the Neumann series. Let $m : [a,b] \times [a,b] \to \mathbb{K}$ be a *continuous* function. We are interested in the unique solvability of the **Volterra integral equation**[4]

$$u(t) - \int_a^t m(t,s)u(s)\,\mathrm{d}s = f(t) \qquad (t \in [a,b])$$

for given $f \in \mathrm{C}[a,b]$. To this end, we consider the abstract **Volterra operator** $V : \mathrm{C}[a,b] \to \mathrm{C}[a,b]$ given by

$$(Vf)(t) = \int_a^t m(t,s)f(s)\,\mathrm{d}s \qquad (t \in [a,b],\ f \in \mathrm{C}[a,b]). \tag{11.10}$$

That means, V is an integral operator with kernel function

$$k(t,s) = \mathbf{1}_{[a,t]}(s)\, m(t,s).$$

One can show that Vf is indeed a continuous function; see Exercise 11.24.

Lemma 11.14. *Let $m : [a,b] \times [a,b] \to \mathbb{K}$ be continuous and let V be the associated Volterra operator, given by (11.10). Then*

$$|V^n f(t)| \le \frac{\|m\|_\infty^n (t-a)^n}{n!} \|f\|_\infty$$

for all $f \in \mathrm{C}[a,b]$, $t \in [a,b]$, $n \in \mathbb{N}$. Consequently,

$$\|V^n\|_{\mathcal{L}(\mathrm{C}[a,b])} \le \|m\|_\infty^n \frac{(b-a)^n}{n!} \qquad \textit{for every } n \in \mathbb{N}.$$

Ex.11.12 **Proof.** The proof is an easy induction and left as an exercise. □

If V is an abstract Volterra operator, then by the previous lemma

$$\sum_{n=0}^\infty \|V^n\| \le \sum_{n=0}^\infty \frac{\|m\|_\infty^n (b-a)^n}{n!} = \mathrm{e}^{\|m\|_\infty (b-a)} < \infty.$$

[4]Vito Volterra (1860–1940), Italian mathematician and physicist.

Hence by Theorem 11.13 the operator $\mathrm{I} - V$ is invertible. This leads to the following corollary.

Corollary 11.15. *If $m : [a, b] \times [a, b] \to \mathbb{K}$ is continuous, then for every $f \in \mathrm{C}[a, b]$ the Volterra integral equation*

$$u(t) - \int_a^t m(t, s)u(s)\,\mathrm{d}s = f(t) \qquad (t \in [a, b])$$

has a unique solution $u \in \mathrm{C}[a, b]$. Ex.11.13

Exercises 11A

Exercise 11.1. a) Using Theorem 11.2 show that for $p = \{1, 2\}$,

$$\mathrm{span}\,\{f \otimes g \mid f \in \mathrm{C}[a, b],\, g \in \mathrm{C}[c, d]\} \quad \text{is } \|\cdot\|_p\text{-dense in} \quad \mathrm{L}^p((a, b) \times (c, d)).$$

b) Show that if $f \in \mathrm{L}^\infty(X)$ and $g \in \mathrm{L}^\infty(Y)$, then $f \otimes g \in \mathrm{L}^\infty(X \times Y)$ with

$$\|f \otimes g\|_{\mathrm{L}^\infty(X\times Y)} = \|f\|_{\mathrm{L}^\infty(X)}\, \|g\|_{\mathrm{L}^\infty(Y)}\,.$$

Exercise 11.2. Prove Lemma 11.3.

Exercise 11.3. Show that for each $f \in \mathrm{L}^2(a, b)$ there is exactly one solution $u \in \mathrm{H}^2(a, b)$ of the problem

$$u'' = -f, \quad u(a) = 0 = u'(b).$$

Determine the corresponding Green's function.

Exercise 11.4. Let $k \in \mathrm{L}^2(X \times Y)$ and $A = A_{[k]}$ be the associated Hilbert–Schmidt integral operator. Show that

$$\langle k, f \otimes \overline{g}\rangle_{\mathrm{L}^2(X\times Y)} = \langle Ag, f\rangle_{\mathrm{L}^2(X)}$$

for $f \in \mathrm{L}^2(X), g \in \mathrm{L}^2(Y)$. Show that if $A = 0$, then $k = 0$ almost everywhere. [Hint: Theorem 11.2.]

Exercise 11.5. Show that the operator

$$\Delta_D^{-1} : \mathrm{L}^2(a, b) \longrightarrow \mathrm{H}^2(a, b)$$

is bounded, i.e., there is a constant $c > 0$ such that

$$\left\|\Delta_D^{-1} f\right\|_{\mathrm{H}^2} \le c\,\|f\|_{\mathrm{L}^2} \qquad \text{for all } f \in \mathrm{L}^2(a, b).$$

[Hint: Confer also Exercise 10.10.]

Exercise 11.6. Show that the Laplace transform (Example 11.7) is not a Hilbert–Schmidt operator.

Exercise 11.7. Determine the operator norm $\|J^n\|$ of J^n, $n \in \mathbb{N}$, acting on $\mathrm{C}[a, b]$ with the supremum norm.

Exercise 11.8. Explain the sentence "This is a valid definition, since …" from the main text.

Exercise 11.9. For $n \in \mathbb{N}$ and $1 \le j \le n$ define

$$e_j := \sqrt{n}\,\mathbf{1}_{(\frac{j-1}{n},\frac{j}{n})}.$$

Then $\{e_1, \dots, e_n\}$ is an orthonormal system in $\mathrm{L}^2(0,1)$. Let

$$P_n : \mathrm{L}^2(0,1) \longrightarrow \operatorname{span}\{e_1, \dots, e_n\}$$

be the associated orthogonal projection. Show that P_n is a Hilbert–Schmidt operator and compute $\|P_n\|_{\mathrm{HS}}$. Conclude that the operator norm and the Hilbert–Schmidt norm are not equivalent norms on the space of Hilbert–Schmidt operators.

Exercise 11.10. Compute $\|\Delta_D^{-1}\|_{\mathrm{HS}}$ for the Dirichlet Laplacian Δ_D on (a,b).

Exercise 11.11. Determine the Hilbert–Schmidt norm $\|J^n\|_{\mathrm{HS}}$ of the operator J^n, $n \in \mathbb{N}$ acting on $\mathrm{L}^2(a,b)$. Show that $\mathrm{I} - J$ is invertible and show that $(\mathrm{I} - J)^{-1} - \mathrm{I}$ is again a Hilbert–Schmidt operator. Determine its integral kernel.

Exercise 11.12. Prove Lemma 11.14.

Exercise 11.13. a) Let $\alpha, \beta \in \mathbb{K}$ and $p \in \mathrm{C}[a,b]$ be given. Show that for each $f \in \mathrm{C}[a,b]$ there is a unique solution u of the initial-value problem

$$u'' - pu = f, \qquad u \in \mathrm{C}^2[a,b],\ u(a) = \alpha,\ u'(a) = \beta.$$

[Hint: Convert the problem into a Volterra integral equation.]

b) Suppose now that $p \ge 0$ on $[a,b]$ and let $u \in \mathrm{C}[a,b]$ be the solution of $u'' = pu$ with $u(a) = 0$ and $u'(a) = 1$. Show that $u(t) \ge 0$ and $u'(t) \ge 1$ for all $t \in [a,b]$.

Exercises 11B

Exercise 11.14. Determine a Green's function for the problem

$$u'' = -f, \qquad u \in \mathrm{H}^2(0,1),\ u(0) = u(1),\ \int_0^1 u(s)\,\mathrm{d}s = 0.$$

Exercise 11.15. Determine a Green's function for the problem

$$u'' = -f, \qquad u \in \mathrm{H}^2(0,1),\ u(0) = u'(0),\ u(1) = u'(1).$$

Exercise 11.16. Determine a Green's function for the problem

$$u'' = -f, \qquad u \in \mathrm{H}^2(0,1),\ u(0) = u'(1),\ u(1) = u'(0).$$

Exercise 11.17. Let $P : \mathrm{L}^2(0,1) \to \{\mathbf{1}\}^\perp$ be the orthogonal projection. Determine the integral kernel of the Hilbert–Schmidt operator

$$A := PJ^2P.$$

Exercise 11.18 (Neumann Laplacian). The **Neumann Laplacian** is the operator

$$\Delta_N : \operatorname{dom}(\Delta_N) \longrightarrow \mathrm{L}^2(a,b), \qquad \Delta_N u := u''$$

with $\operatorname{dom}(\Delta_N) := \{u \in \mathrm{H}^2(a,b) \mid u'(a) = u'(b) = 0\}$.

a) Show that the operator

$$\mathrm{I} - \Delta_N : \mathrm{dom}(\Delta_N) \longrightarrow \mathrm{L}^2(a,b), \qquad (\mathrm{I} - \Delta_N)u = u - u''$$

is bijective. [Hint: Exercise 10.14.]

b) Let $A := (\mathrm{I} - \Delta_N)^{-1}$ be the inverse operator. Show that for $f, u \in \mathrm{L}^2(a,b)$ we have

$$Af = u \quad \text{if and only if} \quad \langle u, v\rangle_{\mathrm{H}^1} = \langle f, v\rangle_{\mathrm{L}^2} \quad \text{for all } v \in \mathrm{H}^1(a,b).$$

Then show that $\|A\| \le 1$ if we consider A as an operator $\mathrm{L}^2(a,b) \to \mathrm{H}^1(a,b)$.

c) Let $T : \mathrm{H}^1(a,b) \to \mathrm{L}^2(a,b)$ be a bounded linear operator. Show that for given $f \in \mathrm{L}^2(a,b)$ the problem

$$u \in \mathrm{H}^2(a,b), \quad u - u'' - Tu = f, \quad u'(a) = u'(b) = 0$$

has a unique solution u, provided $\|T\|$ is sufficiently small. (How small?)

Exercise 11.19. Let the operator A be defined by

$$(Af)(t) := \int_{\mathbb{R}} \arctan(ts) f(s)\,\mathrm{d}s \qquad (f \in \mathrm{L}^1(\mathbb{R}),\ t \in \mathbb{R}).$$

a) Show that the linear operator $A : \mathrm{L}^1(\mathbb{R}) \to \mathrm{C_b}(\mathbb{R})$ has norm $\|A\| = \pi/2$.

b) Let $h \in \mathrm{L}^1(\mathbb{R})$ be another function with $\|h\|_1 = 5/8$. Based on the theory developed in Section 11.4 can we be sure that the equation

$$f(t) = g(t) + h(t) \int_{\mathbb{R}} \arctan(ts) f(s)\,\mathrm{d}s \qquad (t \in \mathbb{R})$$

has a unique solution $f \in \mathrm{L}^1(\mathbb{R})$ for every $g \in \mathrm{L}^1(\mathbb{R})$? Justify your claims.

Exercise 11.20. Let $k : [a,b] \times [a,b] \to \mathbb{C}$ be continuous, and let A be the integral operator with integral kernel k, i.e.,

$$(Af)(t) = \int_a^b k(t,s) f(s)\,\mathrm{d}s \qquad (t \in [a,b]).$$

We consider A as an operator on $(\mathrm{C}[a,b], \|\cdot\|_\infty)$. Show that

$$\|A\| = \sup_{t \in [a,b]} \int_a^b |k(t,s)|\,\mathrm{d}s$$

[Hint: Exercise 2.16.]

Exercise 11.21. Let $a < t < b$ and define $f_n := 2n\mathbf{1}_{(t-1/n,\,t+1/n)}$ for $n \in \mathbb{N}$ large enough. Define

$$A_n(f) := \int_a^b f_n(s) f(s)\,\mathrm{d}s \qquad (f \in \mathrm{C}[a,b]).$$

a) Show that each A_n is a bounded linear functional on $\mathrm{C}[a,b]$ (with respect to the supremum norm, of course) and
$$A_n(f) \to \delta_t(f) = f(t)$$
for each $f \in \mathrm{C}[a,b]$.

b) (more involved) Show that $(A_n)_{n\in\mathbb{N}}$ does not converge in the operator norm of $\mathcal{L}(\mathrm{C}[a,b];\mathbb{C})$.

Exercises 11C

Exercise 11.22 (Discrete Hilbert–Schmidt Operators I)**.** Here is the sequence space analogue of the HS-integral operators: Let $a = (a_{ij})_{i,j\in\mathbb{N}}$ be an infinite matrix such that
$$\|a\|_2 := \Bigl(\sum\nolimits_{i,j\in\mathbb{N}} |a_{ij}|^2\Bigr)^{1/2} < \infty.$$
Show that a induces a linear operator A on ℓ^2 by
$$(Af)(n) := \sum\nolimits_{j=1}^{\infty} a_{nj} f(j) \qquad (n \in \mathbb{N}),$$
and $\|A\|_{\ell^2\to\ell^2} \le \|a\|_2$.

Exercise 11.23. Consider $X = Y = \mathbb{R}_+$ and the kernel function
$$k(x,y) = \frac{1}{x+y} \qquad (x, y > 0).$$
The associated integral operator $\mathcal{H}$ is called the **Hilbert–Hankel**[5] operator. Formally, it is given by
$$(\mathcal{H}f(x) = \int_0^\infty \frac{f(y)}{x+y}\,\mathrm{d}y \qquad (x > 0).$$
Show that $\mathcal{H}$ is a bounded operator on $\mathrm{L}^2(\mathbb{R}_+)$. [Hint: Use the same trick as in Example 11.7.] Then show that $\mathcal{H} = \mathcal{L}^2$. Is $\mathcal{H}$ a Hilbert–Schmidt operator?

Exercise 11.24. Let $m : [a,b]^2 \to \mathbb{K}$ be continuous.

a) Show that if $t_n \to t_0$ in $[a,b]$ then $m(t_n, s) \to m(t_0, s)$ *uniformly* in $s \in [a,b]$. [Hint: m is uniformly continuous (why?).]

b) Use a) to show that if $f \in \mathrm{L}^1(a,b)$ then with
$$(Vf)(t) = \int_a^t m(t,s)f(s)\,\mathrm{d}s \qquad (t \in [a,b])$$
the function Vf is continuous.

c) Use a) to show that if $f \in \mathrm{L}^1(a,b)$ then with
$$(Af)(t) = \int_a^b m(t,s)f(s)\,\mathrm{d}s \qquad (t \in [a,b])$$
the function Af is continuous.

[5]Hermann Hankel (1839–1873), German mathematician.

Exercise 11.25 (Continuity of Multiplications). Let X, Y, Z be normed spaces. Suppose one has defined a "multiplication" $X \times Y \to Z$, i.e., a mapping $(x, y) \mapsto x \cdot y = xy$ satisfying $(x + x')y = xy + x'y$ and $x(y + y') = xy + xy'$. Suppose further that there is a constant $c > 0$ such that

$$\|xy\|_Z \le c\, \|x\|_X\, \|y\|_Y \qquad (x \in X, y \in Y).$$

Show that if $x_n \to x$ in X and $y_n \to y$ in Y, then $x_n y_n \to xy$ in Z.

Show that this applies in the following cases:

a) scalar multiplication $\mathbb{K} \times E \to E, \qquad (\lambda, f) \mapsto \lambda f;$

b) inner product $\langle \cdot, \cdot \rangle : H \times H \to \mathbb{K}, \qquad (f, g) \mapsto \langle f, g \rangle;$

c) operator evaluation $\mathcal{L}(E; F) \times E \to F, \qquad (T, f) \mapsto Tf;$

d) operator composition
$\mathcal{L}(F; G) \times \mathcal{L}(E; F) \to \mathcal{L}(E; G), \qquad (S, T) \mapsto ST := S \circ T;$

e) multiplying functions

$$\mathrm{L}^\infty(X) \times \mathrm{L}^2(X) \longrightarrow \mathrm{L}^2(X), \qquad (f, g) \longmapsto f \cdot g;$$

f) "tensoring" functions

$$\mathrm{L}^2(X) \times \mathrm{L}^2(Y) \longrightarrow \mathrm{L}^2(X \times Y), \quad (f, g) \longmapsto f \otimes g.$$

[Hint: Copy the proof of the cases a) and b) from Theorem 4.13.]

Exercise 11.26. Let $m \in \mathrm{C}[a, b]$ and let $A : \mathrm{L}^1(a, b) \to \mathbb{C}$ be defined by

$$Af := \int_a^b m(s) f(s)\,\mathrm{d}s \qquad (f \in \mathrm{L}^1(a, b)).$$

Show that $\|A\| = \|m\|_\infty$. [Hint: Find $t \in (a, b)$ such that $|m(t)|$ is very close to $\|m\|_\infty$, then consider $f_n = 2n\mathbf{1}_{(t-1/n, t+1/n)}$ for $n \in \mathbb{N}$ large and apply Exercise 11.21.a).]

Exercise 11.27. Let $k : [a, b] \times [a, b] \to \mathbb{C}$ be continuous, and let A be the integral operator with integral kernel k, i.e.,

$$(Af)(t) = \int_a^b k(t, s) f(s)\,\mathrm{d}s \qquad (t \in [a, b]).$$

We consider A as an operator $A : \mathrm{L}^1(a, b) \to \mathrm{C}[a, b]$. Show that $\|A\| = \|k\|_\infty$. [Hint: Exercise 11.26.]

Exercise 11.28. Let $k \in \mathrm{L}^\infty((a, b) \times (a, b))$, and let A be the integral operator with integral kernel k, i.e.,

$$(Af)(t) = \int_a^b k(t, s) f(s)\,\mathrm{d}s \qquad (t \in [a, b]).$$

We consider A as an operator $A : \mathrm{L}^1(a, b) \to \mathrm{L}^\infty(a, b)$. Show that $\|A\| \le \|k\|_{\mathrm{L}^\infty}$. (One actually has identity here. Can you prove that?)

Exercise 11.29. Under the same conditions as in Exercise 11.27 consider A as an operator $A : \mathrm{L}^2(a,b) \to \mathrm{C}[a,b]$. Show that

$$\|A\| = \sup_{t\in[a,b]} \left(\int_a^b |k(t,s)|^2 \, \mathrm{d}s \right)^{1/2}.$$

Exercise 11.30. Let E be a Banach space and let $A \in \mathcal{L}(E)$. Show that

$$\exp(A) := \sum\nolimits_{n=0}^{\infty} \frac{1}{n!} A^n$$

exists in $\mathcal{L}(E)$. Show that if $AB = BA$, then $\exp(A+B) = \exp(A)\exp(B)$. Then show that $\exp(A)$ is invertible for each $A \in \mathcal{L}(E)$.

Exercise 11.31. Let E be a Banach space and let $T \in \mathcal{L}(E)$ be an invertible operator. Show that if $S \in \mathcal{L}(E)$ is such that $\|T - S\| < \|T^{-1}\|^{-1}$, then S is invertible, too. [Hint: Show first that $S = (\mathrm{I} - (T - S)T^{-1})T$.]

Exercise 11.32. Let E be a Banach space. By Exercise 11.31 the set

$$\mathcal{L}(E)^{\times} := \{T \in \mathcal{L}(E) \mid T \text{ is invertible}\}$$

is open. Show that the mapping

$$\mathcal{L}(E)^{\times} \longrightarrow \mathcal{L}(E)^{\times}, \qquad T \longmapsto T^{-1}$$

is continuous.

Chapter 12

Operator Theory II

12.1. Compact Operators

Only finite data sets can be handled numerically. Therefore, in actual computations one must replace a precise solution $u = Af$ by an approximation u_n of it that lies in a pre-described finite-dimensional subspace. Doing this in a uniform manner for all input values f amounts to construct operators A_n with $\dim \operatorname{ran}(A_n) < \infty$ and such that $u_n = A_n f$ is an approximation to $u = Af$.

If H is a Hilbert space with orthonormal basis $(e_j)_{j \in \mathbb{N}}$ we can, as in Example 11.11, consider for each $n \in \mathbb{N}$ the orthogonal projection P_n onto the subspace $\operatorname{span}\{e_1, \dots, e_n\}$. Then $A_n f := P_n A f \to Af$ for each $f \in H$, and $\operatorname{ran}(A_n)$ is finite-dimensional.

The problem with this approximation is that convergence happens only in the *strong* sense (cf. Section 11.3) and we do not have control over the error $\|A_n f - Af\|$ in terms of $\|f\|$ only. This would require convergence in the *operator norm*, whence the following definition.

Definition 12.1. An operator $A : E \to F$ between normed spaces E and F is called of **finite rank** or a **finite-dimensional operator** if $\operatorname{ran}(A)$ has finite dimension. And it is called **finitely approximable** if there is Ex.12.1
a sequence $(A_n)_{n \in \mathbb{N}}$ of finite-dimensional operators in $\mathcal{L}(E; F)$ such that $\|A_n - A\| \to 0$.

For Banach spaces E, F we denote by

$$\mathcal{C}_0(H; K) := \{A \in \mathcal{L}(E; F) \mid A \text{ is finitely approximable}\}$$

the space of all finitely approximable operators. To show that an operator is finitely approximable, one uses the definition or the following useful theorem.

Ex.12.2 **Theorem 12.2.** *Let E, F, G be Banach spaces. Then $\mathcal{C}_0(E; F)$ is a closed linear subspace of $\mathcal{L}(E; F)$. If $A : E \to F$ is finitely approximable, $C \in \mathcal{L}(F; G)$ and $D \in \mathcal{L}(G; E)$, then CA and AD are finitely approximable.*

Proof. The set $\mathcal{C}_0(E, F)$ is the operator norm closure of the linear subspace(!) of $\mathcal{L}(E; F)$ consisting of all finite-rank operators. Hence it is a closed subspace, by Example 4.3.1. Let $A : E \to F$ be finitely approximable. Then by definition there is a sequence of finite rank operators $(A_n)_{n \in \mathbb{N}}$ such that $A_n \to A$ in operator norm. Hence

$$A_n D \to AD \quad \text{and} \quad CA_n \to CA$$

in norm, by (11.6). However, CA_n and $A_n D$ clearly are of finite rank, so AD and CA are finitely approximable. □

Example 12.3 (Hilbert–Schmidt Operators). *Let $X, Y \subseteq \mathbb{R}$ be intervals and $k \in \mathrm{L}^2(X \times Y)$. Then the associated Hilbert–Schmidt integral operator*

$$A = A_{[k]} : \mathrm{L}^2(Y) \longrightarrow \mathrm{L}^2(X), \qquad (Ah)(x) = \int_Y k(x, y)h(y)\, \mathrm{d}y$$

is finitely approximable.

Proof. By Theorem 11.2, the space

$$E := \operatorname{span}\{f \otimes g \mid f \in \mathrm{L}^2(X),\ g \in \mathrm{L}^2(Y)\}$$

is dense in $\mathrm{L}^2(X \times Y)$. If $k = f \otimes g$, then $Ah = \langle h, \overline{g} \rangle \cdot f$ and so $\operatorname{ran}(A)$ is one-dimensional. Hence if $k \in E$, then $A_{[k]}$ is of finite rank. If $k \in \mathrm{L}^2(X \times Y)$ we can find $k_n \in E$ with $\|k - k_n\|_{\mathrm{L}^2} \to 0$. Abbreviating $A_n := A_{[k_n]}$ we estimate

$$\|A - A_n\|_{\mathcal{L}} \le \|A - A_n\|_{\mathrm{HS}} = \left\|A_{[k - k_n]}\right\|_{\mathrm{HS}} = \|k - k_n\|_{\mathrm{L}^2} \to 0$$

and so A is finitely approximable. □

An alternative proof of Theorem 12.3 that avoids Theorem 11.2 is given in Section 12.4.

The following theorem is often useful to show that an operator is not finitely approximable.

Theorem 12.4. *Let E, F be Banach spaces, let $A : E \to F$ be a finitely approximable operator, and let $(f_n)_{n \in \mathbb{N}}$ be a bounded sequence in E. Then the sequence $(Af_n)_{n \in \mathbb{N}} \subseteq F$ has a convergent subsequence.*

Recall that a subsequence of a sequence $(x_n)_{n\in\mathbb{N}}$ is determined by a strictly increasing map $\pi : \mathbb{N} \to \mathbb{N}$, the subsequence then being $(x_{\pi(n)})_{n\in\mathbb{N}}$. See Appendix A.1.

Proof. First of all we note that the theorem is true if A is of finite rank. Indeed, by the boundedness of A, the sequence $(Af_n)_{n\in\mathbb{N}}$ is a bounded sequence in the finite-dimensional space $\operatorname{ran}(A)$. By the Bolzano-Weierstrass theorem, it must have a convergent subsequence.

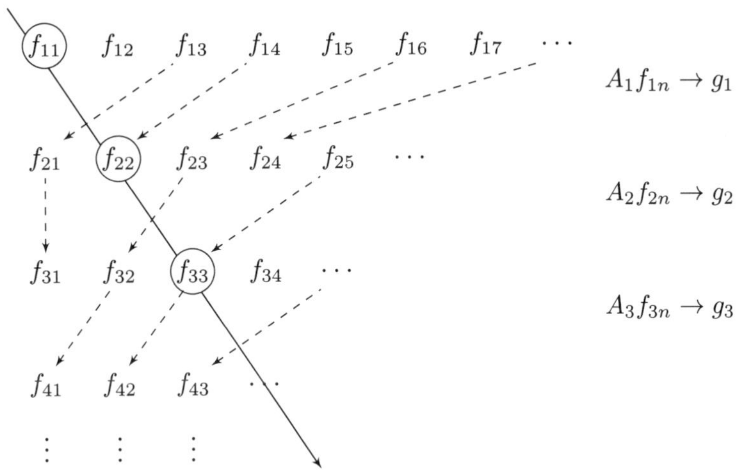

Figure 21. Each row is a subsequence of all its predecessors. And the diagonal sequence is (eventually) a subsequence of every row.

In the general case we shall employ a so-called **diagonal argument**. Find a sequence $(A_m)_{m\in\mathbb{N}}$ of finite-rank operators such that $\|A_m - A\| \to 0$ as $m \to \infty$. Now take the original sequence $(f_n)_{n\in\mathbb{N}}$ and pick a subsequence $f_{11}, f_{12}, f_{13}, \dots$ of it such that

$$A_1 f_{1n} \to g_1, \quad \text{say.}$$

Then pick a subsequence $f_{21}, f_{22}, f_{23}, \dots$ of the first subsequence such that

$$A_2 f_{2n} \to g_2, \quad \text{say.}$$

Continuing in this way, we obtain a "nested" sequence of subsequences $(f_{kn})_{n\in\mathbb{N}}$, each a subsequence of all its predecessors. The "diagonal sequence" $(f_{nn})_{n\in\mathbb{N}}$ is therefore eventually a subsequence of *every* subsequence constructed before. More precisely,

$$\text{for each } k \in \mathbb{N}, \quad (f_{nn})_{n\geq k} \quad \text{is a subsequence of} \quad (f_{kn})_{n\geq k}.$$

Consequently, $A_m f_{nn} \to g_m$ for *every* $m \in \mathbb{N}$. Ex.12.3

Finally, we show that $(Af_{nn})_{n\in\mathbb{N}}$ converges. Since F is complete, it suffices to show that the sequence is Cauchy. The usual estimate yields

$$\begin{aligned}&\|Af_{nn} - Af_{mm}\| \\ &\quad \le \|Af_{nn} - A_k f_{nn}\| + \|A_k f_{nn} - A_k f_{mm}\| + \|A_k f_{mm} - Af_{mm}\| \\ &\quad \le 2M\,\|A - A_k\| + \|A_k f_{nn} - A_k f_{mm}\|\end{aligned}$$

where $M := \sup_{j\in\mathbb{N}} \|f_j\|$. Given $\epsilon > 0$ we can find an index k so large that $2M\,\|A - A_k\| < \epsilon$ and for that k we can find $N \in \mathbb{N}$ so large that $\|A_k f_{nn} - A_k f_{mm}\| < \epsilon$ for $m, n \ge N$. Hence

$$\|Af_{nn} - Af_{mm}\| \le \epsilon + \epsilon = 2\epsilon \qquad \text{for } m, n \ge N. \qquad \square$$

The last step in the proof of Theorem 12.4 is an instance of the "Fundamental Principle of Analysis" in Exercise 5.21 (why?).

The previous theorem gives rise to a new concept.

Definition 12.5. A linear operator $A : E \to F$ between Banach spaces E, F is called **compact** if it has the property that whenever $(f_n)_{n\in\mathbb{N}}$ is a bounded sequence in E, then the sequence $(Af_n)_{n\in\mathbb{N}} \subseteq F$ has a convergent subsequence.

We denote by

$$\mathcal{C}(E;F) := \{A : E \to F \mid A \text{ is a compact operator}\}$$

Ex.12.4 the set of compact operators.

Remarks 12.6. 1) Each compact operator is bounded, i.e., $\mathcal{C}(E;F) \subseteq$
Ex.12.5 $\mathcal{L}(E;F)$.

2) A linear operator $A : E \to F$ is compact if and only if A maps the unit
Ex.12.6 ball of E into a compact subset of F; see Exercise 12.6.

3) The set $\mathcal{C}(E;F)$ of compact operators is a closed linear subspace of $\mathcal{L}(E;F)$; see Exercise 12.27.

4) Theorem 12.4 states that *each finitely approximable operator is compact*, i.e., $\mathcal{C}_0(E;F) \subseteq \mathcal{C}(E;F)$.

The next (optional) result states that *on Hilbert spaces* the concepts of compact and finitely approximable operators coincide.

***Theorem 12.7.** *Let H be a Hilbert space, E a Banach space, and $A : E \to H$ a linear operator. Then A is finitely approximable if and only if it is compact.*

Proof. One implication is simply Theorem 12.4. For the converse suppose that $A : E \to H$ is compact. Then, by Exercise 12.6, the closure of the set $\{Af \mid \|f\| \leq 1\}$ is compact, whence separable (Lemma 4.39). It follows that $K := \overline{\operatorname{ran}}(A)$ is a separable Hilbert space. By Lemma 8.20, one can find an orthonormal basis $(e_n)_{n\in\mathbb{N}}$ of K. Let P_n be the orthonormal projection of H onto $\operatorname{span}\{e_1, \dots, e_n\}$, and $A_n := P_n A$. We claim that $\|A_n \to A\| \to 0$. To establish this, suppose towards a contradiction that $\|A_n - A\| \not\to 0$. After passing to a subsequence there is $\epsilon > 0$ such that $\|A_n - A\| > \epsilon$ for all $n \in \mathbb{N}$. By definition of the operator norm, there is $f_n \in E$ with $\|f_n\| \leq 1$ and $\|A_n f_n - Af_n\| \geq \epsilon$. Passing to another subsequence, by the compactness of A we may suppose that $Af_n \to g$. Note that $g \in K$ and hence $P_n g \to g$. Hence

$$\begin{aligned}\epsilon \leq \|A_n f_n - Af_n\| &\leq \|A_n f_n - P_n g\| + \|P_n g - g\| + \|g - Af_n\| \\ &\leq 2\|Af_n - g\| + \|P_n g - g\| \to 0,\end{aligned}$$

a contradiction. (We used that $\|P_n\| \leq 1$ for all $n \in \mathbb{N}$.) □

By Theorem 12.7, on Hilbert spaces the class of compact operators coincides with the class of finitely approximable operators. It was a long standing open problem — strongly promoted by Grothendieck[1] as one of the central problems in Banach space theory — whether this is true for general Banach spaces. One old problem posed by Mazur[2] in the legendary "Scottish Book" was known to be equivalent to the question, and a living goose was the set prize for the person who would solve it. Finally, long after Grothendieck had given up functional analysis, the Swedish mathematician Enflo[3] solved the problem in the negative [**Enf73**]. He received the goose in a solemn ceremony in 1972.

The following example is central to understand the difference between compact and noncompact operators.

[1]Alexander Grothendieck (1928–), stateless mathematician.

[2]Stanisław Mazur (1905–1981), Polish mathematician.

[3]Per Enflo (1944–), Swedish mathematician.

Example 12.8. Let $(e_j)_{j\in\mathbb{N}}$ be an orthonormal basis for the Hilbert space H, and let $A \in \mathcal{L}(H)$ be given by

$$Af = \sum_{j=1}^{\infty} \lambda_j \langle f, e_j\rangle e_j \qquad (f \in H)$$

for some sequence $\lambda = (\lambda_j)_{j\in\mathbb{N}} \in \ell^\infty$. *Then A is compact if and only if* $\lim_{j\to\infty} \lambda_j = 0$.

Proof. If $\lambda_j \not\to 0$, there is $\epsilon > 0$ and a subsequence $(\lambda_{j_n})_{n\in\mathbb{N}}$ such that $|\lambda_{j_n}| \ge \epsilon$. Define $f_n := \lambda_{j_n}^{-1} e_{j_n}$. Then $(f_n)_{n\in\mathbb{N}}$ is a bounded sequence and $Af_n = e_{j_n}$ for each $n \in \mathbb{N}$. Since this has no convergent subsequence (see Example 4.22), the operator A cannot be compact, by Theorem 12.4.

Now suppose that $\lim_{j\to\infty} \lambda_j = 0$. Define the "truncation"

$$A_n f := \sum_{j=1}^{n} \lambda_j \langle f, e_j\rangle e_j \qquad (f \in H).$$

Then by Parseval and Bessel

$$\begin{aligned}\|Af - A_n f\|^2 &= \Big\| \sum_{j=n+1}^{\infty} \lambda_j \langle f, e_j\rangle e_j \Big\|^2 = \sum_{j=n+1}^{\infty} |\lambda_j|^2 \, |\langle f, e_j\rangle|^2 \\ &\le \Big(\sup_{j>n} |\lambda_j|^2\Big) \sum_{j=n+1}^{\infty} |\langle f, e_j\rangle|^2 \le \Big(\sup_{j>n} |\lambda_j|^2\Big) \|f\|^2 .\end{aligned}$$

Taking the supremum over all f from the unit ball of H yields

$$\|A - A_n\|^2 \le \Big(\sup_{j>n} |\lambda_j|^2\Big) \to 0$$

as $n \to \infty$, since $\lambda_j \to 0$. So A is a norm-limit of finite rank operators,
Ex.12.7 hence compact. □

12.2. Adjoints of Hilbert Space Operators

If $A = (a_{ij})_{ij}$ is a real $d \times d'$-matrix, then its **transposed** matrix is

$$A^* := A^t = (a_{lk})_{kl}.$$

A short computation (that you probably have seen in a linear algebra course) yields

$$\langle Ax, y\rangle = \langle x, A^* y\rangle \qquad \text{for all } x \in \mathbb{R}^{d'}, y \in \mathbb{R}^d, \tag{12.1}$$

where $\langle \cdot, \cdot \rangle$ denotes the standard inner product on $\mathbb{R}^d$ (left-hand side) and $\mathbb{R}^{d'}$ (right-hand side), respectively. The matrix A is called **symmetric** if $d = d'$ and $A^* = A$, and one of the central theorems of linear algebra states that such a matrix is diagonalizable.

The present section and the whole Chapter 13 is devoted to a generalization of these facts to infinite-dimensional Hilbert spaces. First we have to introduce the infinite-dimensional version of a transposed matrix. So let

H, K be Hilbert spaces and let $A : H \to K$ be a bounded linear operator. Guided by (12.1) we require A^* to be an operator

$$A^* : K \longrightarrow H \quad \text{satisfying} \quad \langle Af, g\rangle = \langle f, A^*g\rangle \quad \text{for all } f \in H, g \in K.$$

Can we use this for a *definition* of A^*? We need the following result.

Theorem 12.9. *Let H, K be Hilbert spaces, and let $b : H \times K \to \mathbb{K}$ be a sesquilinear form which is bounded, i.e., there is $c > 0$ such that*

$$|b(f, g)| \le c\, \|f\|\, \|g\| \quad \text{for all } f \in H,\ g \in K.$$

Then there is a unique linear operator $B : K \to H$ such that

$$b(f, g) = \langle f, Bg\rangle \quad \text{for all } f \in H,\ g \in K. \tag{12.2}$$

The operator B is bounded and $\|B\| \le c$.

Proof. Uniqueness: if B and B' are two operators with (12.2) then for each $g \in K$ we have $(Bg - B'g) \perp H$, i.e., $Bg - B'g = 0$. Hence, $B = B'$.

Existence: Fix $g \in K$. Then the mapping

$$\varphi : H \to \mathbb{K}, \qquad f \longmapsto b(f, g)$$

is a linear functional on H. By (12.2), φ is bounded, with $\|\varphi\| \le c\,\|g\|$. The Riesz–Fréchet theorem yields an element $h \in H$ that induces this functional, i.e., such that

$$b(f, g) = \varphi(f) = \langle f, h\rangle \quad \text{for all } f \in H.$$

The element h is unique with this property, and depends only on g, so we are allowed to write $Bg := h$. This defines $Bg \in H$ for each $g \in K$.

For the linearity of B we observe that for given $g, h \in K$ and $\lambda \in \mathbb{K}$,

$$\begin{aligned}\langle f, B(\lambda g + h)\rangle &= b(f, \lambda g + h) = b(f, g)\overline{\lambda} + b(f, h)\\ &= \langle f, Bg\rangle\, \overline{\lambda} + \langle f, Bh\rangle = \langle f, \lambda Bg + Bh\rangle\end{aligned}$$

for all $f \in H$, whence $B(\lambda g + h) = \lambda Bg + Bh$.

The boundedness of B follows from

$$\|Bg\| = \sup_{\|f\|\le 1} |\langle f, Bg\rangle| = \sup_{\|f\|\le 1} |b(f, g)| \le \sup_{\|f\|\le 1} c\,\|f\|\,\|g\| = c\,\|g\|$$

(cf. Example 2.23.) □

Let us return to our original question. Suppose that $A : H \to K$ is a bounded operator. Then

$$b : H \times K \to \mathbb{K}, \qquad b(f, g) := \langle Af, g\rangle$$

is a well-defined sesquilinear form. It is bounded since

$$|b(f, g)| = |\langle Af, g\rangle| \le \|Af\|\,\|g\| \le \|A\|\,\|f\|\,\|g\| \quad \text{for all } f \in H,\ g \in K.$$

Hence, Theorem 12.9 yields an operator $B : K \to H$ such that

$$\langle Af, g\rangle = \langle f, Bg\rangle \qquad \text{for all } f \in H,\ g \in K.$$

We write $A^* := B$ and call it the **adjoint** of A.

Corollary 12.10. *Let H, K be Hilbert spaces, and let $A : H \to K$ be a bounded linear operator. Then there is a unique bounded linear operator $A^* : K \to H$ such that*

$$\langle Af, g\rangle_K = \langle f, A^*g\rangle_H \quad \textit{for all} \quad f \in H,\ g \in K.$$

Furthermore, one has $(A^*)^* = A$ and $\|A^*\| = \|A\|$.

Proof. Only the last two statements have not been proved yet. Observe that

$$\langle g, (A^*)^*f\rangle = \langle A^*g, f\rangle = \overline{\langle f, A^*g\rangle} = \overline{\langle Af, g\rangle} = \langle g, Af\rangle$$

for all $f \in H$, $g \in K$. This implies that $(A^*)^* = A$. But then $\|A\| = \|(A^*)^*\| \le \|A^*\| \le \|A\|$ and the theorem is completely proved. □

The formal properties of adjoints are

$$(A+B)^* = A^* + B^*, \quad (\alpha A)^* = \overline{\alpha}A^*, \quad (AB)^* = B^*A^* \tag{12.3}$$

(where we suppose that $H = K$ in the last identity in order to render the composition meaningful). The proof of these identities is left as Exercise
Ex.12.8 12.8.

Corollary 12.11. *Let $A \in \mathcal{L}(H; K)$. If A is of finite rank or compact, then so is A^*.*

Ex.12.9 **Proof.** The finite rank case is Exercise 12.9. If A is compact then by Theorem 12.7 there is a sequence of finite rank operators A_n such that $\|A_n - A\| \to 0$. But then $\|A_n^* - A^*\| = \|(A_n - A)^*\| = \|A_n - A\| \to 0$, and the claim follows. □

Before we turn to examples, let us note the following result.

Lemma 12.12. *If $A \in \mathcal{L}(H; K)$ then $\ker(A)^\perp = \overline{\operatorname{ran}}(A^*)$, i.e.,*

$$H = \overline{\operatorname{ran}}(A^*) \oplus \ker(A)$$

as an orthogonal decomposition.

Proof. Observe that

$$\begin{aligned} f \perp \overline{\operatorname{ran}}(A^*) &\iff f \perp \operatorname{ran}(A^*) \iff \langle f, A^*g\rangle = 0 \text{ for all } g \in K \\ &\iff \langle Af, g\rangle = 0 \text{ for all } g \in K \iff Af = 0 \iff f \in \ker(A). \end{aligned}$$

Hence $\ker(A)$ is the orthogonal complement of $\overline{\operatorname{ran}}(A^*)$. □

Examples of Adjoints. 1) As already noted above, the adjoint of a real matrix $A = (a_{ij})_{i,j}$ is its transpose $A^* = A^t$. In the case of complex scalars, it is the **conjugate transposed** matrix

$$A^* = (a^*_{kl})_{k,l} = (\overline{a_{lk}})_{k,l}. \tag{12.4}$$

2) Consider the shifts L, R on $H = \ell^2$. Then

$$\langle Re_n, e_k\rangle = \langle e_{n+1}, e_k\rangle = \delta_{n+1,k} = \delta_{n,k-1} = \langle e_n, Le_k\rangle$$

for all $n, k \in \mathbb{N}$. Since $\operatorname{span}\{e_m \mid m \in \mathbb{N}\}$ is dense in H, by sesquilinearity and continuity we conclude that $\langle Rf, g\rangle = \langle f, Lg\rangle$ for all $f, g \in \ell^2$, whence $L^* = R$ and $R^* = L$.

3) Suppose $X, Y \subseteq \mathbb{R}$ are intervals, $k \in \mathrm{L}^2(X \times Y)$ is square-integrable and $A = A_{[k]}$ is the associated Hilbert–Schmidt operator. Then if $f \in \mathrm{L}^2(Y)$, $g \in \mathrm{L}^2(X)$,

$$\begin{aligned}\int_{X\times Y} \left|k(x,y)f(y)\overline{g(x)}\right| \,\mathrm{d}\lambda^2(x,y) &= \int_{X\times Y} |k\cdot(\overline{g}\otimes f)| \,\mathrm{d}\lambda^2 \\ &\le \|k\|_{\mathrm{L}^2(X\times Y)} \|f\|_{\mathrm{L}^2(Y)} \|g\|_{\mathrm{L}^2(X)} < \infty\end{aligned}$$

by Cauchy–Schwarz. So one can apply Fubini's theorem:

$$\begin{aligned}\langle f, A^*g\rangle = \langle Af, g\rangle &= \int_X \int_Y k(x,y)f(y)\,\mathrm{d}y\,\overline{g(x)}\,\mathrm{d}x \\ &= \int_Y f(y)\int_X k(x,y)\overline{g(x)}\,\mathrm{d}x\,\mathrm{d}y = \int_Y f(y)\overline{\left(\int_X \overline{k(x,y)}g(x)\,\mathrm{d}x\right)}\,\mathrm{d}y \\ &= \int_Y f(y)\overline{\left(\int_X k^*(y,x)g(x)\,\mathrm{d}x\right)}\,\mathrm{d}y\end{aligned}$$

where $k^* \in \mathrm{L}^2(Y \times X)$ — the **adjoint kernel function** — is

$$k^*(y,x) := \overline{k(x,y)} \qquad (x \in X, y \in Y).$$

The computation shows that the adjoint of $A_{[k]}$ is

$$A^*_{[k]} = A_{[k^*]},$$

the Hilbert–Schmidt operator corresponding to the adjoint kernel function. This is a continuous analogue to the formula (12.4) for matrices. Ex.12.10 Ex.12.11

12.3. *The Lax–Milgram Theorem

In this (optional) section we describe an abstract framework to treat boundary value problems like the Poisson problem.

Suppose that $(H, \langle\cdot,\cdot\rangle_H)$ is a Hilbert space, $V \subseteq H$ is a linear subspace of H, and $\langle\cdot,\cdot\rangle_V$ is an inner product on V that turns V into a Hilbert space.

Moreover, we suppose that the embedding $V \hookrightarrow H$ is continuous, i.e., there is a constant $C \geq 0$ such that

$$\|v\|_H \leq C \|v\|_V \quad \text{for all } v \in V. \tag{12.5}$$

Our model for this situation is $V = \mathrm{H}_0^1(a,b)$ with inner product $\langle u, v\rangle_V := \langle u', v'\rangle_{\mathrm{L}^2}$; then (12.5) is just the Poincaré inequality.

Let $a : V \times V \to \mathbb{K}$ be a sesquilinear mapping with the following properties:

1) a is **bounded**, i.e., there is $c > 0$ such that

$$|a(u,v)| \leq c \|u\|_V \|v\|_V \qquad (u, v \in V). \tag{12.6}$$

2) a is **coercive**, i.e., there is $\delta > 0$ such that

$$|a(u,u)| \geq \delta \|u\|^2 \qquad (u \in V).$$

(The number δ is called the **coercivity constant**.)

Then we have the following theorem.

Theorem 12.13 (Lax–Milgram[4,5]). *In the situation described above, for each $f \in H$ there is a unique $u \in V$ such that*

$$a(u,v) = \langle f, v\rangle_H \quad \textit{for all } v \in V.$$

Moreover, the operator $A : H \to V$ defined by $Af := u$ has norm $\|A\| \leq C/\delta$.

Proof. Let us show uniqueness first: if $a(u,v) = 0$ for all $v \in V$ then $0 = a(u,u) \geq \delta \|u\|_V^2$ and hence $u = 0$.

To see existence we combine two facts. First, by (12.6), the sesquilinear form $b : V \times V \to \mathbb{C}$ defined by $b(v,u) := \overline{a(u,v)}$ is bounded. Hence Theorem 12.9 yields an operator $B : V \to V$ such that $b(v,u) = \langle v, Bu\rangle_V$, i.e.,

$$a(u,v) = \langle Bu, v\rangle_V \quad \text{for all } u, v \in V$$

and $\|B\| \leq c$. By Exercise 8.23, B is invertible and $\|B^{-1}\| \leq 1/\delta$.

Next, given $f \in H$, the functional $\varphi : V \to \mathbb{K}$ defined by $\varphi(v) := \langle v, f\rangle_H$ for $v \in V$ is bounded:

$$|\varphi(f)| = |\langle v, f\rangle_H| \leq \|f\|_H \|v\|_H \leq C \|f\|_H \|v\|_V \qquad (v \in V)$$

by (12.5). Hence the Riesz–Fréchet theorem yields a unique $Sf \in V$ with

$$\langle v, Sf\rangle_V = \langle v, f\rangle_H \quad \text{for all } v \in V.$$

[4]Peter David Lax (1926–), Hungarian-American mathematician, Abel Prize 2005.

[5]Arthur Norton Milgram (1912–1961), American mathematician.

It is easy to see that $S : H \to V$ is linear and, by virtue of (2.6), that $\|S\| \le C$. But then $A := B^{-1}S$ is the required solution operator since with $u := Af = B^{-1}Sf$ we obtain

$$a(u,v) = \langle Bu, v\rangle_V = \left\langle BB^{-1}Sf, v\right\rangle_V = \langle Sf, v\rangle_V = \langle f, v\rangle_H$$

for all $v \in V$. It follows that $\|A\| = \left\|B^{-1}S\right\| \le \left\|B^{-1}\right\| \|S\| \le C/\delta$. □

Since the space V is a subspace of H, we may consider the solution operator A from Theorem 12.13 as an operator on H.

Corollary 12.14. *In the situation of Theorem* 12.13, *if the sesquilinear form a is symmetric, then A is self-adjoint, i.e. $A^* = A$.*

Proof. Take $f, g \in H$ and write $u = Af$. Then

$$\langle Af, g\rangle_H = \langle u, g\rangle_H = \overline{\langle g, u\rangle_H} = \overline{a(Ag, u)} = a(u, Ag) = a(Af, Ag) = \langle f, Ag\rangle_H$$

by construction. The claim follows. □

Application. As an application we consider the **Sturm–Liouville**[6,7] **problem**

$$-(pu')' + qu = f \quad u(0) = u(1) = 0. \tag{12.7}$$

Here $p \in \mathrm{L}^\infty(0,1)$ and $\alpha > 0$ are such that $p(s) \ge \alpha$ almost everywhere, $q \in \mathrm{L}^\infty(0,1)$ satisfies $\operatorname{Re} q \ge 0$ almost everywhere, and $f \in \mathrm{L}^2(0,1)$. A *solution* to the problem (12.7) is any $u \in \mathrm{H}^1_0(0,1)$ that satisfies the differential equation $-(pu')' + qu = f$ in the weak sense. One can show that this happens if and only if

$$u \in \mathrm{H}^1_0(0,1) \quad \text{and} \quad a(u,v) = \langle f, v\rangle_{\mathrm{L}^2} \quad \text{for all } v \in \mathrm{H}^1_0(0,1),$$

where $a(u,v) := \langle pu', v'\rangle_{\mathrm{L}^2} + \langle qu, v\rangle_{\mathrm{L}^2}$. This form turns out to be bounded (by Poincaré's inequality) and coercive. By the Lax-Milgram theorem, for every $f \in \mathrm{L}^2(0,1)$ the problem (12.7) has a unique solution $u = Af$. If q is real-valued, then a is symmetric, whence by Corollary 12.14, A is self-adjoint. We leave the details as Exercise 12.25.

12.4. *Abstract Hilbert–Schmidt Operators

We have introduced Hilbert–Schmidt operators as a class of integral operators between Hilbert spaces of the form $\mathrm{L}^2(X)$. In this section we present a purely operator-theoretic description of these operators. We begin with a simple, but very useful lemma.

[6]Charles Sturm (1803–1855), Swiss-French mathematician.
[7]Joseph Liouville (1809–1882), French mathematician and physicist.

Lemma 12.15. *Let $f, h \in \mathrm{L}^2(X)$ and $g, k \in \mathrm{L}^2(Y)$. Then*

$$\langle f \otimes \overline{g}, h \otimes \overline{k} \rangle_{\mathrm{L}^2(X\times Y)} = \langle f, h\rangle_{\mathrm{L}^2(X)} \cdot \overline{\langle g, k\rangle_{\mathrm{L}^2(Y)}}.$$

Proof. This is left as Exercise 12.21. □

With the help of the lemma we can show that Hilbert–Schmidt integral operators have a very special property.

Theorem 12.16. *Let $k \in \mathrm{L}^2(X\times Y)$ and $A = A_{[k]}$ be the associated Hilbert–Schmidt integral operator. Then*

$$\sum\nolimits_{j=1}^{\infty} \|Ae_j\|^2 < \infty \tag{12.8}$$

for each orthonormal basis $(e_j)_{j\in\mathbb{N}}$ of $\mathrm{L}^2(Y)$.

Proof. Fix orthonormal bases $(e_j)_{j\in\mathbb{N}}$ of $\mathrm{L}^2(Y)$ and $(f_m)_{m\in\mathbb{N}}$ of $\mathrm{L}^2(X)$. By Lemma 12.15, $(f_m \otimes \overline{e_j})_{j,m}$ is an orthonormal system in $\mathrm{L}^2(X \times Y)$. Now

$$\langle k, f_m \otimes \overline{e_j}\rangle_{\mathrm{L}^2(X\times Y)} = \langle Ae_j, f_m\rangle_{\mathrm{L}^2(X)}$$

and hence

$$\sum_j \|Ae_j\|^2 = \sum_j \sum_k |\langle Ae_j, f_m\rangle|^2 = \sum_{j,k} |\langle k, f_m \otimes \overline{e_j}\rangle|^2 \le \|k\|^2_{\mathrm{L}^2(X\times Y)} < \infty$$

by Parseval's identity in $\mathrm{L}^2(X)$ and Bessel's inequality in $\mathrm{L}^2(X \times Y)$. □

The proof of Theorem 12.16 yields the inequality $\sum_{j=1}^{\infty} \|Ae_j\|^2 \le \|A\|^2_{\mathrm{HS}}$. Actually one has equality here; see Exercise 12.22. In the next step we show that (12.8) even characterizes Hilbert–Schmidt integral operators.

Theorem 12.17. *Let $(e_j)_{j\in\mathbb{N}}$ be an orthonormal basis of $\mathrm{L}^2(Y)$ and let $A : \mathrm{L}^2(Y) \to \mathrm{L}^2(X)$ be a bounded linear operator satisfying $\sum_{j=1}^{\infty} \|Ae_j\|^2 < \infty$. Then the series*

$$k := \sum_{j=1}^{\infty} Ae_j \otimes \overline{e_j}$$

converges in $\mathrm{L}^2(X \times Y)$, and $A = A_{[k]}$.

Proof. By Lemma 12.15 the vectors $Ae_j \otimes \overline{e_j}$ are pairwise orthogonal in the Hilbert space $\mathrm{L}^2(X \times Y)$. By Theorem 8.13 it suffices to test square-summability, and indeed

$$\sum_{j=1}^{\infty} \|Ae_j \otimes \overline{e_j}\|^2 = \sum_{j=1}^{\infty} \|Ae_j\|^2 \, \|\overline{e_j}\|^2 = \sum_{j=1}^{\infty} \|Ae_j\|^2 < \infty,$$

by Theorem 11.2 and the hypothesis. Define $B := A_{[k]}$, and fix $f \in \mathrm{L}^2(X)$ and $m \in \mathbb{N}$. Then

$$\begin{aligned}\langle Be_m, f\rangle &= \langle k, f \otimes \overline{e_m}\rangle = \sum_j \langle Ae_j \otimes \overline{e_j}, f \otimes \overline{e_m}\rangle \\ &= \sum_j \langle Ae_j, f\rangle \langle e_m, e_j\rangle = \langle Ae_m, f\rangle .\end{aligned}$$

It follows that $Be_m = Ae_m$ for all $m \in \mathbb{N}$, and by linearity and the maximality of the orthonormal system $(e_j)_{j\in\mathbb{N}}$ we obtain $A = B$ as claimed. □

As a corollary we obtain a new proof that a Hilbert–Schmidt operator is compact (Example 12.3), avoiding the use of the density statement in Theorem 11.2. Indeed, define

$$k_n := \sum\nolimits_{j=1}^{n} Ae_j \otimes \overline{e_j}.$$

Then $\left\|A_{[k_n]} - A_{[k]}\right\| = \left\|A_{[k_n-k]}\right\| \le \|k_n - k\|_{\mathrm{L}^2} \to 0$. But

$$\operatorname{ran}(A_{[k_n]}) \subseteq \operatorname{span}\{Ae_1, \dots, Ae_n\}$$

is finite-dimensional and hence $A_{[k]}$ is compact.

Let us take a step towards further generality.

Definition 12.18. Let H, K be Hilbert spaces. A bounded linear operator $A : H \to K$ is called an **(abstract) Hilbert–Schmidt operator** if

$$\sum\nolimits_j \|Ae_j\|^2 < \infty \tag{12.9}$$

for some orthonormal basis $(e_j)_j$ of H.

It follows from Theorems 12.16 and 12.17 that the abstract Hilbert–Schmidt operators between $\mathrm{L}^2(X)$-spaces are exactly the Hilbert–Schmidt integral operators.

Lemma 12.19. *Let H, K be two Hilbert spaces, let $(e_j)_j$ and $(f_m)_m$ be orthonormal bases in H and K, respectively. Then for a bounded operator $A : H \to K$ the following assertions are equivalent:*

(i) $\sum_j \|Ae_j\|^2 < \infty$.

(ii) $\sum_{j,m} |\langle Ae_j, f_m\rangle|^2 < \infty$.

(iii) $\sum_m \|A^* f_m\|^2 < \infty$.

Proof. This is left as Exercise 12.23. □

Lemma 12.19 shows that an abstract Hilbert–Schmidt operator satisfies (12.9) for *each* orthonormal basis $(e_j)_j$ (why?). Moreover, it follows that A is Hilbert–Schmidt if and only if A^* is Hilbert–Schmidt.

Suppose $A \in \mathcal{L}(H; K)$. Then by the ususal abstract Fourier expansion

$$Af = A\Big(\sum\nolimits_{j=1}^{\infty} \langle f, e_j\rangle\, e_j\Big) = \sum\nolimits_{j=1}^{\infty} \langle f, e_j\rangle\, Ae_j.$$

In general, this series is convergent, but not absolutely convergent. However, in the case that A is an abstract Hilbert–Schmidt operator,

$$\begin{aligned}\sum\nolimits_{j=1}^{\infty} \|\langle f, e_j\rangle\, Ae_j\| &= \sum\nolimits_{j=1}^{\infty} |\langle f, e_j\rangle|\, \|Ae_j\| \\ &\le \Big(\sum\nolimits_{j=1}^{\infty} |\langle f, e_j\rangle|^2\Big)^{1/2} \Big(\sum\nolimits_{j=1}^{\infty} \|Ae_j\|^2\Big)^{1/2} \\ &= \Big(\sum\nolimits_{j=1}^{\infty} \|Ae_j\|^2\Big)^{1/2} \|f\| < \infty\end{aligned}$$

by Cauchy–Schwarz in ℓ^2 and Parseval's identity in H.

Theorem 12.20. *Every abstract Hilbert–Schmidt operator is compact.*

Proof. Let A_n be the truncation defined by

$$A_n f := \sum\nolimits_{j=1}^{n} \langle f, e_j\rangle\, Ae_j \qquad (f \in H).$$

Then the computation from above yields $\|Af - A_n f\|^2 \le \sum_{j=n+1}^{\infty} \|Ae_j\|^2$ for all $f \in H$ i.e., $\|A_n - A\|^2 \le \sum_{j=n+1}^{\infty} \|Ae_j\|^2 \to 0$. □

Recall that we recognized the operator $-\Delta_D^{-1}$ — the solution operator for the Poisson problem — as a Hilbert–Schmidt operator by explicitly determining its kernel function. In more general situations this is not possible any more, and then the next result is of great help.

Theorem 12.21. *Let H be a Hilbert space and $A : H \to \mathrm{L}^\infty(a, b)$ be a bounded operator. Then A, considered as an operator $A : H \to \mathrm{L}^2(a, b)$ is an abstract Hilbert–Schmidt operator.*

Proof. The proof needs a little experience with measure theory. For the special case that $\operatorname{ran}(A) \subseteq \mathrm{C}[a, b]$, an alternative proof is in Exercise 12.18.

We suppose that H has a countable orthonormal basis $(e_j)_{j=1}^{\infty}$ (but the proof can be adapted to cover the general case). Fix $d \in \mathbb{N}$ and $\lambda \in \mathbb{C}^d$.

Then

$$\left\|\sum_{j=1}^d \lambda_j Ae_j\right\|_{\mathrm{L}^\infty} = \left\|A\left(\sum_{j=1}^d \lambda_j e_j\right)\right\|_{\mathrm{L}^\infty} \le \|A\|_{H\to\mathrm{L}^\infty} \left\|\sum_{j=1}^d \lambda_j e_j\right\|_H$$
$$= \|A\|_{H\to\mathrm{L}^\infty} \|\lambda\|_2 .$$

This means that

$$\left|\sum_{j=1}^d \lambda_j (Ae_j)(x)\right| \le \|A\|_{H\to\mathrm{L}^\infty} \|\lambda\|_2 \quad \text{for almost all } x \in (a,b).$$

Now we take the supremum over all $\lambda \in \mathbb{Q}^d$ with $\|\lambda\|_2 \le 1$, which are only countably many, and obtain

$$\sum_{j=1}^d |(Ae_j)(x)|^2 \le \|A\|_{H\to\mathrm{L}^\infty}^2 \quad \text{for almost all } x \in (a,b).$$

Integrating yields $\sum_{j=1}^d \|Ae_j\|_{\mathrm{L}^2}^2 \le (b-a)\|A\|_{H\to\mathrm{L}^\infty}^2$. Since $d \in \mathbb{N}$ was arbitrary, we obtain

$$\sum_{j=1}^\infty \|Ae_j\|_{\mathrm{L}^2}^2 \le (b-a)\|A\|_{H\to\mathrm{L}^\infty}^2 < \infty. \qquad \square$$

Remark 12.22. One can weaken the assumptions of Theorem 12.21 by requiring only that $A : H \to \mathrm{L}^2(a,b)$ is bounded and $\operatorname{ran}(A) \subseteq \mathrm{L}^\infty(a,b)$. This follows from the closed graph theorem; see Corollary 15.14.

Example 12.23 (Sturm–Liouville Problem)**.** Recall the Sturm–Liouville problem

$$-(pu')' + qu = f, \qquad u(0) = u(1) = 0$$

treated at the end of Section 12.3. Under the conditions on p and q formulated there, for each $f \in \mathrm{L}^2(0,1)$ the problem has a unique solution $u = Af \in \mathrm{H}_0^1(0,1)$. Because we have

$$\mathrm{H}_0^1(0,1) \subseteq \mathrm{C}[0,1] \subseteq \mathrm{L}^\infty(0,1) \subseteq \mathrm{L}^2(0,1)$$

with continuous inclusions, we can view A as a bounded operator $\mathrm{L}^2(0,1) \to \mathrm{L}^\infty(0,1)$. Hence, by Theorem 12.21, the operator A, considered as an operator on $\mathrm{L}^2(0,1)$ is Hilbert-Schmidt, whence compact. As a consequence, A is an integral operator with an L^2-kernel, i.e., the solution u is computed from f by means of an integral

$$u(t) = \int_0^1 k(t,s) f(s)\,\mathrm{d}s$$

and $k \in \mathrm{L}^2((0,1)\times(0,1))$. This function k is the **Green's function** for the problem.

Exercises 12A

Exercise 12.1. Let H, K be Hilbert spaces and let $A : H \to K$ be of finite rank. Show that there are vectors $g_1, \ldots, g_n \in H$ and an orthonormal system $e_1, \ldots, e_n \in K$ such that

$$Af = \sum_{j=1}^{n} \langle f, g_j \rangle e_j$$

for all $f \in H$.

Exercise 12.2. Let $A : E \to F$ be a finite rank/finitely approximable operator such that $\mathrm{I} - A$ is invertible. Show that $\mathrm{I} - (\mathrm{I} - A)^{-1}$ is finite rank/finitely approximable, too.

Exercise 12.3. Show that in the proof of Theorem 12.4, the sequence $(f_{nn})_{n \ge k}$ is a subsequence of $(f_{kn})_{n \ge k}$, for each $k \in \mathbb{N}$.

Exercise 12.4. Let $A : E \to F$ be a compact operator such that $\mathrm{I} - A$ is invertible. Show that $\mathrm{I} - (\mathrm{I} - A)^{-1}$ is compact, too.

Exercise 12.5. Show that a compact operator $A : E \to F$ is necessarily bounded.

Exercise 12.6. Let E, F be Banach spaces and $A \in \mathcal{C}(E; F)$. Show that

$$\overline{\{Af \mid f \in H,\ \|f\| \le 1\}}$$

is a sequentially compact subset of F; cf. Exercise 4.22.

Exercise 12.7. Show that if H is an infinite-dimensional Hilbert space and $A \in \mathcal{L}(H)$ is invertible, then A cannot be a compact operator. [Hint: Example 4.22.]

Exercise 12.8. Let H, K be Hilbert spaces, and let $A, B \in \mathcal{L}(H; K)$. Show that

$$(A + B)^* = A^* + B^* \quad \text{and} \quad (\alpha A)^* = \overline{\alpha} A^*$$

where $\alpha \in \mathbb{K}$. If $K = H$, show that $(AB)^* = B^* A^*$.

Exercise 12.9. Let H, K be Hilbert spaces, and let $A \in \mathcal{L}(H; K)$ be of finite rank. Show that A^* is of finite rank, too. [Hint: Exercise 12.1.]

Exercise 12.10. Let J be the integration operator on $H = \mathrm{L}^2(a, b)$. Determine J^*. Show that

$$(J + J^*)f = (Jf)(b) \cdot \mathbf{1} = \langle f, \mathbf{1} \rangle \cdot \mathbf{1}$$

for $f \in \mathrm{L}^2(a, b)$. (See also Exercise 12.16.)

Exercise 12.11. For the following operators A on ℓ^2 determine the adjoint A^* and decide whether A is compact or not. (Justify your answer.)

1) $A : (x_1, x_2, \ldots) \mapsto (x_2, x_1 + x_3, x_2 + x_4, \ldots)$.
2) $A : (x_1, x_2, \ldots) \mapsto (x_1, \frac{x_1 + x_2}{2}, \frac{x_2 + x_3}{3}, \ldots)$.
3) $A : (x_1, x_2, \ldots) \mapsto (x_1, x_3, x_5, \ldots)$.

Exercises 12B

Exercise 12.12. Let H, K be Hilbert spaces and $A : H \to K$ a bounded and invertible linear operator. Show that $A^* : K \to H$ is invertible, too, and that $(A^*)^{-1} = (A^{-1})^*$.

Exercise 12.13. Show that JJ^* is a Hilbert–Schmidt integral operator on $\mathrm{L}^2(a,b)$, and determine its kernel function.

Exercise 12.14. Show that

$$\langle J^2 f, g\rangle = \langle f, J^2 g\rangle$$

for all $f, g \in \mathrm{L}^2(a,b)$ such that $f, g \in \{\mathbf{1}\}^\perp$.

Exercise 12.15. Determine the integral kernel of the Hilbert–Schmidt operator

$$A := \frac{1}{2}(J^2 + J^{*2}).$$

on $\mathrm{L}^2(0,1)$.

Exercise 12.16. Let J be the integration operator, considered as an operator $J : \mathrm{L}^2(a,b) \to \mathrm{H}^1(a,b)$. Determine a formula for $J^* : \mathrm{H}^1(a,b) \to \mathrm{L}^2(a,b)$. [Attention: This J^* is different from the J^* computed in Exercise 12.10! Do you understand, why?]

Exercise 12.17. Let $F := \{u \in \mathrm{H}^1(0,1) \mid u(0) = 0\}$. Determine $F^\perp$, the orthogonal complement of F in the Hilbert space $\mathrm{H}^1(0,1)$. [Hint: $F = \operatorname{ran} J$.]

Exercise 12.18. Let H be a Hilbert space with orthonormal basis $(e_j)_{j\in\mathbb{N}}$ and $A : H \to \mathrm{C}[a,b]$ a bounded linear operator.

a) Let $x \in [a,b]$. Show that there is a unique $g_x \in H$ such that

$$\langle f, g_x\rangle = (Af)(x) \qquad \text{for all } f \in H.$$

b) Show that $\|g_x\| \le \|A\|_{H\to\mathrm{C}[a,b]}$.

c) Use a) and b) to show that for every $x \in [a,b]$,

$$\sum\nolimits_{j=1}^{\infty} |(Ae_j)(x)|^2 \le \|A\|^2_{H\to\mathrm{C}}\,.$$

d) Show that $\sum_{j=1}^{\infty} \|Ae_j\|^2_{\mathrm{L}^2} < \infty$.

e) Show that for $f \in H$,

$$(Af)(x) = \sum_{j=1}^{\infty} \langle f, e_j\rangle\,(Ae_j)(x) \qquad (x \in [a,b])$$

the series being uniformly convergent on $[a,b]$, and absolutely convergent for each $x \in [a,b]$.

Exercise 12.19. Let $A : H \to H$ be a bounded linear operator, and $F \subseteq H$ a closed subspace. Suppose that

$$\langle Af, g\rangle = \langle f, Ag\rangle \qquad (f, g \in F).$$

Find an operator $B \in \mathcal{L}(H)$ such that $B^* = B$ and $A - B$ maps F into $F^\perp$. (There are several possibilities.)

Exercise 12.20. Show that the integration operator $J : \mathrm{L}^1(0,1) \to \mathrm{L}^\infty(0,1)$ is not compact. [Hint: Look at the functions $f_n = n\mathbf{1}_{(0,1/n)}$; recall that on $\mathrm{C}[0,1]$ the L^∞-norm coincides with the supremum norm.]

Conclude with the help of Exercise 11.28 that the function $\mathbf{1}_A$, $A = \{(s,t) \mid s \le t\}$, is not contained in the L^∞-closure of $\mathrm{L}^\infty(0,1) \otimes \mathrm{L}^\infty(0,1)$.

Exercises 12C

Exercise 12.21. Let $f, h \in \mathrm{L}^2(X)$ and $g, k \in \mathrm{L}^2(Y)$. Show that

$$\langle f \otimes \overline{g}, h \otimes \overline{k} \rangle_{\mathrm{L}^2(X \times Y)} = \langle f, h \rangle_{\mathrm{L}^2(X)} \cdot \overline{\langle g, k \rangle_{\mathrm{L}^2(Y)}}.$$

Draw the following conclusions:

a) Let $(e_j)_j, (f_m)_m$ be orthonormal systems in $\mathrm{L}^2(Y)$ and $\mathrm{L}^2(X)$, respectively. Then $(f_m \otimes \overline{e_j})_{m,j}$ is an orthonormal system in $\mathrm{L}^2(X \times Y)$.

b) If $(e_j)_j$ is an orthogonal system in $\mathrm{L}^2(Y)$ and $(f_j)_j$ is sequence of vectors in $\mathrm{L}^2(X)$, then $(f_j \otimes e_j)_j$ is an orthogonal system in $\mathrm{L}^2(X \times Y)$.

Exercise 12.22. Let $(e_j)_j, (f_m)_m$ are orthonormal *bases* in $\mathrm{L}^2(Y)$ and $\mathrm{L}^2(X)$, respectively. Show that $(f_m \otimes \overline{e_j})_{m,j}$ is an orthonormal basis in $\mathrm{L}^2(X \times Y)$. [Hint: Exercise 12.21 and Theorem 11.2.] Prove that $\|A\|_{\mathrm{HS}}^2 = \sum_j \|Ae_j\|^2$ in Theorem 12.16.

Exercise 12.23. Prove Lemma 12.19.

Exercise 12.24 (Discrete Hilbert–Schmidt Operators II)**.** Let $a = (a_{ij})_{i,j}$ be an infinite matrix such that $\sum_{i,j} |a_{ij}|^2 < \infty$. Let $A : \ell^2 \to \ell^2$ be the discrete Hilbert–Schmidt operator associated with the matrix a, i.e., A is given by

$$(Af)(n) = \sum_{j=1}^{\infty} a_{nj} f(j) \qquad (f \in \ell^2,\ n \in \mathbb{N}).$$

(See Exercise 11.22.) Show that A is compact.

Exercise 12.25 (Sturm–Liouville Problem, Variational Approach)**.** In the following we work over the field $\mathbb{K} = \mathbb{C}$ of complex numbers. Let $p \in \mathrm{L}^\infty(0,1)$ and $\alpha > 0$ be such that $p(s) \ge \alpha$ almost everywhere. Furthermore, let $q \in \mathrm{L}^\infty(0,1)$ with $\operatorname{Re} q \ge 0$ almost everywhere. We want to discuss the problem

$$-(pu')' + qu = f \quad u(0) = u(1) = 0 \tag{12.10}$$

for given $f \in \mathrm{L}^2(0,1)$. A *solution* to the problem (12.10) is any $u \in \mathrm{H}_0^1(0,1)$ that satisfies the differential equation $-(pu')' + qu = f$ in the weak sense.

On the space $\mathrm{H}_0^1(0,1)$ we use the inner product $\langle u, v \rangle_{\mathrm{H}_0^1} := \langle u', v' \rangle_{\mathrm{L}^2}$. Recall that it was shown in Section 10.4 that this is indeed an inner product that turns $\mathrm{H}_0^1(0,1)$ into a Hilbert space.

a) Show that the sesqulinear form $a : \mathrm{H}^1_0(0,1) \times \mathrm{H}^1_0(0,1) \to \mathbb{C}$ given by

$$a(u,v) := \langle pu', v' \rangle_{\mathrm{L}^2} + \langle qu, v \rangle_{\mathrm{L}^2} \qquad (u, v \in \mathrm{H}^1_0(0,1))$$

is bounded and coercive with coercivity constant α.

b) Show that u is a solution to (12.10) if and only if

$$a(u,v) = \langle f, v \rangle_{\mathrm{L}^2} \quad \text{for all } v \in \mathrm{H}^1_0(0,1).$$

c) Apply the Lax-Milgram theorem to show that there is a unique solution $u =: Af$ of the problem (12.7), and that $\|u\|_{\mathrm{H}^1} \le (C/\alpha) \|f\|_{\mathrm{L}^2}$, where C is the constant in the Poincaré inequality.

d) Show that A, considered as an operator on $\mathrm{L}^2(0,1)$, satisfies $A^* = A$ if q is real-valued.

Exercise 12.26. Under the hypotheses of Exercise 12.25, let $r \in \mathrm{L}^\infty(0,1)$. Show that if $\|r\|_{\mathrm{L}^\infty}$ is small enough, then for each $f \in \mathrm{L}^2(0,1)$ the problem

$$-(pu')' + qu + ru' = f \qquad u \in \mathrm{H}^1_0(0,1) \tag{12.11}$$

has a unique solution. (How small should $\|r\|_{\mathrm{L}^\infty}$ be, in terms of the data?)

Exercise 12.27 (Compact Operators on Banach Spaces). Let E, F, G be Banach spaces.

a) Show that $\mathcal{C}(E;F)$ is a closed linear subspace of $\mathcal{L}(E;F)$.

b) Suppose that $A : E \to F$ is compact, and $C \in \mathcal{L}(F;G)$, $D \in \mathcal{L}(G;E)$. Show that CA and AD are compact.

Chapter 13

Spectral Theory of Compact Self-Adjoint Operators

One of the most important results of finite-dimensional linear algebra states that a symmetric real square matrix A is orthogononally diagonalizable. Equivalently, each such matrix has an orthonormal basis consisting of eigenvectors of A. In this chapter we shall derive an infinite-dimensional version of this result.

13.1. Approximate Eigenvalues

Recall that if $A : E \to F$ is a linear operator between vector spaces E, F, then an **eigenvalue** of A is a scalar $\lambda \in \mathbb{K}$ such that the **eigenspace** $\ker(\lambda \mathrm{I} - A)$ satisfies

$$\ker(\lambda \mathrm{I} - A) \neq \{0\},$$

and every $0 \neq f$ such that $Af = \lambda f$ is called an associated **eigenvector**.

In finite dimensions the eigenvalues tell a great deal about the operator. Indeed, the theory of the so-called "Jordan canonical form" says that in the case $\mathbb{K} = \mathbb{C}$ a square matrix is determined up to similarity (change of basis) by the dimensions of the **generalized eigenspaces**

$$\ker(\lambda \mathrm{I} - A)^k$$

where λ runs through the eigenvalues and k the natural numbers.

There is no analogue of such a result for operators on infinite-dimensional spaces. In fact, there are relatively simple operators having no eigenvalues at all.

Example 13.1. Let $H := \mathrm{L}^2(0,1)$ and let $A : H \to H$ be the multiplication
Ex.13.1 operator $(Af)(t) := t \cdot f(t)$, $t \in (0,1)$. Then A has no eigenvalues.

So let us generalize the notion of an eigenvalue.

Definition 13.2. Let E be a normed space and $A : E \to E$ a bounded operator. A scalar $\lambda \in \mathbb{K}$ is called an **approximate eigenvalue** of A if there is a sequence $(f_n)_{n\in\mathbb{N}}$ in E such that $\|f_n\| = 1$ for all $n \in \mathbb{N}$ and $\|Af_n - \lambda f_n\| \to 0$.

Note that an eigenvalue is also an approximate eigenvalue. The next example shows that the converse does not hold.

Example 13.3. In Example 13.1, every $\lambda \in [0,1]$ is an approximate eigenvalue. Indeed, for $n \in \mathbb{N}$ we can find f_n such that

$$\|f_n\|_2 = 1 \quad \text{and} \quad f_n(t) = 0 \quad (|t - \lambda_0| \geq 1/n).$$

(Choose $f_n := c_n \mathbf{1}_{[\lambda_0 - 1/n, \lambda_0 + 1/n]}$ with a suitable constant c_n.) Then

$$\|Af_n - \lambda_0 f_n\|_2^2 = \int_0^1 |t - \lambda_0|^2 \, |f_n(t)|^2 \, dt \leq \frac{1}{n^2} \int_0^1 |f_n(t)|^2 \, dt = \frac{1}{n^2} \to 0.$$

Ex.13.2
Ex.13.3
Ex.13.4

For a matrix A, the collection of its eigenvalues is also called its 'spectrum'. By Example 13.1, this notion of spectrum is not very reasonable beyond finite dimensions. It turns out that a "good" definition of the **spectrum** of a general operator A on a Banach space E is

$$\sigma(A) := \{\lambda \in \mathbb{K} \mid \lambda \mathrm{I} - A \text{ is not invertible}\}.$$

(For matrices, this coincides of course with the collection of eigenvalues.) We shall not pursue this topic further in the main text, and refer to Exercises 13.12-13.17 and books on operator theory for further information.

Lemma 13.4. *Let A be a bounded operator on the Banach space E. If $\lambda\mathrm{I} - A$ is invertible, then λ cannot be an approximate eigenvalue. If $|\lambda| > \|A\|$, then $\lambda\mathrm{I} - A$ is invertible.*

Proof. If $\|Af_n - \lambda f_n\| \to 0$ and $\lambda\mathrm{I} - A$ is invertible, then

$$f_n = (\lambda\mathrm{I} - A)^{-1}(\lambda f_n - Af_n) \to 0$$

which contradicts the requirement that $\|f_n\| = 1$ for all $n \in \mathbb{N}$. Take $|\lambda| > \|A\|$. Then $\|\lambda^{-1}A\| < 1$ and hence

$$\lambda\mathrm{I} - A = \lambda(\mathrm{I} - \lambda^{-1}A)$$

is invertible with

$$(\lambda\mathrm{I} - A)^{-1} = \lambda^{-1}\sum\nolimits_{n=0}^{\infty}(\lambda^{-1}A)^n = \sum\nolimits_{n=1}^{\infty}\lambda^{-(n+1)}A^n$$

(Theorem 11.13). □ Ex.13.5

Example 13.5. In Example 13.1 (the multiplication operator $(Af)(t) = t \cdot f(t)$ on $\mathrm{L}^2(0,1)$), for every $\lambda \in \mathbb{C} \setminus [0,1]$ the operator $\lambda\mathrm{I} - A$ is invertible and the inverse is given by

$$\big[(\lambda\mathrm{I} - A)^{-1}f\big](t) = \frac{1}{\lambda - t}f(t) \qquad (t \in (0,1)).$$

Ex.13.6

The following result gives a hint why we can expect good results for compact operators.

Theorem 13.6. *Let A be a compact operator on a Banach space E and let $\lambda \neq 0$ be an approximate eigenvalue of A. Then λ is an eigenvalue and* $\ker(A - \lambda\mathrm{I})$ *is finite-dimensional.*

Proof. By definition, there is a sequence $(f_n)_{n\in\mathbb{N}}$ in E such that $\|f_n\| = 1$ for all n and $\|Af_n - \lambda f_n\| \to 0$. As A is compact, by passing to a subsequence we may suppose that $g := \lim_n Af_n$ exists. Consequently,

$$\|\lambda f_n - g\| \le \|\lambda f_n - Af_n\| + \|Af_n - g\| \to 0.$$

Thus $\|g\| = \lim_n \|\lambda f_n\| = \lim_n |\lambda|\,\|f_n\| = |\lambda| \neq 0$. Moreover,

$$Ag = A(\lim_n \lambda f_n) = \lambda \lim_n Af_n = \lambda g$$

which shows that λ is an eigenvalue with eigenvector g. Suppose that $F := \ker(A - \lambda\mathrm{I})$ is infinite-dimensional. Then there is a sequence of unit vectors $(e_n)_{n\in\mathbb{N}}$ in F with $\|e_n - e_m\| \ge 1$ for $n \neq m$. (See Example 4.22 for the Hilbert space and Corollary 4.34 for the Banach space case.)
For $n \neq m$,

$$\|Ae_n - Ae_m\| = \|\lambda e_n - \lambda e_m\| = |\lambda|\,\|e_n - e_m\| \ge |\lambda|\,.$$

Since $\lambda \neq 0$, the sequence $(Ae_n)_{n\in\mathbb{N}}$ does not have a convergent subsequence, which is in contradiction to the compactness of A. □

13.2. Self-Adjoint Operators

A bounded operator A on a Hilbert space H is called **self-adjoint** or **Hermitian**, if $A^* = A$. By definition of the adjoint, A is self-adjoint if and only if

$$\langle Af, g\rangle = \langle f, Ag\rangle$$

for all $f, g \in H$.

Examples 13.7. a) Each orthogonal projection is self-adjoint; see Exercise 8.4.

b) Let $\lambda \in \ell^\infty$, then the multiplication operator A_λ on ℓ^2 (Example 2.25) is self-adjoint if and only if λ is a real sequence.

c) A Hilbert–Schmidt integral operator $A = A_{[k]}$ on $\mathrm{L}^2(a, b)$ is self-adjoint if $\overline{k(x, y)} = k(y, x)$ for almost all $x, y \in (a, b)$. This is true for instance for the Green's function for the Poisson problem (see page 184).

We shall need the following (technical) result.

Theorem 13.8. *Let A be a bounded self-adjoint operator on a Hilbert space A. Then $\langle Af, f\rangle \in \mathbb{R}$ for all $f \in H$ and*

$$\|A\| = \|\!|A|\!\| := \sup\{|\langle Af, f\rangle| \mid f \in H,\ \|f\| = 1\}.$$

The quantity $\|\!|A|\!\|$ is called the **numerical radius** of A.

Proof. One has $\langle Af, f\rangle = \langle f, Af\rangle = \overline{\langle Af, f\rangle}$ so $\langle Af, f\rangle$ is real. By Cauchy–Schwarz,

$$|\langle Af, f\rangle| \le \|Af\|\,\|f\| \le \|A\|\,\|f\|^2 = \|A\|\,,$$

if $\|f\| = 1$. This proves that $\|\!|A|\!\| \le \|A\|$.
For the converse we start from the identity

$$\langle A(u+v), u+v\rangle - \langle A(u-v), u-v\rangle = 2\langle Au, v\rangle + 2\langle Av, u\rangle = 4\,\mathrm{Re}\,\langle Au, v\rangle$$

valid for all $u, v \in H$. (This is the polarization identity for the symmetric sesquilinear form $a(u, v) := \langle Au, v\rangle$; see Lemma A.8.) Hence

$$\begin{aligned}
4\,\mathrm{Re}\,\langle Au, v\rangle &\le |\langle A(u+v), u+v\rangle| + |\langle A(u-v), u-v\rangle| \\
&\le \|\!|A|\!\|\left(\|u+v\|^2 + \|u-v\|^2\right) \\
&= \|\!|A|\!\|\left(2\|u\|^2 + 2\|v\|^2\right).
\end{aligned}$$

We replace v be cv for $c \in \mathbb{C}$ with $|c| = 1$ and apply (A.1) to obtain

$$2\,|\langle Au, v\rangle| \le \|\!| A |\!\|\,(\,\|u\|^2 + \|v\|^2\,).$$

Now taking the supremum over all u, v with $\|u\|$, $\|v\| \le 1$ yields $2\,\|A\| \le 2\,\|\!| A |\!\|$ and the proof is complete. □

Finally, we collect the spectral-theoretic facts of self-adjoint operators. Recall that a subspace F is called **A-invariant** if $A(F) \subseteq F$.

Lemma 13.9. *Let A be a self-adjoint operator on a Hilbert space. Then the following assertions hold.*

a) *Every eigenvalue of A is real.*

b) *Eigenvectors with respect to different eigenvalues are orthogonal to each other.*

c) *If F is an A-invariant subspace of H, then $F^\perp$ is also A-invariant.*

Proof. a) If $Af = \lambda f$ and $\|f\| = 1$ then

$$\lambda = \lambda\,\|f\|^2 = \langle \lambda f, f\rangle = \langle Af, f\rangle \in \mathbb{R}$$

by Theorem 13.8. Ex.13.7

b) Suppose that $\lambda, \mu \in \mathbb{R}$ and $f, g \in H$ such that $Af = \lambda f$ and $Ag = \mu g$. Then

$$(\lambda - \mu)\,\langle f, g\rangle = \langle \lambda f, g\rangle - \langle f, \mu g\rangle = \langle Af, g\rangle - \langle f, Ag\rangle = 0$$

since $A = A^*$. Hence $\lambda \ne \mu$ implies that $f \perp g$.

c) Finally, let $f \in F$, $g \in F^\perp$. Then $Af \in F$ and hence $\langle f, Ag\rangle = \langle Af, g\rangle = 0$. As $f \in F$ was arbitrary, $Ag \in F^\perp$. □

Example 13.1 shows that even self-adjoint operators need not have eigenvalues. If A is compact, however, this is different.

Lemma 13.10. *Let A be a compact self-adjoint operator on a Hilbert space. Then A has an eigenvalue λ such that $|\lambda| = \|A\|$.*

Proof. We may suppose that $A \ne 0$. By definition, we can find a sequence $(f_n)_n$ in H such that $\|f_n\| = 1$ and $|\langle Af_n, f_n\rangle| \to \|\!| A |\!\|$. By passing to a subsequence we may suppose that $\langle Af_n, f_n\rangle \to \lambda$ with $|\lambda| = \|\!| A |\!\| = \|A\|$. Then, since λ and $\langle Af_n, f_n\rangle$ are both real,

$$\begin{aligned} 0 &\le \|Af_n - \lambda f_n\|^2 = \|Af_n\|^2 - 2\lambda \operatorname{Re}\langle Af_n, f_n\rangle + \lambda^2\,\|f_n\|^2 \\ &\le \|A\|^2 - 2\lambda\,\langle Af_n, f_n\rangle + \lambda^2 \to \|A\|^2 - \lambda^2 = 0. \end{aligned}$$

Hence λ is an approximate eigenvalue, and Theorem 13.6 concludes the proof. □

13.3. The Spectral Theorem

We are now in the position to state and prove the main result.

Theorem 13.11 (Spectral Theorem). *Let A be a compact self-adjoint operator on a Hilbert space H. Then A is of the form*

$$Af = \sum\nolimits_j \lambda_j \langle f, e_j \rangle e_j \qquad (f \in H) \tag{13.1}$$

for some (finite or countably infinite) orthonormal system $(e_j)_j$ and real numbers $\lambda_j \neq 0$ satisfying $\lim_{j\to\infty} \lambda_j = 0$. Moreover, $Ae_j = \lambda e_j$ for each j.

More precisely, the orthonormal system is either $(e_j)_{j=1}^N$ for some $N \in \mathbb{N}$ or $(e_j)_{j\in\mathbb{N}}$. Of course, the condition $\lim_{j\to\infty} \lambda_j = 0$ is only meaningful in the second case.

Proof. We shall find (e_j, λ_j) step by step. If $A = 0$, then there is nothing to show. So let us assume that $\|A\| > 0$.

Write $H_1 = H$. By Lemma 13.10, A has an eigenvalue λ_1 such that $|\lambda_1| = \|A\|$. Let $e_1 \in H$ be such that $\|e_1\| = 1$ and $Ae_1 = \lambda_1 e_1$.

Now $F_1 := \operatorname{span}\{e_1\}$ is clearly an A-invariant linear subspace of H_1. By Lemma 13.9.c), $H_2 := F_1^\perp$ is also A-invariant. Hence we can consider the restriction $A|H_2$ of A on H_2 and iterate. If $A|H_2 = 0$, the process stops. If
Ex.13.8 not, since $A|H_2$ is a compact self-adjoint operator on H_2, we can find a unit vector e_2 and a scalar λ_2 such that $Ae_2 = \lambda_2 e_2$ and

$$|\lambda_2| = \|A|H_2\|_{\mathcal{L}(H_2)} \leq \|A|H_1\|_{\mathcal{L}(H_1)}.$$

After n steps we have constructed an orthonormal system $e_1, \dots, e_n$ and a sequence $\lambda_1, \dots, \lambda_n$ such that

$$Ae_j = \lambda_j e_j, \quad |\lambda_j| = \|A|H_j\|_{\mathcal{L}(H_j)} \quad \text{where} \quad H_j = \{e_1, \dots, e_{j-1}\}^\perp$$

for all $j = 1, \dots, n$. In the next step define $H_{n+1} := \{e_1, \dots, e_n\}^\perp$, note that it is A-invariant and consider the restriction $A|H_{n+1}$ thereon. This is again a compact self-adjoint operator. If $A|H_{n+1} = 0$, the process stops, otherwise one can find a unit eigenvector associated with an eigenvalue λ_{n+1} such that $|\lambda_{n+1}| = \|A|H_{n+1}\|_{\mathcal{L}(H_{n+1})}$.

Suppose that the process stops after the n-th step. Then $A|H_{n+1} = 0$. If $f \in H$, then

$$f - \sum\nolimits_{j=1}^n \langle f, e_j \rangle e_j \in \{e_1, \dots, e_n\}^\perp = H_{n+1}$$

and so A maps it to 0; this means that

$$Af = A\sum\nolimits_{j=1}^n \langle f, e_j \rangle e_j = \sum\nolimits_{j=1}^n \langle f, e_j \rangle Ae_j = \sum\nolimits_{j=1}^n \lambda_j \langle f, e_j \rangle e_j,$$

i.e., (13.1). Now suppose that the process does not stop, i.e., $|\lambda_n| = \|A|H_n\| > 0$ for all $n \in \mathbb{N}$. We claim that $|\lambda_n| \to 0$, and suppose towards a contradiction that this is not the case. Then there is $\epsilon > 0$ such that $|\lambda_n| \geq \epsilon$ for all $n \in \mathbb{N}$. But then

$$\|Ae_j - Ae_k\|^2 = \|\lambda_j e_j - \lambda_k e_k\|^2 = |\lambda_j|^2 + |\lambda_k|^2 \geq 2\epsilon^2$$

for all $j \neq k$. So $(Ae_j)_{j\in\mathbb{N}}$ cannot have a convergent subsequence, contradicting the compactness of A.

Now let $f \in H$ and define

$$y_n := f - \sum\nolimits_{j=1}^{n-1} \langle f, e_j \rangle \, e_j \;\in\; \{e_1, \dots, e_{n-1}\}^{\perp} = H_n.$$

Note that y_n is the orthogonal projection of f onto H_n, and so $\|y_n\| \leq \|f\|$. Hence

$$\|Ay_n\| \leq \|A|H_n\|_{\mathcal{L}(H_n)} \|y_n\| \leq |\lambda_n| \, \|f\| \to 0.$$

This implies

$$Af - \sum\nolimits_{j=1}^{n-1} \lambda_j \langle f, e_j \rangle \, e_j = Ay_n \to 0,$$

which proves (13.1). □

The spectral theorem (Theorem 13.11) contains additional information. Let us denote by J the index set for the orthonormal system in the spectral theorem. So $J = \{1, \dots, N\}$ or $J = \mathbb{N}$. Moreover, let

$$P_0 : H \longrightarrow \ker(A)$$

be the orthogonal projection onto the kernel of A and $P_r := \mathrm{I} - P_0$ its complementary projection. Then we can write

$$Af = 0 \cdot P_0 f + \sum\nolimits_{j\in J} \lambda_j \langle f, e_j \rangle \, e_j$$

for all $f \in H$. This formula is called the **spectral decomposition** of A.

Corollary 13.12. *Let A be as in the spectral theorem (Theorem 13.11). Then the following assertions hold.*

a) $\overline{\operatorname{ran}}(A) = \overline{\operatorname{span}}\{e_j \mid j \in J\}$ *and* $\ker(A) = \{e_j \mid j \in J\}^{\perp}$.

b) $P_r f = \sum\nolimits_{j\in J} \langle f, e_j \rangle \, e_j$ *for all $f \in H$.*

c) *Every nonzero eigenvalue of A occurs in the sequence $(\lambda_j)_{j\in J}$, and its geometric multiplicity is*

$$\dim \ker(\lambda \mathrm{I} - A) = \operatorname{card}\{j \in J \mid \lambda = \lambda_j\} < \infty.$$

Proof. a) By the formula (13.1) it is obvious that $\operatorname{ran}(A) \subseteq \overline{\operatorname{span}}\{e_j \mid j \in J\}$. For the converse, simply note that since each $\lambda_j \neq 0$, each $e_j = A(\lambda_j^{-1} e_j)$ is contained in $\operatorname{ran}(A)$.

Since $A = A^*$, Lemma 12.12 yields the orthogonal decomposition $H = \ker(A) \oplus \overline{\operatorname{ran}}(A)$. Hence $P_r = \mathrm{I} - P_0$ is the orthogonal projection onto $\overline{\operatorname{ran}}(A)$, and so the formula in b) follows from the abstract theory in Chapter 8. Moreover, by a)

$$\ker(A) = \overline{\operatorname{ran}}(A)^\perp = \{e_j \mid j \in J\}^\perp,$$

and this is the second identity in a).

c) Suppose that $Af = \lambda f$. If $\lambda \neq \lambda_j$ for every j, then by Lemma 13.9.b) $f \perp e_j$ for every j and so $Af = 0$, by a). This proves the first assertion of
Ex.13.9 c). The remaining statement is left as an exercise. □

The Eigenvalue Equation. We employ the spectral theorem to discuss the solvability of the equation

$$Au - \lambda u = f, \tag{13.2}$$

where $f \in H$ and $\lambda \in \mathbb{K}$ are given, and A is a compact self-adjoint operator on H. It turns out that we have a complete overview about existence and uniqueness of solutions u. Take $(e_j)_{j\in J}$ and $(\lambda_j)_{j\in J}$ as in the spectral theorem.

Theorem 13.13 (Fredholm Alternative[1]). *In the situation above, precisely one of the following cases holds:*

1) *If $\lambda \neq 0$ is different from every λ_j, then $(\lambda\mathrm{I} - A)$ is invertible and*

$$u := (A - \lambda\mathrm{I})^{-1} f = -\frac{1}{\lambda} P_0 f + \sum\nolimits_{j\in J} \frac{1}{\lambda_j - \lambda} \langle f, e_j\rangle\, e_j$$

is the unique solution to (13.2).

2) *If $\lambda \neq 0$ is an eigenvalue of A, then* (13.2) *has a solution if and only if $f \perp \ker(\lambda\mathrm{I} - A)$. In this case a particular solution is*

$$u := -\frac{1}{\lambda} P_0 f + \sum\nolimits_{j\in J_\lambda} \frac{1}{\lambda_j - \lambda} \langle f, e_j\rangle\, e_j,$$

where $J_\lambda := \{j \in J \mid \lambda_j \neq \lambda\}$.

[1] Erik Ivar Fredholm (1866–1927), Swedish mathematician and physicist.

3) *If* $\lambda = 0$, *then* (13.2) *is solvable if and only if* $f \in \operatorname{ran}(A)$; *in this case one particular solution is*

$$u := \sum\nolimits_{j \in J} \frac{1}{\lambda_j} \langle f, e_j \rangle \, e_j,$$

this series being indeed convergent.

Proof. We first show uniqueness in 1). Suppose that u_1, u_2 satisfy (13.2). Then $u := u_1 - u_2$ satisfies $Au = \lambda u$, and since every eigenvalue of A appears in the sequence $(\lambda_j)_j$ we have $u = 0$.

Now we show that in 2) the condition $f \perp \ker(A - \lambda \mathrm{I})$ is necessary for $f \in \operatorname{ran}(A - \lambda \mathrm{I})$. Indeed, if $f = Au - \lambda u$ and $Ag = \lambda g$, then

$$\langle f, g \rangle = \langle Au - \lambda u, g \rangle = \langle u, Ag - \lambda g \rangle = 0$$

since λ is an eigenvalue of $A = A^*$, hence a real number.

To prove existence in 1) and 2) simultaneously, take $0 \neq \lambda$ and define $J_\lambda := \{j \in J \mid \lambda \neq \lambda_j\}$. (In the situation 1), $J_\lambda = J$.) Take $f \in H$ such that $f \perp \ker(A - \lambda \mathrm{I})$. (In the situation 1), this is always satisfied.) Then we can write

$$f = P_0 f + P_r f = P_0 f + \sum\nolimits_{j \in J} \langle f, e_j \rangle \, e_j = P_0 f + \sum\nolimits_{j \in J_\lambda} \langle f, e_j \rangle \, e_j$$

because for $j \notin J_\lambda$ we have $f \perp e_j$. Now note that

$$c := \sup_{j \in J_\lambda} \left| \frac{1}{\lambda_j - \lambda} \right| < \infty,$$

because $\lambda_j \to 0 \neq \lambda$. Hence

$$\sum\nolimits_{j \in J_\lambda} \left| \frac{\langle f, e_j \rangle}{\lambda_j - \lambda} \right|^2 \le c^2 \sum\nolimits_{j \in J} \left| \langle f, e_j \rangle \right|^2 \le c^2 \left\| f \right\|^2 < \infty,$$

by Bessel's inequality. So by Theorem 8.13 and since we are in a Hilbert space, the series

$$v := \sum\nolimits_{j \in J_\lambda} \frac{1}{\lambda_j - \lambda} \langle f, e_j \rangle \, e_j$$

converges. Define $u := (-1/\lambda) P_0 f + v$. Then

$$\begin{aligned} Au - \lambda u &= Av - \lambda v + P_0 f \\ &= P_0 f + \sum\nolimits_{j \in J_\lambda} \frac{\lambda_j}{\lambda_j - \lambda} \langle f, e_j \rangle \, e_j - \sum\nolimits_{j \in J_\lambda} \frac{\lambda}{\lambda_j - \lambda} \langle f, e_j \rangle \, e_j \\ &= P_0 f + \sum\nolimits_{j \in J_\lambda} \langle f, e_j \rangle \, e_j = f \end{aligned}$$

as claimed.

In 3), $f \in \operatorname{ran}(A)$ is certainly necessary for the solvability of $Au = f$. Now suppose that $f = Av$ for some $v \in H$. Then by (13.1)

$$\langle f, e_j \rangle = \lambda_j \langle v, e_j \rangle$$

for all $j \in J$. In particular, $\sum_{j \in J} \left| \lambda_j^{-1} \langle f, e_j \rangle \right|^2 < \infty$; since H is a Hilbert space, the series

$$u := \sum\nolimits_{j \in J} \frac{\langle f, e_j \rangle}{\lambda_j} e_j$$

is convergent. Note that actually $u = P_r v$. Hence

$$Au = AP_r v = AP_0 v + AP_r v = A(P_0 v P_r v) = Av = f.$$

This concludes the proof. □

13.4. *The General Spectral Theorem

It is natural to ask whether a "spectral theorem" holds under less restrictive hypotheses than the ones we have considered. It turns out that for a compact operator A to admit a spectral decomposition of H into pairwise orthogonal eigenspaces it is necessary and sufficient that the operator is **normal**, i.e., satisfies $A^*A = AA^*$. (For a proof see Exercises 13.23 and 13.24.)

The situation changes when one leaves the realm of compact operators. If A is self-adjoint (or normal) but not compact, then possibly it has no eigenvalues at all, so what should a "spectral decomposition" be in the first place?

To answer this question we re-interpret the compact case. Suppose for simplicity that $\dim(H) = \infty$, $\ker(A) = \{0\}$, and $A = \sum_{j=1}^{\infty} \lambda_j \langle \cdot, e_j \rangle e_j$ is the spectral decomposition of A. Then $(e_j)_{j \in \mathbb{N}}$ is an orthonormal *basis* of H and as such induces an isometric isomorphism $U : H \to \ell^2$ mapping $e_j \in H$ onto $\tilde{e_j}$, the j-th standard unit vector in ℓ^2. The operator $\tilde{A} := UAU^{-1} : \ell^2 \to \ell^2$ is then simply the multiplication operator with multiplier sequence $(\lambda_j)_{j \in \mathbb{N}} \in c_0$. One says that the pair (H, A) is **unitarily equivalent** to $(\ell^2, \tilde{A})$.

The last step then requires measure-theoretic integration theory. Basically, instead of Lebesgue measure on subsets $\mathbb{R}^d$ one considers general "measure spaces" (X, Σ, μ). Here X is a set, Σ is a σ-algebra of subsets of X (see Theorem 7.2.b) and $\mu : \Sigma \to [0, \infty]$ is countably additive (see Theorem 7.2.c). The map μ is then called a **measure** and one can build up the integration theory as for the Lebesgue measure (Chapter 7). This leads to the spaces $\mathrm{L}^p(X, \Sigma, \mu)$ for $1 \le p \le \infty$.

For the special case $X = \mathbb{N}$ and μ the "counting measure" (i.e., $\mu(A) = \operatorname{card}(A)$ for $A \subseteq \mathbb{N}$), the corresponding L^2-space is nothing other than ℓ^2. So the spectral theorem for a compact normal operator A just says that A is unitarily equivalent to a multiplication operator on a very specific L^2-space, namely ℓ^2. Now we can formulate the general result.

Theorem 13.14. *Let A be a normal operator on a Hilbert space H. Then A is unitarily equivalent to a multiplication operator on an L^2-space.*

More precisely, there exists a measure space (X, Σ, μ), a function $m \in L^\infty(X, \Sigma, \mu)$ and an isometric isomorphism $U : H \to L^2(X, \Sigma, \mu)$ such that

$$Af = U^{-1}(m \cdot Uf) \quad \text{for all } f \in H.$$

The common proofs are developed in the framework on so-called C^*-algebras; see, e.g., [**Con00**, Theorem 11.5] of [**Haa06**, Appendix D].

Theorem 13.14 is just *one* version of the general spectral theorem. In fact, there is a wealth of different formulations to be found in the literature. The one that is most often used, in particular in the context of mathematical physics, uses the notion of a **spectral measure**; see. e.g. [**Con90**, IX,§2].

The multiplier version we stated here is, despite its elegance and simplicity, often neglected in standard functional analysis textbooks. (Werner's excellent [**Wer00**] is a notable exception, but that book has not been translated into English yet.) To the effect that one eminent 20th century mathematician, Paul R. Halmos[2], felt the need to write an article with the title "What does the spectral theorem say?" [**Hal63**] in which he promoted the multiplier formulation and provided a clear and relatively elementary proof.

The history of the spectral theorem from its first rudimentary appearance in the early texts on analytic geometry in the 16th century until its modern forms within the theory of commutative Banach algebras developed mainly by Gelfand[3] is recorded in Steen's article [**Ste73**].

Exercises 13A

Exercise 13.1. Let $H := L^2(0,1)$ and $A : H \to H$ is the multiplication operator $(Af)(t) := t \cdot f(t)$, $t \in (0,1)$. Show that A has no eigenvalues.

Exercise 13.2. Show that 0 is not an eigenvalue but an approximate eigenvalue of the integration operator J on $C[a,b]$.

Exercise 13.3. Consider the left shift L on ℓ^2. Show that $\lambda \in \mathbb{K}$ is an eigenvalue of L if and only if $|\lambda| < 1$.

[2]Paul Richard Halmos (1916–2006), Hungarian-American mathematician.

[3]Israel Moissejewitsch Gelfand (1913–2009), Russian mathematician.

Exercise 13.4. Let $(e_n)_{n\in\mathbb{N}}$ be the standard unit vectors on ℓ^2 and let $\lambda \in \mathbb{K}$ with $|\lambda| = 1$. Define $f_n := (1/\sqrt{n})(\lambda e_1 + \lambda^2 e_2 + \cdots + \lambda^n e_n)$. Use the sequence $(f_n)_{n\in\mathbb{N}}$ to show that λ is an approximate eigenvalue of the left shift L on ℓ^2.

Exercise 13.5. Consider the integration operator J on $\mathrm{L}^2(a,b)$. Show that for every $\lambda \neq 0$, $\lambda\mathrm{I} - J$ is invertible and compute a (Hilbert–Schmidt) integral kernel for $(1/\lambda)\mathrm{I} - (\lambda\mathrm{I} - J)^{-1}$.

Exercise 13.6. Let A be the multiplication operator $(Af)(t) = t \cdot f(t)$ on $\mathrm{L}^2(0,1)$. Show that for every $\lambda \in \mathbb{C} \setminus [0,1]$ the operator $\lambda\mathrm{I} - A$ is invertible and the inverse is given by

$$\big[(\lambda\mathrm{I} - A)^{-1} f\big](t) = \frac{1}{\lambda - t} f(t) \qquad (t \in (0,1)).$$

Exercise 13.7. Let A be a bounded self-adjoint operator and let λ be an approximate eigenvalue of A. Show that $\lambda \in \mathbb{R}$.

Exercise 13.8. Let H be a Hilbert space, let $A : H \to H$ be a bounded linear operator, and let $F \subseteq H$ be an A-invariant closed subspace of H. Let $B := A\big|_F$ be the restriction of A to F. Show that the following assertions hold.

a) $\|B\|_{\mathcal{L}(F)} \le \|A\|_{\mathcal{L}(H)}$.

b) If A is self-adjoint, then B is self-adjoint.

c) If A is compact, then B is compact.

Exercise 13.9. In the situation of the spectral theorem, show that

$$\ker(\lambda\mathrm{I} - A) = \operatorname{span}\{e_j \mid \lambda_j = \lambda\}$$

for each $\lambda \neq 0$.

Exercises 13B

Exercise 13.10. Let E be a Banach space and let $A \in \mathcal{L}(E)$. Show that λ is not an approximate eigenvalue if and only if there is $\delta > 0$ such that

$$\|\lambda f - Af\| \ge \delta \, \|f\| \qquad (f \in H).$$

By Exercises 5.15 and 15.5, this happens if and only if $\lambda\mathrm{I} - A$ is injective and $\operatorname{ran}(\lambda\mathrm{I} - A)$ is a closed subspace of E.

Exercise 13.11 (Positive Operators). Let H be a Hilbert space. A bounded self-adjoint operator A on H is called **positive** if $\langle Af, f\rangle \ge 0$ for all $f \in H$. We write $A \ge 0$ if A is positive.

a) Show that if $A, B \in \mathcal{L}(H)$ are self-adjoint and $\alpha, \beta \in \mathbb{R}$, then $\alpha A + \beta B$ is self-adjoint. Then show that if $A, B \ge 0$ and $\alpha, \beta \ge 0$, then also $\alpha A + \beta B \ge 0$.

b) Let A be self-adjoint such that $A \ge 0$, and let $C \in \mathcal{L}(H)$ be arbitrary. Show that C^*AC is selfadjoint and $C^*AC \ge 0$.

c) Show that if A is self-adjoint and $\alpha \ge \|A\|$, then $\alpha\mathrm{I} \pm A \ge 0$.

d) Let A be a positive self-adjoint bounded operator on H. Show that

(13.3) $$\|Af\|^2 \le \|A\| \langle Af, f\rangle$$

for all $f \in H$. [Hint: Let $\alpha := \|A\|$ and $C := \alpha \mathrm{I} - A$ and show that $\alpha^2 A - \alpha A^2 = ACA + CAC$.]

For a bounded operator A on a Banach space E we let

$$\sigma(A) := \{\lambda \in \mathbb{K} \mid (\lambda \mathrm{I} - A) \text{ is not invertible}\}$$

be its **spectrum**. The complement $\rho(A) := \mathbb{C} \setminus \sigma(A)$ is called its **resolvent set**. By Lemma 13.4, $\sigma(A)$ contains all approximate eigenvalues and is itself contained in the ball/interval $\{\lambda \in \mathbb{K} \mid |\lambda| \le \|A\|\}$. For $\lambda \in \rho(A)$ we introduce the notation

$$R(\lambda, A) := \lambda \mathrm{I} - A$$

and call it the **resolvent** of A at λ.

Exercise 13.12. Let E be a Banach space, $A \in \mathcal{L}(E)$, $\lambda, \mu \in \mathbb{K}$ such that $\lambda, \mu \in \rho(A)$. Show that

$$R(\lambda, A) - R(\mu, A) = (\mu - \lambda)\, R(\lambda, A)\, R(\mu, A).$$

(This identity is called the **resolvent identity**.)

Exercise 13.13. Let H be a Hilbert space, $A \in \mathcal{L}(H)$ and $\lambda \in K$. Show that if $\lambda \in \rho(A)$ then $\overline{\lambda} \in \rho(A^*)$ and

$$R(\overline{\lambda}, A^*) = R(\lambda \mathrm{I}, A)^*.$$

Exercise 13.14. Let E be a Banach space, and let $A \in \mathcal{L}(E)$ be an invertible operator. Let $\lambda \in \rho(A) \setminus \{0\}$. Show that $\lambda^{-1} \in \rho(A^{-1})$ with

$$R(\lambda^{-1}, A^{-1}) = \lambda \mathrm{I} - \lambda^2 R(\lambda, A).$$

Exercise 13.15. Let $A \in \mathcal{L}(E)$ and $\mu \in \rho(A)$. Show that if $|\lambda - \mu| < \|R(\mu, A)\|^{-1}$, then $\lambda \in \rho(A)$, too, and

$$R(\lambda, A) = \sum_{n=0}^{\infty} (\lambda - \mu)^n R(\mu, A)^{n+1}.$$

Conclude that $\sigma(A)$ is compact, $\rho(A)$ is open and the resolvent is a continuous mapping $R(\cdot, A) : \rho(A) \to \mathcal{L}(E)$. Moreover, $\|R(\mu, A)\| \operatorname{dist}(\mu, \sigma(A)) \ge 1$ for $\mu \in \rho(A)$. [See also Exercises 11.31 and 11.32.]

By the previous exercise, $\|R(\mu, A)\| \to \infty$ as μ approaches a point of the spectrum. The following shows that this necessarily means that *boundary points of the spectrum are approximate eigenvalues.*

Exercise 13.16. Let E be a Banach space and let $A \in \mathcal{L}(E)$. Let $(\lambda_n)_{n\in\mathbb{N}}$ be a sequence of scalars in $\rho(A)$ such that

$$\lambda_n \to \lambda \quad \text{and} \quad \|R(\lambda_n, A)\| \to \infty.$$

Show that λ is an approximate eigenvalue of A. [Hint: By definition of the operator norm, find $g_n \in E$ such that $\|g_n\| \le 1$ and $\|R(\lambda_n, A)g_n\| \to \infty$. Define $f_n := {}^{1}\!/\!_{\|R(\lambda_n, A)g_n\|} R(\lambda_n, A)g_n$ and show that $\|f_n\| = 1$ and $(\lambda \mathrm{I} - A)f_n \to 0$.]

Exercise 13.17. Let $T : E \to E$ be an isometry, i.e., an operator satisfying $\|Tf\| = \|f\|$ for every $f \in E$. Prove the following assertions:

a) If $\lambda \in \mathbb{C}$ is an approximate eigenvalue of the isometry T, then $|\lambda| = 1$.

b) The set $\{\lambda \in \sigma(T) \mid |\lambda| < 1\}$ is open in $\mathbb{C}$.

Conclude that *either* T is invertible and $\sigma(T)$ is contained in the unit circle, *or* T is not surjective and $\sigma(T)$ is the full closed unit disc. (In the latter case, $\lambda = 0$ is not an approximate eigenvalue of T.)

Exercises 13C

Exercise 13.18 (Minmax Principle). Let A be a compact self-adjoint operator on the Hilbert space H, and denote the *positive* eigenvalues (with multiplicities) by $(\lambda_n)_{n<N}$ in decreasing order $\lambda_1 \geq \lambda_2 \geq \cdots > 0$. (As in the spectral theorem $N \in \mathbb{N}_0 \cup \{\infty\}$.)

a) Let $\alpha > 0$ and $n \in \mathbb{N}$. Show that the following assertions are equivalent:

   (i) $n < N$ and $\lambda_n \geq \alpha$.

   (ii) There is an n-dimensional subspace $F \subseteq H$ such that $\alpha \|f\|^2 \leq \langle Af, f\rangle$ for all $f \in F$.

b) Show that

$$\lambda_n = \max_F \min_{0 \neq f \in F} \frac{\langle Af, f\rangle}{\|f\|^2}$$

where F ranges over all n-dimensional subspaces F of H, in the sense that the left-hand side is defined if and only if the right-hand is defined and strictly positive.

c) Show that

$$\lambda_n = \min_G \max_{0 \neq f \in F^\perp} \frac{\langle Af, f\rangle}{\|f\|^2},$$

where the minimum is taken over all $(n-1)$-dimensional subspaces G of H (again in the sense that the left-hand side is defined if and only if the right-hand side is defined and strictly positive).

Exercise 13.19 (Singular Value Decomposition). Let H, K be Hilbert spaces.

a) Suppose that $(e_j)_j$ and $(f_j)_j$ are orthonormal systems in H and K, respectively, and $(s_j)_j$ is a decreasing null sequence of positive real numbers. Show that the series

$$A := \sum\nolimits_j s_j \langle \cdot, e_j\rangle f_j$$

converges in the operator norm to a compact operator $A : H \to K$. Show further that $(s_j^2)_j$ is the sequence of eigenvalues of A^*A, counted with multiplicities.

b) Conversely, let $A : H \to K$ be a compact operator between Hilbert spaces H, K. Show that there is a decreasing sequence $(s_j)_j$ of positive real numbers and orthonormal systems $(e_j)_j$ of H and $(f_j)_j$ of K such that

$$A = \sum\nolimits_j s_j \langle \cdot, e_j\rangle f_j \tag{13.4}$$

and $(s_j)_j$ is either finite or a null sequence. [Hint: Apply the spectral theorem to A^*A to obtain its spectral decomposition $A^*A = \sum_j \lambda_j \langle \cdot, e_j \rangle e_j$; show that all $\lambda_j > 0$; then define $f_j = Ae_j / \|Ae_j\|$ and $s_j = \sqrt{\lambda_j}$.]

By a), the sequence $(s_j)_j$ is uniquely determined by A; it is called the sequence of **singular values** of A. The series representation (13.4) is called the **singular value decomposition** of A.

Exercise 13.20. Show that the following statements are equivalent for a linear operator $A : H \to K$:

(i) There are sequences $(u_n)_n$ in H and $(v_n)_n$ in K such that

$$\sum_n \|u_n\| \cdot \|v_n\| < \infty \quad \text{and} \quad A = \sum_n \langle \cdot, u_n \rangle v_n.$$

(ii) A is compact and $\sum_j s_j < \infty$, where $(s_j)_j$ is the sequence of singular values of A.

Such operators are called **nuclear** or of **trace class**.

Exercise 13.21. A bounded operator A on a Hilbert space H is called **normal** if $A^*A = AA^*$. Establish the following claims for any normal operator $A \in \mathcal{L}(H)$:

a) $\|Af\| = \|A^*f\|$ for every $f \in H$.

b) $\lambda \mathrm{I} - A$ is normal for every $\lambda \in \mathbb{C}$.

c) $\ker(\lambda \mathrm{I} - A) = \ker(\overline{\lambda} \mathrm{I} - A^*)$ for all $\lambda \in \mathbb{C}$.

d) $\ker(\lambda \mathrm{I} - A) \perp \ker(\mu \mathrm{I} - A)$ whenever $\lambda \neq \mu$.

e) If $F \subseteq H$ is such that $A(F) \subseteq F$, then $A(F^\perp) \subseteq F$.

Exercise 13.22. Suppose A is a normal operator on a *finite-dimensional* complex Hilbert space H. Show that H has an orthonormal basis of eigenvectors of A. [Hint: Define F to be the subspace of H generated by the eigenvectors of A, and consider $F^\perp$. Can it happen that $\dim(F^\perp) \geq 1$?]

Exercise 13.23 (Spectral Theorem for Compact Normal Operators)**.** Let A be a compact normal operator on a Hilbert space H, with singular value decomposition $A = \sum_j s_j \langle \cdot, e_j \rangle f_j$ as in Exercise 13.19.

a) Show that $A^*Ae_j = s_j^2 e_j$ and $AA^*f_k = s_k^2 f_k$.

b) Conclude from a) that for given $s > 0$ one has

$$H_s := \operatorname{span}\{e_j \mid s_j^2 = s^2\} = \operatorname{span}\{f_k \mid s_k^2 = s^2\}.$$

c) Show that A restricts to a normal operator on the finite-dimensional space H_s and hence (by Exercise 13.22) that

$$H_s = \bigoplus_{|\lambda| = s} \ker(\lambda \mathrm{I} - A).$$

d) Conclude that A has a spectral decomposition

$$A = \sum_j \lambda_j \langle \cdot, \tilde{e}_j \rangle \tilde{e}_j$$

for some finite or infinite null sequence $(\lambda_j)_j$ and an orthonormal system $(\tilde{e}_j)_j$.

Exercise 13.24. Give another proof of the spectral theorem for a compact normal operator A avoiding the singular value decomposition, e.g., by applying the spectral theorem to the self-adjoint operators $B := \frac{1}{2}(A + A^*)$ and $C := \frac{1}{2\mathrm{i}}(A - A^*)$.

Chapter 14

Applications of the Spectral Theorem

In this chapter we discuss some applications and consequences of the spectral theorem.

14.1. The Dirichlet Laplacian

Recall from Section 11.2 that the **Dirichlet Laplacian** is Δ_D defined as

$$\Delta_D u = u'' \quad \text{with domain} \quad \mathrm{dom}(\Delta_D) = \mathrm{H}^2(a,b) \cap \mathrm{H}^1_0(a,b).$$

It is a version of the Laplace operator (= second derivative) with Dirichlet boundary conditions $u(a) = u(b) = 0$.

In Section 11.2 we have shown that Δ_D is bijective, and its negative inverse is given by the Hilbert–Schmidt operator

$$-\Delta_D^{-1} f = Af := \int_a^b g(\cdot, s) f(s)\,\mathrm{d}s$$

where

$$g(t,s) := \begin{cases} (s-a)(b-t) & a \le s \le t \le b, \\ (t-a)(b-s) & a \le t \le s \le b. \end{cases}$$

As a Hilbert–Schmidt operator, A is compact. Since k is symmetric and real-valued, A is self-adjoint, and $\ker(A) = \{0\}$, by construction. To apply the spectral theorem to A, we need to find its nonzero eigenvalues and corresponding eigenvectors.

Lemma 14.1. *Let $\lambda \neq 0$, $\mu = -1/\lambda$. Then*

$$f \in \mathrm{L}^2(a,b) \quad \text{and} \quad Af = \lambda f \quad \Longleftrightarrow \quad f \in \mathrm{dom}(\Delta_D) \quad \text{and} \quad \Delta_D f = \mu f.$$

Moreover, in this case either $f = 0$ or $\lambda > 0$.

Proof. The "if"-part is straightforward. For the "only if" part suppose that $Af = \lambda f$. Then $f = -\mu A f = \mu \Delta_D^{-1} f \in \mathrm{dom}(\Delta_D)$, and hence $\Delta_D f = \mu \Delta_D \Delta_D^{-1} f = f$. If $f \neq 0$, then $\lambda \in \mathbb{R}$ (since A is self-adjoint) and hence

$$0 < \|f\|_2 = \langle f, f \rangle = -\lambda \left\langle f'', f \right\rangle = \lambda \left\langle f', f' \right\rangle$$

since $f \in \mathrm{C}_0^1[a,b]$. But $\langle f', f' \rangle > 0$ (why?) and thus $\lambda > 0$. □

Lemma 14.1 yields that an eigenvector f of A must satisfy $f \in \mathrm{H}^2(a,b)$, in particular $f \in \mathrm{C}[a,b]$. But then $f'' = \Delta_D f = \mu f \in \mathrm{C}[a,b]$, and this shows that actually $f \in \mathrm{C}^2[a,b]$. (Iterating this argument one can prove that $f \in \mathrm{C}^\infty[a,b]$.)

For simplicity, let $(a,b) = (0,1)$ from now on. Employing the classical theory of differential equations we can conclude that $Au = \lambda u$ with $\lambda > 0$ if and only if

$$u(t) = \alpha \cos\left({}^{t}\!/\!_{\sqrt{\lambda}} \right) + \beta \sin\left({}^{t}\!/\!_{\sqrt{\lambda}} \right) \qquad (0 \le t \le 1)$$

for some constants α, β. The boundary condition $u(0) = 0$ forces $\alpha = 0$ and so $\ker(\lambda \mathrm{I} - A)$ is at most one-dimensional. On the other hand, $u(1) = 0$ and $\beta \neq 0$ forces

$$\sin({}^{1}\!/\!_{\sqrt{\lambda}}) = 0.$$

This yields the sequence of eigenvalues

$$\lambda_n = \frac{1}{n^2 \pi^2} \qquad (n \ge 1)$$

with associated (normalized) eigenfunctions

$$e_n(t) = \frac{\sin(n\pi t)}{\sqrt{2}} \qquad (n \ge 1,\ t \in [0,1]).$$

Since A is injective, $\overline{\mathrm{ran}}(A) = H$ and, according to the spectral theorem, the system $(e_n)_{n=1}^\infty$ must be an orthonormal basis for $\mathrm{L}^2(0,1)$. Moreover, the operator A can be written as

$$(Af)(t) = \int_0^1 g(t,s) f(s)\,\mathrm{d}s = \sum_{n=1}^\infty \left(\frac{1}{2n^2\pi^2} \int_0^1 f(s) \sin(n\pi s)\,\mathrm{d}s \right) \sin(n\pi t)$$

with $t \in (0,1)$ and $f \in \mathrm{L}^2(0,1)$. The general theory yields convergence of this series only in the L^2-norm but *not pointwise*. However, in this particular
Ex.14.1 case, we see that the series converges even *uniformly* in $t \in [0,1]$.

It is even true that

$$g(t,s) = \sum_{n=1}^\infty \frac{\sin(n\pi \cdot t)\, \sin(n\pi \cdot s)}{2n^2\pi^2}$$

Ex.14.2 as an absolutely convergent series in $\mathrm{C}([0,1] \times [0,1])$.

14.2. The Schrödinger Operator

We now perturb the Dirichlet Laplacian by a multiplication operator. More precisely, we let $q \in \mathrm{C}[0,1]$ be a fixed *positive* continuous function ($q \geq 0$), and consider the differential operator

$$L : \mathrm{H}^2(0,1) \cap \mathrm{H}^1_0(a,b) \longrightarrow \mathrm{L}^2(a,b) \qquad Lu = -u'' + qu.$$

It is called a one-dimensional **Schrödinger operator**[1] with **potential function** q. We write $\mathrm{dom}(L) := \mathrm{H}^2(0,1) \cap \mathrm{H}^1_0(0,1)$ and call it the **domain** of L.

This operator is a special case of a so-called *Sturm–Liouville* operator; see [**You88**, Chap. 9] and Exercise 12.25.

Let us first look at eigenvalues of L.

Lemma 14.2. *If $u \in \mathrm{dom}(L)$ and $Lu = \lambda u$, then $u \in \mathrm{C}^2[0,1]$ and either $u = 0$ or $\lambda < 0$. In particular, L is injective.*

Proof. If $Lu = \lambda u$, then $u'' = qu + \lambda u \in \mathrm{C}[0,1]$, hence $u \in \mathrm{C}^2[0,1]$. Integration by parts yields

$$\begin{aligned}\lambda \|u\|_2^2 &= \langle \lambda u, u\rangle = \langle qu - u'', u\rangle = \langle pu, u\rangle - \langle u'', u\rangle \\ &= \int_0^1 q(s)\,|u(s)|^2 \,\mathrm{d}s + \|u'\|_2^2 \geq \|u'\|_2^2 \geq 0,\end{aligned}$$

since $q \geq 0$ by assumption. Also, $\lambda > 0$ or $u' = 0$, which implies $u = 0$. □

We shall now show that the **Sturm–Liouville problem**

$$u'' - qu = -f \qquad u \in \mathrm{H}^2(a,b) \cap \mathrm{H}^1_0(a,b)$$

is well-posed for $f \in \mathrm{L}^2(0,1)$. That is to say, $L : \mathrm{dom}(L) \to \mathrm{L}^2(0,1)$ is bijective and its inverse is bounded. To this aim we consider the symmetric sesquilinear form

$$a(u,v) := \langle u', v'\rangle_{\mathrm{L}^2} + \langle qu, v\rangle_{\mathrm{L}^2} \qquad (u, v \in \mathrm{H}^1_0(0,1)).$$

This is an inner product on $\mathrm{H}^1_0(0,1)$ and the associated norm $\|\cdot\|_a$, say, is equivalent to the standard norm on $\mathrm{H}^1_0(0,1)$. In particular, $\mathrm{H}^1_0(0,1)$ is a Hilbert space with respect to the inner product a (Exercise 14.3).

Given $f \in \mathrm{L}^2(0,1)$ the functional $v \mapsto \langle v, f\rangle_{\mathrm{L}^2}$ is bounded on $\mathrm{H}^1_0(0,1)$. Ex.14.3

[1] Erwin Schrödinger (1887–1961), Austrian physicist.

Hence, by the Riesz–Fréchet theorem there is a unique $u \in \mathrm{H}_0^1(0,1)$ such that

$$\langle u', v' \rangle_{\mathrm{L}^2} + \langle qu, v \rangle_{\mathrm{L}^2} = a(u,v) = \langle f, v \rangle \quad \text{for all } v \in \mathrm{H}_0^1(0,1).$$

Specializing $v \in \mathrm{C}_0^1[0,1]$ we see that $u \in \mathrm{H}^2(0,1)$ and $Lu = f$. This shows
Ex.14.4 that L is bijective as claimed. We leave it as Exercise 14.4 to show that

$$L^{-1} : \mathrm{L}^2(0,1) \longrightarrow \mathrm{H}^2(0,1)$$

is bounded.

We now claim that L^{-1}, considered as an operator on $H = \mathrm{L}^2(0,1)$, is compact and self-adjoint. The self-adjointness is straightforward (Exercise
Ex.14.5 14.5). For the compactness there are two alternative proofs. The first uses the theory of abstract Hilbert–Schmidt operators developed in Section 12.4: the operator L^{-1} maps L^2 boundedly into $\mathrm{C}[0,1]$, whence by Theorem 12.21 it is Hilbert–Schmidt on $\mathrm{L}^2(0,1)$. A fortiori, it is compact.

The second proof uses classical tools from differential equations to construct a *Green's function* k for the problem; see the next (optional) section. It turns out that k is continuous on $[0,1] \times [0,1]$, and this implies that L^{-1} is Hilbert-Schmidt, whence compact.

***The Green's Function.** We construct a Green's function for the operator L^{-1}. The method of doing this is classical: first pass to the linear system

$$x' = Ax + y, \quad \text{where} \quad x = \begin{bmatrix} u \\ v \end{bmatrix}, \; y = \begin{bmatrix} 0 \\ -f \end{bmatrix} \quad \text{and} \quad A = \begin{bmatrix} 0 & 1 \\ q & 0 \end{bmatrix}.$$

Second, pick a fundamental system of solutions Ψ for the homogeneous system $x' = Ax$ and try to find the final solution by variation of parameters, i.e., in the form $x = \Psi z$ where z is a $\mathbb{C}^2$-valued function. Finally, take the boundary values into account to determine the free parameters.

As the fundamental system Ψ must satisfy $\Psi' = A\Psi$, it is of the form

$$\Psi = \begin{bmatrix} u & v \\ u' & v' \end{bmatrix}$$

for some particular functions u, v satisfying $u'' = qu$ and $v'' = qv$. (As such, $u, v \in \mathrm{C}^2[0,1]$ and hence we are really in the realm of classical differential equations.) Any linear independent choice of such u, v will do, and it turns out that it is wise to choose them in such a way that

$$\begin{aligned} u(0) &= 0, \quad u'(0) = 1, \\ v(1) &= 0, \quad v'(1) = 1. \end{aligned} \tag{14.1}$$

(See Exercise 11.13 for the construction of u; v can be found analogously.
Ex.14.6 A different method is sketched in Exercise 14.6.)

Lemma 14.3. *With this choice of u and v, the* **Wronskian**[2] $w := u'v - uv'$ *is a constant nonzero function.*

Proof. One easily computes $(u'v - uv')' = u''v - uv'' = puv - upv = 0$, so w is indeed a constant. If $w = 0$, then $u(1) = (uv')(1) = (u'v)(1) = 0$, and this means that $u \in \operatorname{dom}(L)$. Since $u'' = pu$, we would have $Lu = 0$, contradicting the injectivity of S (Lemma 14.2). □

We now make the "Ansatz" $x = \Psi\, z$. Differentiating and equating yields

$$Ax + y = x' = \Psi' z + \Psi z' = A\Psi z + \Psi z' = Ax + \Psi z'.$$

Hence we have a solution if $\Psi z' = y$, i.e., if

$$z' = \Psi^{-1} y = \frac{1}{w}\begin{bmatrix} v' & -v \\ -u' & u \end{bmatrix} \cdot \begin{bmatrix} 0 \\ -f \end{bmatrix} = \frac{1}{w}\begin{bmatrix} vf \\ -uf \end{bmatrix}.$$

This is satisfiable with

$$z = \frac{1}{w}\begin{bmatrix} J(vf) \\ -J(uf) \end{bmatrix} + \begin{bmatrix} c \\ d \end{bmatrix}$$

for $c, d \in \mathbb{C}$ and J, as usual, the integration operator. We multiply with Ψ and take the first component to obtain

$$L^{-1} f = \frac{1}{w}\big(uJ(vf) - vJ(uf)\big) + cu + dv,$$

where c, d have to be such that the right-hand side vanishes at the boundary. This is the case if and only if $d = 0$ and

$$0 = u(1)\int_0^1 v(s) f(s)\, \mathrm{d}s + cu(1).$$

As $u(1) \neq 0$ (otherwise $u = 0$) we have $c = -\int_0^1 v(s) f(s)\, \mathrm{d}s$, and putting this back into the identity above yields

$$(L^{-1} f)(t) = \frac{-1}{w}\Big(\int_0^t v(t) u(s) f(s)\, \mathrm{d}s + \int_t^1 u(t) v(s)\, \mathrm{d}s\Big).$$

In effect, we have proved the following result.

Lemma 14.4. *The operator L^{-1} is a Hilbert–Schmidt operator with (continuous and symmetric) kernel function*

$$k(t,s) = \frac{-1}{w}\begin{cases} u(s)v(t) & \text{if } 0 \le s \le t \le 1, \\ v(s)u(t) & \text{if } 0 \le t \le s \le 1. \end{cases}$$

One can show that $u, u', v \geq 0$ and $v \leq 0$ on $[0,1]$, whence $w > 0$ and $k \geq 0$ on $[0,1] \times [0,1]$. Ex.14.7

[2] Josef Hoëné-Wronski (1776–1853), Polish-French mathematician and philosopher.

The Spectral Decomposition. As a self-adjoint and compact operator, L^{-1} has a spectral decomposition. Similar to Lemma 14.1 for the Dirichlet Laplacian one can show that λ is an eigenvalue for L if and only $1/\lambda$ is an
Ex.14.8 eigenvalue for L^{-1}, with same eigenspaces. Moreover, all eigenvalues are *simple*. For if $Lu = \lambda u$ and $Lv = \lambda v$ with $u \neq 0$, then $u'(0) \neq 0$ and there is $\alpha \in \mathbb{C}$ such that $h := \alpha u - v$ satisfies $h(0) = h'(0) = 1$. This implies that $h = 0$ on $[0, 1]$ since a solution of $u'' - (q + \lambda)u = 0$ is uniquely determined by its two initial values.

The spectral theorem now yields a decomposition

$$L^{-1}f = \sum_{n=1}^{\infty} \lambda_n \langle f, e_n \rangle e_n \qquad (f \in \mathrm{L}^2(0,1)), \tag{14.2}$$

where the $(e_n)_{n\in\mathbb{N}}$ form an orthonormal basis of $\mathrm{L}^2(0,1)$ and $\lambda_1 > \lambda_2 > \cdots \searrow 0$. Since L^{-1} factors through $\mathrm{C}[0,1]$, we obtain $\sum_{n=1}^{\infty} \lambda_n^2 < \infty$ and
Ex.14.9 the series (14.2) converges in the supremum norm.

One can say even more. Since we already know that the Green's function k is continuous and all eigenvalues are positive, we can apply a theorem of Mercer[3] [**Lax02**, Chap. 30, Thm. 11] which states that under these conditions

$$k = \sum_{n=1}^{\infty} \lambda_n e_n \otimes \overline{e_n}$$

is uniformly and absolutely convergent on $[0,1] \times [0,1]$.

But even without Mercer's theorem one can tell a little more. Namely, by passing through the procedure that has led us to the Green's function, one can explicitly determine the eigenvalues of L, and hence of L^{-1}; see Exercise 14.16.

14.3. An Evolution Equation

In this section we look at the following partial differential equation (initial-boundary value problem) on $[0, \infty) \times [0, 1]$:

$$\begin{cases} \partial_t u(t,x) = \partial_{xx} u(t,x) - q(x) u(t,x) & (t,x) \in (0,\infty) \times (0,1), \\ u(t,0) = u(t,1) = 0 & (t > 0), \\ u(0,x) = f(x) & (x \in (0,1)), \end{cases} \tag{14.3}$$

where $0 \le q \in \mathrm{C}[0,1]$ is the potential and $f : (0,1) \to \mathbb{K}$ is a given initial data. For $q = 0$ it is a one-dimensional inhomogeneous **heat equation** with Dirichlet boundary conditions. If f is continuous, it is reasonable to speak

[3] James Mercer (1883–1932), English mathematician.

of a so-called "classical" solution, i.e., a function $u \in \mathrm{C}([0,\infty) \times [0,1]) \cap \mathrm{C}^{1,2}((0,\infty) \times (0,1))$ that solves the PDE in the ordinary sense. However, the most successful strategy is to allow for a very weak notion of solution (in order to make it easy to find one) and then in a second step investigate under which conditions on f this solution is a classical one.

To find a candidate for a solution, one shifts the problem from PDEs to functional analysis. We want our solution u to be a function $u : [0,\infty) \to \mathrm{L}^2(0,1)$ satisfying

$$u(0) = f \in \mathrm{L}^2(0,1) \quad \text{and} \quad u(t) \in \mathrm{dom}(L), \quad u_t(t) = -Lu(t) \quad (t > 0).$$

Here, the time derivative u_t is to be understood in a "weak sense", i.e., it is a function $u_t : [0,\infty) \to \mathrm{L}^2(0,1)$ such that

$$\frac{d}{dt} \langle u(t), v\rangle = \langle u_t(t), v\rangle \qquad (t > 0)$$

for all $v \in \mathrm{L}^2(0,1)$. Equivalently, $L^{-1}u_t = -u$ for all $t > 0$, and writing the operator A in its associated Fourier expansion (see Section 14.2) yields

$$u(t) = -L^{-1}(u_t(t)) = \sum\nolimits_{n=1}^{\infty} -\lambda_j \langle u_t(t), e_n\rangle\, e_n.$$

Since $(e_n)_{n\geq 1}$ is an orthonormal basis for $\mathrm{L}^2(0,1)$, we can replace $u(t)$ on the left by its Fourier expansion to obtain

$$\sum\nolimits_{j=1}^{\infty} \langle u(t), e_n\rangle\, e_n = \sum\nolimits_{j=n}^{\infty} -\lambda_n \langle u_t(t), e_n\rangle\, e_n$$

and comparing Fourier coefficients we arrive at

$$\langle u(t), e_n\rangle = -\lambda_n \langle u_t(t), e_n\rangle \qquad (n \in \mathbb{N}, t > 0).$$

Hence we have transformed the original *partial* differential equation into an infinite system of *ordinary* differential equations

$$\frac{d}{dt} \langle u(t), e_n\rangle = {}^{-1}\!/\!_{\lambda_n} \langle u(t), e_n\rangle\,, \quad \langle u(0), e_n\rangle = \langle f, e_n\rangle \qquad (n \in \mathbb{N}).$$

This is clearly satisfiable by letting

$$u(t) := T(t)f := \sum\nolimits_{n=1}^{\infty} e^{-t/\lambda_n} \langle f, e_n\rangle\, e_n \quad (t \geq 0).$$

It is now a quite tedious but manageable exercise in analysis to prove that the series actually defines a smooth function on $(0,\infty) \times [0,1]$ which satisfies the heat equation. Moreover, the initial condition is met in the sense that $\lim_{t \searrow 0} u(t) = f$ in $\mathrm{L}^2(0,1)$, but one can say more depending on whether f is continuous or has even higher regularity.

The Operator Semigroup. For each $t \geq 0$ we can consider the operator $T(t)$, which maps the initial datum f to the solution $u(t)$ at time t. It is clearly bounded, and a little calculation shows that

$$T(0) = \mathrm{I} \quad \text{and} \quad T(t+s) = T(t)T(s), \quad (t, s \geq 0).$$

Moreover, for fixed $f \in \mathrm{L}^2(0,1)$ the mapping

$$[0, \infty) \longrightarrow \mathrm{L}^2(0,1), \qquad t \longmapsto T(t)f$$

is continuous. Hence the operator family $(T(t))_{t \geq 0}$ is what is called a **strongly continuous operator semigroup**. Because of the similarity with the scalar exponential function one sometimes writes

$$T(t) = \mathrm{e}^{-tL} \qquad (t \geq 0).$$

The method sketched here is a step into the field of *Evolution Equations*. There one transforms finite-dimensional PDEs into ODEs in diverse Banach spaces and applies functional analytic methods in order to solve them or to study the asymptotic behaviour or other properties of their solutions; see [**EN06**] or [**Eva98**, Sec. 7.4].

14.4. *The Norm of the Integration Operator

Several times we have encountered the operator J given by integration:

$$(Jf)(t) := \int_a^t f(s)\,\mathrm{d}s \qquad (t \in [a, b]).$$

This operator is often called *the* **Volterra operator**. It is quite easy to compute its norm when considered as acting on $\mathrm{C}[a, b]$ with the supremum norm; see Exercise 11.7. But what is the norm of J when considered as an operator on $\mathrm{L}^2(a, b)$? Of course, J is an integral operator with kernel

$$k(t, s) = \mathbf{1}_{[a,t]}(s).$$

Hence, one can estimate

$$\begin{aligned} \|J\|^2 \leq \|k\|_{HS}^2 &= \int_a^b \int_a^b |k(t,s)|^2 \,\mathrm{d}s\,\mathrm{d}t \\ &= \int_a^b \int_a^t \mathrm{d}s\,\mathrm{d}t = \int_a^b (t-a)\,\mathrm{d}t = \frac{(b-a)^2}{2}, \end{aligned}$$

which gives $\|J\|_{\mathrm{L}^2 \to \mathrm{L}^2} \leq \sqrt[1]{\!/\sqrt{2}}\,(b-a)$. But we shall see that we do not have equality here.

The idea is to use the spectral theorem. However, J is not a self-adjoint operator and so one has to use a little trick, based on the following lemma.

Lemma 14.5. *Let A be an arbitrary bounded operator on a Hilbert space. Then A^*A and AA^* are (positive) self-adjoint operators with norm*

$$\|A^*A\| = \|AA^*\| = \|A\|^2 .$$

Proof. One has $(A^*A)^* = A^*A^{**} = A^*A$, so A^*A is self-adjoint. It is also positive since $\langle A^*Ax, x\rangle = \langle Ax, Ax\rangle = \|Ax\|^2 \geq 0$ for each $x \in H$. Clearly

$$\|A^*A\| \leq \|A^*\| \, \|A\| = \|A\|^2$$

by Lemma 12.12. But, on the other hand,

$$\|Ax\|^2 = \langle Ax, Ax\rangle = \langle A^*Ax, x\rangle \leq \|A^*Ax\| \, \|x\| \leq \|A^*A\| \, \|x\|^2$$

for all $x \in H$, by Cauchy–Schwarz. Hence $\|A\|^2 \leq \|A^*A\|$, by definition of the norm. For the statements about AA^* just replace A by A^* in these results. □

To apply the lemma we recall from Exercise 12.10 that

$$J^*f = \langle f, \mathbf{1}\rangle \, \mathbf{1} - Jf \qquad (f \in \mathrm{L}^2(a,b)).$$

By the above lemma, the operator $A := JJ^*$ is given by

$$\begin{aligned} Af(t) = JJ^*f(t) &= \langle f, \mathbf{1}\rangle \, (t-a) - J^2 f(t) \\ &= \int_a^b (t-a) f(s) \, \mathrm{d}s - \int_a^t (t-s) \, f(s) \, \mathrm{d}s \\ &= \int_a^b \min(t-a, s-a) f(s) \, \mathrm{d}s \end{aligned}$$

for $f \in \mathrm{L}^2(a,b)$, hence is induced by the kernel $k(t,s) := \min(t-a, s-a)$. Since A is a compact self-adjoint operator, by Lemma 13.10 its norm is equal to the largest modulus of an eigenvalue of A. The following lemma shows that our operator A is associated with the **Laplacian with mixed boundary conditions**.

Lemma 14.6. *For $f, u \in \mathrm{L}^2(a,b)$ we have*

$$Af = u \qquad \Longleftrightarrow \qquad u \in \mathrm{H}^2(a,b), \quad u'' = -f, \quad u(a) = 0 = u'(b).$$

Proof. We leave the proof as Exercise 14.10. □ Ex.14.10

The operator A is self-adjoint and compact as it is Hilbert–Schmidt. Consequently, we know from Lemmas 13.4 and 13.10 that $\|A\| = |\lambda|$, where λ is the eigenvalue of A with largest modulus. To determine this eigenvalue we suppose that $\lambda \neq 0$ and there is $f \neq 0$ and $Af = \lambda f$. Then as in the proof of Lemma 14.1 we find for $\mu := -1/\lambda$ that $f \in \mathrm{H}^2(a,b)$, $f'' = \mu f$, and therefore even $f \in \mathrm{C}^2[a,b]$. Moreover, integration by parts, together with the boundary condition $f(a) = 0 = f'(b)$ yields

$$\|f\|_2^2 = \langle f, f\rangle = -\lambda \langle f'', f\rangle = \lambda \langle f', f'\rangle .$$

Since $f \neq 0$ and f is not a constant function, it follows that $\lambda > 0$. Then the classical theory of differential equations yields that

$$\lambda = \left(\frac{2(b-a)}{(2n-1)\pi}\right)^2 \quad \text{and} \quad f(t) = \cos\left(\frac{(2n-1)\pi(t-a)}{2(b-a)}\right)$$

for some $n \in \mathbb{N}$. The largest choice here is for $n = 1$, and hence

$$\|J\|^2 = \|JJ^*\| = \|A\| = \left(\frac{2(b-a)}{\pi}\right)^2 .$$

Ex.14.11 We find $\|J\| = {}^2\!/\!_\pi(b-a)$, which is slightly smaller than ${}^1\!/\!_{\sqrt{2}}(b-a) = \|J\|_{HS}$.

14.5. *The Best Constant in the Poincaré Inequality

In Lemma 10.10 we encountered the Poincaré inequality

$$\|u\|_2 \le c\,\|u'\|_2 \qquad (u \in \mathrm{H}_0^1(a,b)), \tag{14.4}$$

and have seen there that one can choose $c = \|J\|$. We know now that $\|J\| = {}^2\!/\!_\pi(b-a)$, but it is still not clear what the *best constant*

$$c_0 := \inf\left\{c \ge 0 \mid \|u\|_2 \le c\,\|u'\|_2 \quad \text{for all} \quad u \in \mathrm{H}_0^1(a,b)\right\}$$

actually is. The aim of this section is to determine c_0 as

$$c_0 = {}^1\!/\!_\pi(b-a). \tag{14.5}$$

As a first step we note the following.

Lemma 14.7. *The space* $\mathrm{H}^2(a,b) \cap \mathrm{H}_0^1(a,b)$ *is dense in* $\mathrm{H}_0^1(a,b)$.

Proof. We sketch the proof and leave details as an exercise. Note that $\mathrm{H}_0^1(a,b) = \{Jf \mid f \in \mathrm{L}^2(a,b),\ f \perp \mathbf{1}\}$. Take $f \in \mathrm{L}^2(a,b)$ with $f \perp \mathbf{1}$ and find $f_n \in \mathrm{C}^1[a,b]$ such that $f_n \to f$ in $\|\cdot\|_2$. Then $Jf_n \to Jf$ in $\|\cdot\|_{\mathrm{H}^1}$. Let $g_n(t) := (Jf_n)(t) - \langle f_n, \mathbf{1}\rangle \cdot t$. Then $g \in \mathrm{C}^2[a,b] \cap \mathrm{H}_0^1(a,b)$ and $g_n \to Jf$ in
Ex.14.12 $\|\cdot\|_{\mathrm{H}^1}$. □

The lemma shows that the Poincaré inequality (14.4) is equivalent to

$$\|u\|_2 \le c\,\|u'\|_2 \qquad (u \in \mathrm{H}^2(a,b) \cap \mathrm{H}_0^1(a,b)).$$

By the product rule for H^1 (Exercise 10.11), it is also equivalent to

$$\|u\|_2^2 \le c^2 \langle u, -u'' \rangle \qquad (u \in \mathrm{H}^2(a,b) \cap \mathrm{H}_0^1(a,b)).$$

(Note that the continuous(!) function uu' vanishes at the boundary.) But now recall that with $A := (-\Delta_D)^{-1}$ as in Section 14.1 above,

$$\mathrm{H}^2(a,b) \cap \mathrm{H}_0^1(a,b) = \mathrm{dom}(\Delta_D) = \{Af \mid f \in \mathrm{L}^2(a,b)\}$$

is the domain of the Dirichlet Laplacian. So if we write $-f = u''$ and $u = Af$ in the inequality above, it is equivalent to

$$\|Af\|^2 \le c^2 \langle Af, f \rangle \qquad (f \in \mathrm{L}^2(a,b)). \tag{14.6}$$

By Cauchy–Schwarz,

$$\langle Af, f \rangle \le \|Af\| \, \|f\| \le \|A\| \, \|f\|^2$$

and so (14.6) implies that $\|A\|^2 \le c^2 \|A\|$, whence $c^2 \ge \|A\|$. Hence, the optimal constant satisfies

$$c_0 \ge \sqrt{\|A\|}.$$

On the other hand, A is a *positive* self-adjoint operator in the sense of Exercise 13.11. By part d) of that exercise we have

$$\|Af\|^2 \le \|A\| \langle Af, f \rangle \qquad (f \in \mathrm{L}^2(a,b)).$$

Hence the Poincaré inequality (14.4) holds true with $c^2 = \|A\|$. To sum up, we have shown that

$$c_0 = \sqrt{\|A\|}.$$

Now, the spectral theorem tells us that $\|A\|$ equals the largest absolute value of an eigenvalue of A, which is $1/\pi^2 (b-a)^2$, with corresponding eigenfunction

$$e_1(t) = \sqrt{\frac{b-a}{2}} \sin\left(\frac{\pi(t-a)}{b-a} \right) \qquad (t \in [a,b]).$$

(Adapt the considerations about the operator on $(0,1)$ from above.) Hence indeed $c_0 = 1/\pi(b-a)$ as claimed in (14.5), with the function e_1 as an extremal case. □

Exercises 14A

Exercise 14.1. Show that for every $f \in \mathrm{L}^2(0,1)$ the series

$$\sum_{n=1}^{\infty} \left(\frac{1}{2n^2\pi^2} \int_0^1 f(s) \sin(n\pi s) \, \mathrm{d}s \right) \sin(n\pi t)$$

converges uniformly in $t \in [0,1]$.

Exercise 14.2. Suppose that $(k_n)_n$ is a Cauchy sequence in $\mathrm{L}^2(X \times Y)$, suppose that $k \in \mathrm{L}^2(X \times Y)$ is such that $A_{k_n} f \to A_k f$ in $\mathrm{L}^2(X)$ for every $f \in \mathrm{L}^2(Y)$. Show that $k_n \to k$ in $\mathrm{L}^2(X \times Y)$.

Apply this to prove that

$$g(t, s) = \sum_{n=1}^{\infty} \frac{\sin(n\pi \cdot t)\sin(n\pi s)}{2n^2\pi^2}$$

as an absolutely convergent series in $\mathrm{C}([0,1] \times [0,1])$.

Exercise 14.3. Let $q \in \mathrm{C}[0,1]$ with $q \geq 0$. Show that

$$a(u, v) := \langle u', v' \rangle_{\mathrm{L}^2} + \langle qu, v \rangle_{\mathrm{L}^2} \qquad (u, v \in \mathrm{H}^1_0(0,1)).$$

is an inner product on $\mathrm{H}^1_0(0,1)$ and the associated norm (call it $\|\cdot\|_a$) is equivalent to the standard norm on $\mathrm{H}^1_0(0,1)$. Furthermore, show that for $f \in \mathrm{L}^2(0,1)$ the functional

$$(\mathrm{H}^1_0(0,1), \|\cdot\|_a) \to \mathbb{C}, \qquad v \longmapsto \langle v, f \rangle_{\mathrm{L}^2}$$

is bounded by a constant times $\|f\|_{\mathrm{L}^2}$.

Exercise 14.4. Show that the operator $L^{-1} : \mathrm{L}^2(0,1) \to \mathrm{H}^2(0,1)$ considered in Section 14.2 is bounded.

Exercise 14.5. Show that the operator L^1 considered in Section 14.2 is self-adjoint on $\mathrm{L}^2(0,1)$.

Exercise 14.6. Let $q \in \mathrm{C}[0,1]$ and let $A = \left[\begin{smallmatrix} 0 & 1 \\ q & 0 \end{smallmatrix}\right]$. Define $\tilde{A}(t) := \left[\begin{smallmatrix} 0 & t \\ \int_0^t q(s)\,\mathrm{d}s & 0 \end{smallmatrix}\right]$ and $\Psi(t) := \mathrm{e}^{\tilde{A}(t)}$. Show that $\Psi' = A\Psi$ on $[0,1]$; then show that there is an invertible (constant) matrix T such that $\Psi T = \left[\begin{smallmatrix} u & v \\ u' & v' \end{smallmatrix}\right]$, where u, v satisfy the boundary conditions (14.1).

Exercise 14.7. In the construction of the Green's function k for the Schrödinger operator S on $[0,1]$ with positive potential $q \geq 0$, show that $u, u', v' \geq 0$ and $v \leq 0$ on $[0,1]$. Conclude that the Green's function k is also positive.

Exercise 14.8. Let L be the Schrödinger operator considered in Section 14.2, and let $\lambda \neq 0$. Show that λ is an eigenvalue for L^{-1} with eigenfunction $u \neq 0$ if and only if $1/\lambda$ is an eigenvalue of L with same eigenfunction u.

Exercise 14.9. Let $(e_j)_{j \in \mathbb{N}}$ be an orthonormal system in $\mathrm{L}^2(a,b)$, let $\lambda \in \ell^\infty$ such that $\lambda_j \neq 0$ for all $j \in \mathbb{N}$. and let $A : \mathrm{L}^2(a,b) \to \mathrm{L}^2(a,b)$ be given by

$$Af := \sum_{j=1}^{\infty} \lambda_j \langle f, e_j \rangle e_j$$

for $f \in \mathrm{L}^2(a,b)$. Suppose further that $\operatorname{ran}(A) \subseteq \mathrm{C}[a,b]$ and $A : \mathrm{L}^2(a,b) \to \mathrm{C}[a,b]$ is bounded. Show that

a) Each $e_j \in \mathrm{C}[a,b]$.

b) The series defining A converges uniformly, i.e., in the supremum norm.

c) Show that

$$\sum_{j=1}^{\infty} |\lambda_j|^2 < \infty$$

[Hint: Use Exercise 12.18 or Theorem 12.16.]

Exercise 14.10. Prove Lemma 14.6.

Exercise 14.11. Discuss the spectral decomposition of the operator $A = JJ^*$ on $L^2(a, b)$. In which sense do the appearing series converge?

Exercise 14.12. Fill in the details in the proof of Lemma 14.7.

Exercises 14B

Exercise 14.13. Let $Bu = u''$ defined on $\mathrm{dom}(B) := \{u \in H^2(a, b) \mid u'(a) = 0 = u(b)\}$. Show that

$$B : \mathrm{dom}(B) \longrightarrow L^2(a, b)$$

is bijective. Compute its inverse operator $A := B^{-1}$, show that it is a self-adjoint Hilbert–Schmidt operator mapping $L^2(a, b)$ boundedly into $C[a, b]$. Compute the eigenvalues and the corresponding eigenfunctions and discuss the spectral representation.

Exercise 14.14. Let $P : L^2(0, 1) \to \{\mathbf{1}\}^\perp$ be the orthogonal projection. Show that the operator $A := PJ^2P$ is a self-adjoint Hilbert–Schmidt operator and compute its integral kernel. (See also Exercise 12.14.) Determine the eigenvalues of A and find an orthonormal basis of $L^2(0, 1)$ consisting of eigenfunctions of A.

Exercise 14.15. Consider the operator $A = (1/2)(J^2 + J^{*2})$ on $L^2(a, b)$. Determine its integral kernel. Show that $\mathrm{ran}(A) \subseteq H^2(a, b)$ and $(Af)'' = f$. Find the right boundary conditions to characterize $\mathrm{ran}(A)$. Then determine the spectral decomposition of A.

Exercise 14.16. Let L be the Schrödinger operator as considered in Section 14.2, i.e., $Lu = -u'' + qu$, with $0 \le q \in C[0, 1]$. Show that the eigenvalues of L are precisely the numbers $\lambda > \int_0^1 q(s)\, ds$ that solve the equation

$$\sin\Big(\sqrt{\lambda - \textstyle\int_0^1 q(s)\, ds}\Big) = 0,$$

i.e., $\lambda_n = \int_0^1 q(s)\, ds + \pi^2 n^2$ for $n \ge 1$. Compute a formula for the corresponding eigenfunctions.

Exercise 14.17. We consider the Poisson problem

$$-u'' = f, \qquad u'(0) = u'(1) = u(1) - u(0).$$

Show that this has a solution if and only if $f \in H := \{\mathbf{1}, J\mathbf{1}\}^\perp$. Then show that this solution is unique in the space $H^2(0, 1) \cap H$. Show that the solution operator is self-adjoint, determine the Green's function and the spectral decomposition.

Chapter 15

Baire's Theorem and Its Consequences

In this chapter we encounter two of the most important and useful results from functional analysis: the uniform boundedness principle and the open mapping theorem. The proofs need nothing more than a little metric space theory and some basics about Banach spaces and bounded operators, and hence the systematic part of this chapter could have been placed quite in the beginning (after Chapter 5, say).

15.1. Baire's Theorem

Both, the uniform boundedness principle and the open mapping theorem rest on a relatively simple but very powerful result about metric spaces called *Baire's theorem*[1]. Recall that an *open ball* in a metric space (Ω, d) is any set of the form

$$\mathrm{B}(x, r) = \{y \in \Omega \mid d(x, y) < r\}$$

for some "center" $x \in \Omega$ and "radius" $r > 0$. The *closed ball* with the same center and radius is

$$\mathrm{B}[x, r] = \{y \in \Omega \mid d(x, y) \le r\}.$$

We have seen in Lemma 4.2 that an open ball is indeed an open, and a closed ball is indeed a closed subset of Ω. Note that

$$0 < s < r \quad \Rightarrow \quad \mathrm{B}[x, s] \subseteq \mathrm{B}(x, r) \subseteq \mathrm{B}[x, r]$$

[1]René-Louis Baire (1874–1932), French mathematician.

for any $x \in \Omega$. So every open ball contains a closed ball with the same
Ex.15.1 center and positive radius.

In its simplest form, Baire's theorem states that if a complete metric space is exhausted by a union of countably many closed subsets, then at least one of these subsets must contain an open ball. More precisely, we have the following.

Theorem 15.1 (Baire). *Let (Ω, d) be a nonempty complete metric space and let $(A_n)_{n\in\mathbb{N}}$ be a sequence of closed subsets of Ω such that*

$$\Omega = \bigcup_{n\in\mathbb{N}} A_n.$$

Then there is $n \in \mathbb{N}$ and $x \in \Omega, r > 0$ with $\mathrm{B}(x, r) \subseteq A_n$.

Theorem 15.1 is not exactly what is known as "Baire's theorem" in the literature, but a little weaker. For the full version see [**Rud87**, 5.6] and Exercise 15.11 below.

Ex.15.2 The proof of Baire's theorem will become quite perspicuous if one knows the following lemma.

Lemma 15.2 (Principle of Nested Balls). *Let (Ω, d) be a complete metric space, and let*

$$\mathrm{B}[x_1, r_1] \supseteq \mathrm{B}[x_2, r_2] \supseteq \mathrm{B}[x_3, r_3] \supseteq \dots$$

be a nested *sequence of closed balls in it. If $r_n \to 0$, then $x := \lim_{n\to\infty} x_n$ exists and*

$$\bigcap_{n\in\mathbb{N}} \mathrm{B}[x_n, r_n] = \{x\}. \tag{15.1}$$

Proof. Note first that if $y \in \mathrm{B}[x_n, r_n]$ for all $n \in \mathbb{N}$, then $d(y, x_n) \le r_n \to 0$. Since limits are unique, the intersection $\bigcap_{n\in\mathbb{N}} \mathrm{B}[x_n, r_n]$ contains at most one point.

Now fix $\epsilon > 0$ and find $N \in \mathbb{N}$ such that $r_N \le \epsilon/2$. If $n, m \ge N$ we have $x_n, x_m \in \mathrm{B}[x_N, r_N]$, and hence

$$d(x_n, x_m) \le d(x_n, x_N) + d(x_N, x_m) \le r_N + r_N = 2r_N \le \epsilon.$$

Hence $(x_n)_{n\in\mathbb{N}}$ is a Cauchy sequence. By completeness, $x := \lim_{n\to\infty} x_n$ exists. For any $N \in \mathbb{N}$, the set $\mathrm{B}[x_N, r_N]$ is closed and contains all x_n with $n \ge N$. Consequently, $x \in \mathrm{B}[x_N, r_N]$, and this proves (15.1). □

Proof of Theorem 15.1. We suppose that no A_n contains an open ball, and claim that then there exists $x \in \Omega$ which is not contained in any A_n. To find that x we are going to construct a sequence of nested balls.

In the first step, pick any $x_1 \in \Omega \setminus A_1$. This must exist, otherwise $A_1 = \Omega$, which trivially contains an open ball. Since A_1 is closed, if $r_1 > 0$ is small enough one has

$$A_1 \cap \mathrm{B}[x_1, r_1] = \emptyset.$$

By hypothesis, A_2 does not contain $\mathrm{B}(x_1, r_1)$, so there is $x_2 \in \mathrm{B}(x_1, r_1)$ but $x_2 \notin A_2$. Since A_2 is closed and $\mathrm{B}(x_1, r_1)$ is open, for $r_2 > 0$ small enough we have

$$A_2 \cap \mathrm{B}[x_2, r_2] = \emptyset \quad \text{and} \quad \mathrm{B}[x_2, r_2] \subseteq \mathrm{B}(x_1, r_1).$$

Again by hypothesis, the set A_3 does not contain the ball $\mathrm{B}(x_2, r_2)$, and hence we find $x_3 \in \mathrm{B}(x_2, r_2)$ but $x_3 \notin A_3$. Since A_3 is closed and $\mathrm{B}(x_2, r_2)$ is open, for small enough $r_3 > 0$ we have

$$A_3 \cap \mathrm{B}[x_3, r_3] = \emptyset \quad \text{and} \quad \mathrm{B}[x_3, r_3] \subseteq \mathrm{B}(x_2, r_2).$$

In this manner we find a nested sequence of balls $\mathrm{B}[x_n, r_n]$ such that

$$\mathrm{B}[x_n, r_n] \cap A_n = \emptyset \qquad \text{for all } n \in \mathbb{N}. \tag{15.2}$$

Since in each step we can make the radius r_n as small as we like, we can arrange it such that $r_n \to 0$. By the principle of nested balls, the centers $(x_n)_{n\in\mathbb{N}}$ converge to $x \in \bigcap_{n\in\mathbb{N}} \mathrm{B}[x_n, r_n]$. By (15.2), $x \notin A_n$ for each $n \in \mathbb{N}$, and the proof is complete. □

Baire's theorem has plenty of interesting applications, in particular to the theory of real functions. For instance, one can employ it to show that the set of nowhere differentiable functions is dense in $\mathrm{C}[a, b]$; see [**Boa96**, Section 10]. However, going deeper here would lead us too far astray, so we restrict ourselves to (some) applications of Baire's theorem in functional analysis.

15.2. The Uniform Boundedness Principle

Recall that a linear mapping $T : E \to F$, where E, F are normed spaces, is *bounded* if there is a number $c \geq 0$ such that

$$\|Tf\| \leq c\,\|f\| \qquad \text{for all } f \in E.$$

In this case, the *operator norm*

$$\|T\| := \sup\{\|Tf\| \mid \|f\| \leq 1\}$$

is a finite number, and one has $\|Tf\| \leq \|T\|\,\|f\|$ for all $f \in E$ (see Chapter 2). In this section we consider whole collections $\mathcal{T}$ of bounded linear mappings.

Definition 15.3. Let E, F be normed linear spaces. A collection $\mathcal{T}$ of linear mappings from E to F is called **uniformly bounded** if there is a $c \ge 0$ such that

$$\|Tf\| \le c\,\|f\| \qquad \text{for all } f \in E \text{ and all } T \in \mathcal{T}.$$

In other words, $\mathcal{T}$ is uniformly bounded if each $T \in \mathcal{T}$ is bounded and

$$\sup\{\|T\| \mid T \in \mathcal{T}\} < \infty,$$

i.e., $\mathcal{T}$ is a bounded subset of the normed space $\mathcal{L}(E; F)$.

We remark that we encountered uniformly bounded *sequences* of operators already in Section 9.6 in connection with the "strong convergence lemma" (Corollary 9.23) and Fejér's theorem.

Suppose that $\mathcal{T} \subseteq \mathcal{L}(E; F)$ is a uniformly bounded collection of linear operators. Then for each $f \in E$ one has

$$\|Tf\| \le \|T\|\,\|f\| \le \big(\sup_{S\in\mathcal{T}} \|S\|\big)\,\|f\|$$

for all $T \in \mathcal{T}$, and hence $\sup_{T\in\mathcal{T}} \|Tf\| < \infty$. We say that the operator family $\mathcal{T}$ is *pointwise bounded.* The uniform boundedness principle asserts that in case that E is complete, i.e., a Banach space, one has the converse implication.

Theorem 15.4 (Uniform Boundedness Principle)**.** *Let E be a Banach space, let F be a normed space, and let $\mathcal{T}$ be a collection of bounded linear operators from E to F. Then $\mathcal{T}$ is uniformly bounded if and only if it is pointwise bounded.*

One implication was already mentioned, and is trivial. For the proof of the nontrivial implication we need a little lemma.

Lemma 15.5. *Let $(E, \|\cdot\|)$ be a normed space and let $K \subseteq E$ be a subset with the following properties:*

1) *K is "midpoint-convex", i.e., if $f, g \in K$, then also $\frac{1}{2}(f+g) \in K$;*
2) *K is "symmetric", i.e., if $f \in K$, then also $-f \in K$.*

Ex.15.3 *Then, if K contains some open ball of radius $r > 0$, it also contains $\mathrm{B}(0, r)$.*
Ex.15.4

Proof. (See also Figure 22.) Suppose that $r > 0$ and $g \in E$ are such that $\mathrm{B}(g, r) \subseteq K$. By 2), K contains also $-\mathrm{B}(g, r)$ and by 1) it then must also contain $\frac{1}{2}(\mathrm{B}(g, r) - \mathrm{B}(g, r))$. But the latter contains $\mathrm{B}(0, r)$, as each $h \in E$ with $\|h\| < r$ can be written as

$$h = \frac{2h + g - g}{2} = \frac{(g+h) - (g-h)}{2} \in \tfrac{1}{2}\big(\mathrm{B}(g, r) - \mathrm{B}(g, r)\big). \qquad \square$$

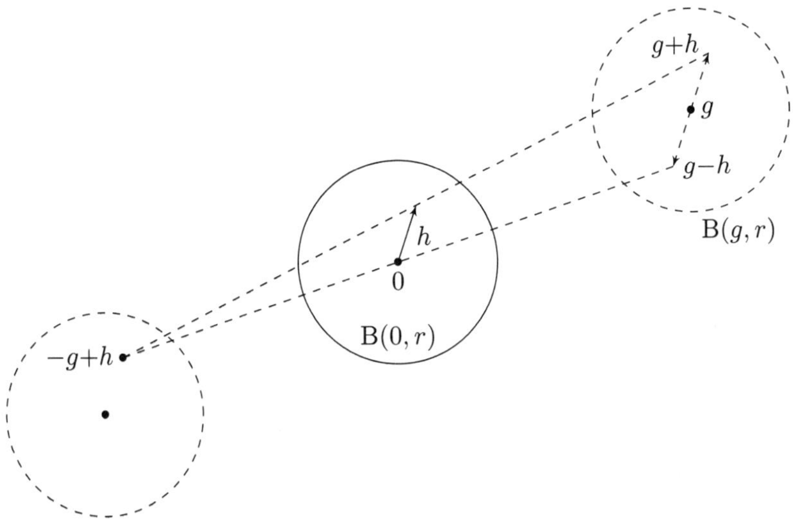

Figure 22. $B(0,r) \subseteq \frac{1}{2}(B(g,r) - B(g,r))$.

Proof of Theorem 15.4. Let $\mathcal{T} \subseteq \mathcal{L}(E;F)$ be pointwise bounded. For $n \in \mathbb{N}$, define

$$K_n := \{f \in E \mid \|Tf\| \le n \text{ for all } T \in \mathcal{T}\}.$$

Since each $T \in \mathcal{T}$ is bounded and the norm mapping is continuous, K_n is a closed subset of E. By hypothesis, every $f \in E$ is contained in at least one K_n, so

$$E = \bigcup\nolimits_{n \in \mathbb{N}} K_n.$$

Since E is complete, Baire's theorem applies and yields $n \in \mathbb{N}$, $r > 0$, and $g \in E$ with $B(g,r) \subseteq K_n$. By straightforward arguments, K_n is midpoint-convex and symmetric. Hence Lemma 15.5 implies that $B(0,r) \subseteq K_n$, and since K_n is closed, we have even $B[0,r] \subseteq K_n$, by Exercise 15.1.

Now take $f \in E$ with $\|f\| \le 1$. Then $rf \in B[0,r] \subseteq K_n$, which means that $r\,\|Tf\| = \|T(rf)\| \le n$ for each $T \in \mathcal{T}$. Dividing by r yields

$$\|Tf\| \le n/r \qquad \text{for all } T \in \mathcal{T},$$

and hence $\|T\| \le n/r$ for all $T \in \mathcal{T}$. □

We continue with an important consequence, complementing the strong convergence lemma (Corollary 9.23).

Theorem 15.6 (Banach–Steinhaus[2]). *Let E, F be Banach spaces, and let $(T_n)_{n\in\mathbb{N}} \subseteq \mathcal{L}(E;F)$ be a sequence such that*

$$Tf := \lim_{n\to\infty} T_n f$$

exists for every $f \in E$. Then T is a bounded operator, $(T_n)_{n\in\mathbb{N}}$ is uniformly bounded, and

$$\|T\| \le \liminf_{n\to\infty} \|T_n\|.$$

Proof. For each $f \in E$, since $(T_n f)_{n\in\mathbb{N}}$ converges, also $(\|T_n f\|)_{n\in\mathbb{N}}$ converges, and therefore $\sup_{n\in\mathbb{N}} \|T_n f\| < \infty$. By the uniform boundedness principle, $\sup_{n\in\mathbb{N}} \|T_n\| < \infty$. If $f \in E$ with $\|f\| \le 1$, then by the continuity of the norm,

$$\|Tf\| = \lim_{n\to\infty} \|T_n f\| = \liminf_{n\to\infty} \|T_n f\| \le \liminf_{n\to\infty} \|T_n\|.$$

Taking the supremum over all such f concludes the proof. □

15.3. Nonconvergence of Fourier Series

Let us turn to a nice application of the uniform boundedness principle to the theory of classical Fourier series.

Theorem 15.7 (Du Bois-Reymond). *There exists a function $f \in \mathrm{C_{per}}[0,1]$ such that its partial Fourier series at $t = 0$,*

$$S_n f(0) = \sum_{k=-n}^{n} \widehat{f}(k) e^{2\pi i k t}\Big|_{t=0} = \sum_{k=-n}^{n} \widehat{f}(k) \qquad (n \in \mathbb{N})$$

does not converge to $f(0)$.

Proof. We consider the linear functionals

$$T_n : \mathrm{C_{per}}[0,1] \longrightarrow \mathbb{C}, \qquad T_n f := (S_n f)(0)$$

for $n \in \mathbb{N}$. Then

$$T_n f = \sum_{k=-n}^{n} \int_0^1 \mathrm{e}^{2\pi i k s} f(s)\,\mathrm{d}s = \int_0^1 \frac{\sin(2n+1)\pi s}{\sin \pi s} f(s)\,\mathrm{d}s$$

for all $f \in \mathrm{C_{per}}[0,1]$. We now consider $E := \mathrm{C_{per}}[0,1]$ with the supremum norm. Then each T_n is bounded and E is a Banach space. By the uniform boundedness principle it remains to prove that $\sup_{n\in\mathbb{N}} \|T_n\| = \infty$ to conclude the existence of a function $f \in \mathrm{C_{per}}[0,1]$ such that $\sup_{n\in\mathbb{N}} |S_n f(0)| = \infty$. In particular, $\lim_{n\to\infty} S_n f(0)$ does not exist.

[2]Hugo Steinhaus (1887–1972), Polish mathematician.

Define the **Dirichlet kernel**

$$D_n(s) := \frac{\sin(2n+1)\pi s}{\sin \pi s},$$

so that $T_n f = \int_0^1 D_n(s) f(s)\,\mathrm{d}s$ for $f \in E$. We claim that

$$\|T_n\| = \int_0^1 |D_n(s)|\,\mathrm{d}s.$$

For $\mathrm{C}[0,1]$ in place of $\mathrm{C_{per}}[0,1]$, this has been done in Example 2.27. But the argument there can be adapted, because D_n is periodic, and hence also the approximants $\overline{D_n}/(|D_n|+\epsilon)$ are periodic, i.e., contained in E.

Finally, we use the inequality $0 \le \sin s \le s$ for $s \in [0,\pi]$ to estimate

$$\begin{aligned}
\int_0^1 |D_n(s)|\,\mathrm{d}s &= \int_0^1 \frac{|\sin(2n+1)\pi s|}{\sin \pi s}\,\mathrm{d}s = \frac{1}{\pi}\int_0^\pi \frac{|\sin(2n+1)s|}{\sin s}\,\mathrm{d}s \\
&\ge \frac{1}{\pi}\int_0^\pi \frac{|\sin(2n+1)s|}{s}\,\mathrm{d}s = \frac{1}{\pi}\int_0^{(2n+1)\pi} \frac{|\sin s|}{s}\,\mathrm{d}s \\
&= \sum_{k=1}^{2n+1} \frac{1}{\pi}\int_{(k-1)\pi}^{k\pi} \frac{|\sin s|}{s}\,\mathrm{d}s \ge \sum_{k=1}^{2n+1} \frac{1}{\pi^2 k}\int_{(k-1)\pi}^{k\pi} |\sin s|\,\mathrm{d}s \\
&= \frac{1}{\pi^2}\left(\int_0^\pi |\sin s|\,\mathrm{d}s\right)\sum_{k=1}^{2n+1}\frac{1}{k} = \frac{2}{\pi^2}\sum_{k=1}^{2n+1}\frac{1}{k}
\end{aligned}$$

for all $n \in \mathbb{N}$. Hence

$$\sup_{n\in\mathbb{N}} \int_0^1 |D_n(s)|\,\mathrm{d}s = \infty$$

since the harmonic series diverges, and this concludes the proof. □

15.4. The Open Mapping Theorem

The second major result based on Baire's theorem is the so-called open mapping theorem. Actually, it is not just one theorem, but a collection of results all of the same flavor. Here is the most central formulation:

Theorem 15.8 (Open Mapping Theorem). *Let E, F be Banach spaces and let $T : E \to F$ be a bounded linear mapping which is surjective. Then there is $a > 0$ such that for each $g \in F$ there is $f \in E$ with $\|f\| \le a\,\|g\|$ and $Tf = g$.*

The name "open mapping theorem" stems from the fact that the conclusion of the theorem is equivalent to saying that T maps open subsets of E onto open subsets of F. But at least for our purposes we can entirely avoid this equivalent formulation.

If T is not only surjective, but also injective, then it is algebraically invertible and Theorem 15.8 yields $\|T^{-1}g\| \le a\|g\|$ for all $g \in F$. But this just means that the inverse T^{-1} is bounded, with operator norm $\|T^{-1}\| \le a$. Hence we have established the following important corollary.

Corollary 15.9 (Bounded Inverse Theorem). *If E, F are Banach spaces, and if $T \in \mathcal{L}(E; F)$ is bijective, then the inverse operator T^{-1} is also*
Ex.15.5 *bounded.*

This theorem is by no means obvious, and it is false if one drops the completeness of one of the spaces. E.g., if $E = (\mathrm{C}[0,1], \|\cdot\|_\infty)$ and $F = (\mathrm{C}[0,1], \|\cdot\|_2)$, then the identity operator

$$\mathrm{I} : (\mathrm{C}[0,1], \|\cdot\|_\infty) \longrightarrow (\mathrm{C}[0,1], \|\cdot\|_2)$$

is bounded and bijective, but its inverse is not bounded.

Since boundedness of an operator is the same as continuity, Corollary 15.9 helps us to understand Theorem 15.8. Mere surjectivity of T just tells us that to any $g \in F$ there is always a pre-image of g, but it tells us nothing about the error we make in picking such a pre-image if g is slightly perturbed. Theorem 15.8 now says that this error can be *controlled.* In fact, it is amplified by a factor of at most a. However, the theorem does not tell how large a is, and thus it is of limited use in those situations when such information matters.

The following corollary sheds new light on our discussions of the supremum norm, L^2-norm and L^1-norm on $\mathrm{C}[a, b]$ in previous chapters.

Corollary 15.10. *Let E be a linear space that is a Banach space with respect to either one of two given norms $\|\cdot\|_1$ and $\|\cdot\|_2$ on E. If there is $M \ge 0$ with*

$$\|f\|_2 \le M\|f\|_1 \qquad \text{for all } f \in E,$$

then the two norms are equivalent.

Proof. The hypothesis just says that $\mathrm{I} : (E, \|\cdot\|_1) \to (E, \|\cdot\|_2)$ is bounded. If both are Banach spaces, Corollary 15.9 applies, and hence $\mathrm{I} = \mathrm{I}^{-1}$:

$(E, \|\cdot\|_2) \to (E, \|\cdot\|_2)$ is bounded, too. This yields a constant $M' \geq 0$ such that $\|f\|_1 \leq M' \|f\|_1$ for every $f \in E$, and hence both norms are equivalent. □

The proof of the open mapping theorem is based on Baire's theorem — but not exclusively. Another ingredient is the following rather technical result, interesting in its own right. Roughly speaking, it says that if an operator is "approximately surjective" and one can pick "approximate pre-images" in a controlled way, then the mapping must be surjective. Here is the precise formulation.

Theorem 15.11. *Let E, F be Banach spaces, and let $T \in \mathcal{L}(E; F)$. Suppose that there exist $0 \leq q < 1$ and $a \geq 0$ such that for every $g \in F$ with $\|g\| \leq 1$ there is $f \in E$ such that*

$$\|f\| \leq a \quad \textit{and} \quad \|Tf - g\| \leq q.$$

Then for each $g \in F$ there is $f \in E$ such that $Tf = g$ and $\|f\| \leq \frac{a}{1-q}\|g\|$.

Proof. It suffices to prove the statement in the case $\|g\| \leq 1$, because for $g = 0$ the assertion is trivial, and if $g \neq 0$ we may replace g by $(1/\|g\|)g$. Let $A := \{Tf \mid f \in E,\ \|f\| \leq a\}$. Then the hypothesis asserts that

$$h \in F,\ \|h\| \leq 1 \quad \Longrightarrow \quad \exists\, h' \in A\ :\ \|h - h'\| \leq q.$$

We shall recursively construct a sequence $(g_n)_{n\in\mathbb{N}}$ in A such that

$$\left\| g - \sum\nolimits_{k=1}^{n} q^{k-1} g_k \right\| \leq q^n \tag{15.3}$$

for $n \geq 1$. The case $n = 1$ is just the hypothesis, since $\|g\| \leq 1$. Suppose that $g_1, \dots, g_n$ are already constructed such that (15.3) holds. Then by hypothesis there is $g_{n+1} \in A$ such that

$$\left\| q^{-n}\Big(g - \sum\nolimits_{k=1}^{n} q^{k-1} g_k\Big) - g_{n+1} \right\| \leq q,$$

and multiplying by q^n yields (15.3) with n replaced by $n + 1$.

By definition of A, for each $n \in \mathbb{N}$ we find $f_n \in E$ with $\|f_n\| \leq a$ and $Tf_n = g_n$. Since $0 \leq q < 1$ and E is a Banach space, the series $f := \sum_{n=1}^{\infty} q^{n-1} f_n$ converges absolutely in E, with

$$\|f\| \leq \sum\nolimits_{k=1}^{\infty} q^{k-1} \|f_k\| \leq \sum\nolimits_{k=1}^{\infty} a q^{k-1} = \frac{a}{1-q}.$$

Since T is bounded, one has

$$Tf = T\Big(\sum\nolimits_{k=1}^{\infty} q^{k-1} f_k\Big) = \sum\nolimits_{k=1}^{\infty} q^{k-1} T f_k = \sum\nolimits_{k=1}^{\infty} q^{k-1} g_k = g,$$

by (15.3). This concludes the proof. □

Proof of Theorem 15.8. Let E, F be Banach spaces and let $T \in \mathcal{L}(E; F)$ be surjective. For $n \in \mathbb{N}$ define

$$B_n := \{g \in F \mid \exists f \in E \text{ s.t. } \|f\| \le n,\ Tf = g\} = T(\mathrm{B}_E[0, n]).$$

Note that B_n is midpoint-convex and symmetric, hence — by Exercise 15.3 — $A_n := \overline{B_n}$ has the same properties. Moreover, $F = \bigcup_{n \in \mathbb{N}} B_n$, by the surjectivity of T, and hence

$$F = \bigcup\nolimits_{n \in \mathbb{N}} A_n.$$

Since all the A_n are closed, and F is complete, Baire's theorem applies and we find an $n \in \mathbb{N}$ such that A_n contains an open ball. But A_n is midpoint-convex and symmetric, whence by Lemma 15.5 there is $r > 0$ such that $\mathrm{B}_E(0, r) \subseteq A_n$. Since A_n is closed, we have even

$$\mathrm{B}_F[0, r] \subseteq A_n = \overline{B_n} = \overline{T(\mathrm{B}_E[0, n])}. \tag{15.4}$$

Dividing by r yields

$$\mathrm{B}_F[0, 1] \subseteq \overline{T(\mathrm{B}_E[0, n/r])},$$

and this means that the hypotheses of Theorem 15.11 are satisfied with $a = n/r$ and any $q \in (0, 1)$. The conclusion of Theorem 15.11 yields exactly what we want. □

The Closed Graph Theorem. Let E, F be two normed spaces. A linear mapping $T : E \to F$ is said to have a **closed graph** if

$$f_n \to f, \quad Tf_n \to g \quad \Longrightarrow \quad Tf = g$$

holds for all sequences $(f_n)_{n \in \mathbb{N}}$ in E and all $f \in E$, $g \in F$. That means, T has closed graph if its graph

$$\operatorname{graph}(T) = \{(f, Tf) \mid f \in E\}$$

is closed in the normed space $E \times F$; see Exercise 4.20.

Theorem 15.12 (Closed Graph Theorem). *If E, F are Banach spaces and $T : E \to F$ is a linear mapping, then T is bounded if and only if it has a closed graph.*

Proof. Define the new norm $|||f||| := \|f\|_E + \|Tf\|_F$ for $f \in E$. Then $\|f\|_E \le |||f|||$ for each $f \in E$. The closedness of the graph of T and since both E, F are complete implies that E is complete with respect to this new
Ex.15.6 norm. Hence by Corollary 15.10 there must be a constant $c > 0$ such that $\|Tf\| \le |||f||| \le c\,\|f\|$ for all $f \in E$. □

15.5. Applications with a Look Back

In this section we sketch some situations where the results of this chapter apply.

Countable Algebraic Bases. A vector space E with a countable algebraic basis $(e_j)_{j\in\mathbb{N}}$ is linearly isomorphic to the space c_{00}, the space of finite sequences. We have seen many norms on it, but none was complete. This is actually a general fact.

Theorem 15.13. *A normed space with a countable algebraic basis is never a Banach space.*

Proof. Apply Baire's theorem to $A_n := \operatorname{span}\{e_1, \dots, e_n\}$. □ Ex.15.7

Boundedness and Convergence. Let $\alpha = (\alpha_j)_{j\in\mathbb{N}}$ be a scalar sequence. If $\alpha \in \ell^\infty$, then clearly

$$\sum\nolimits_{j=1}^{\infty} \alpha_j x_j$$

converges for every $x \in \ell^1$. We claim that the converse holds. Indeed, for each $n \in \mathbb{N}$ the linear functional

$$\varphi_n : \ell^1 \longrightarrow \mathbb{C}, \qquad \varphi_n(x) := \sum\nolimits_{j=1}^{n} \alpha_j x_j$$

is clearly bounded, and it is easy to see that

$$\|\varphi_n\| = \max\{|\alpha_1|, \dots, |\alpha_n|\}.$$

By assumption, $\lim_{n\to\infty} \varphi_n(x)$ exists for each $x \in \ell^1$, whence the sequence $(\varphi_n)_n$ is pointwise bounded. Since ℓ^1 is a Banach space, the uniform boundedness principle applies and yields that

$$\|\alpha\|_\infty = \sup\nolimits_{n\in\mathbb{N}} \max\{|\alpha_1|, \dots, |\alpha_n|\} = \sup\nolimits_{n\in\mathbb{N}} \|\varphi_n\| < \infty$$

as claimed. Can you find a direct proof without using the uniform boundedness principle? Ex.15.8 Ex.15.9

Nonequivalence and Noncompleteness. Recall the trivial inequality

$$\|f\|_2 \le \sqrt{b-a}\,\|f\|_\infty \qquad (f \in \mathrm{C}[a,b])$$

for the L^2-norm and the supremum norm on $\mathrm{C}[a,b]$. Consider the statements

(i) $\|\cdot\|_\infty$ and $\|\cdot\|_2$ are not equivalent.

(ii) $\mathrm{C}[a,b]$ is not complete with respect to $\|\cdot\|_2$.

(iii) $\mathrm{C}[a,b]$ is complete with respect to $\|\cdot\|_\infty$.

Each statement has been proved individually (Example 3.14, Theorem 5.8, Example 5.13). But the open mapping theorem (or better: Corollary 15.10) tells us that (i) and (ii) are equivalent in the presence of (iii). This means we could have saved some work on the expense of having only an abstract argument in place of a concrete counterexample.

Well-Posedness. Recall our discussion in Section 11.2 of the well-posedness of an equation on the basis of the Poisson problem. The solution operator there is $-\Delta_D^{-1} : \mathrm{L}^2(a,b) \to \mathrm{H}^2(a,b)$, and for well-posedness of the original equation one needs the boundedness of this operator, where one has to take the H^2-norm on the target space. This was established in Exercise 11.5, but the closed graph theorem renders this superfluous. Namely, one has the following corollary.

Corollary 15.14. *Let E, F, G be Banach spaces with $F \subseteq G$ and this inclusion is continuous. Let $T : E \to F$ be a bounded linear operator such that* $\operatorname{ran}(T) \subseteq F$. *Then $T : E \to F$ is bounded.*

Proof. That the inclusion is continuous means that $\|\cdot\|_G$ is weaker on F than $\|\cdot\|_F$. By the closed graph theorem it suffices to show that $T : E \to F$ has a closed graph. So suppose that $f_n, f \in E$ and $g \in F$ such that $\|f_n - f\|_E \to 0$ and $\|Tf_n - g\|_F \to 0$. Since T is continuous into G we obtain $\|Tf_n - Tf\|_G \to 0$, and since the inclusion $F \subseteq G$ is continuous we
Ex.15.10 have $\|Tf_n - g\|_G \to 0$. But limits are unique, whence $g = Tf$. □

Now, since we knew that $-\Delta_D^{-1}$ is a Hilbert–Schmidt operator — and hence bounded — on $\mathrm{L}^2(a,b)$, and since $\mathrm{H}^2(a,b)$ is a Banach space, Corollary 15.14 yields that $-\Delta_D^{-1}$ is bounded from L^2 to H^2. Similarly one can weaken the assumptions in Theorem 12.21, the criterion for being an abstract Hilbert–Schmidt operator; see Remark 12.22.

Invertible Operators and Spectral Theory. In our definition of an *invertible operator* $T : E \to F$ apart from the bijectivity of T we required also the boundedness of T^{-1}. If E, F are Banach spaces, this is automatic, by Corollary 15.9. In particular, if $T \in \mathcal{L}(E)$, and E is a Banach space, then $\lambda \in \mathbb{C}$ is in the spectrum of T if $\lambda \mathrm{I} - T$ is not bijective. This is a simpler definition of the spectrum $\sigma(T)$ than the one given on page 232.

The bounded inverse theorem also accounts for another proof of the well-posedness of the Poisson problem. Namely, $\operatorname{dom}(\Delta_D)$ is a closed subspace of $\mathrm{H}^2(a,b)$ (why?), hence a Hilbert space. Since $\mathrm{L}^2(a,b)$ is a Hilbert space and Δ_D is obviously bounded, its inverse must be bounded, which is the well-posedness.

Tietze's Theorem. Let (Ω, d) be a metric space. Any subset $A \subseteq \Omega$ is a metric space with respect to the induced metric, and if $f \in \mathrm{C_b}(\Omega)$ is a bounded continuous function, one can consider its restriction

$$Tf := f|_A \in \mathrm{C_b}(A)$$

to the set A. The operator $T : \mathrm{C_b}(\Omega) \to \mathrm{C_b}(A)$ is linear with $\|T\| \le 1$. Tietze's theorem states that if A is closed, then T is surjective.

Theorem 15.15 (Tietze[3]). *Let (Ω, d) a metric space, $A \subseteq \Omega$ a closed subset and $g \in \mathrm{C_b}(A;\mathbb{R})$. Then there is $h \in \mathrm{C_b}(\Omega;\mathbb{R})$ such that $h|_A = g$ and $\|h\|_\infty = \|g\|_\infty$.*

Note: For the case $\Omega = \mathbb{R}$ and $A = [a, b]$ the result is straighforward. But think of A being a "weird" closed set, e.g., the Cantor set!

Before we enter the actual proof, we need to note a general fact about metric spaces, interesting in its own right. Namely, if A and B are *disjoint* closed subsets of Ω, then the function

$$f(x) := \frac{d(x, A) - d(x, B)}{d(x, A) + d(x, B)} \qquad (x \in \Omega)$$

(see Definition 3.15 and Exercise 8.9) has the property that it is continuous with

$$f(\Omega) \subseteq [-1, 1], \quad f = -1 \text{ on } A \quad \text{and} \quad f = 1 \text{ on } B.$$

By adding constants and scaling one can modify f in such a way that for given $a < b$ one has $f(\Omega) \subseteq [a, b]$, $f = a$ on A and $f = b$ on B.

Proof of Tietze's theorem. We want to apply Theorem 15.11 with $E = \mathrm{C_b}(\Omega)$, $F = \mathrm{C_b}(A)$, T the restriction mapping as above. Let $g : A \to [-1, 1]$ be continuous. Then the sets $g^{-1}[-1, -1/3]$ and $g^{-1}[1/3, 1]$ are closed in Ω (why?). By the remarks above we can find a continuous function $f : \Omega \to [-1/3, 1/3]$ such that $f = -1/3$ on $g^{-1}[-1, -1/3]$ and $f = 1/3$ on $g^{-1}[1/3, 1]$. This implies that

$$\|f\|_\infty \le \frac{1}{3} =: a \quad \text{and} \quad \|f|_A - g\|_\infty \le \frac{2}{3} =: q$$

Theorem 15.11 applies and yields $h \in \mathrm{C_b}(\Omega)$ such that $h|_A = g$ and

$$\|h\|_\infty \le \frac{a}{1-q} \|g\|_\infty = \|g\|_\infty \le \|h\|_\infty$$

as claimed. □

[3]Heinrich Tietze (1880–1964), Austrian mathematician.

Exercises 15A

Exercise 15.1. Let E be a normed space. Show that

$$\overline{\mathrm{B}(f,r)} = \mathrm{B}[f,r]$$

for every $f \in E$ and $r > 0$.

Exercise 15.2 (Reformulation of Baire's Theorem). Let (Ω, d) be a nonempty complete metric space, and let $O_n \subseteq \Omega$ be open in Ω for $n \in \mathbb{N}$. Show, using Baire's theorem 15.1, the implication

$$\overline{O_n} = \Omega \quad \text{for all } n \in \mathbb{N} \quad \Longrightarrow \quad \bigcap_{n\in\mathbb{N}} O_n \neq \emptyset.$$

Exercise 15.3. Let E be a normed space, and let $A \subseteq E$ be any subset. Show that if A is midpoint-convex/symmetric, then $\overline{A}$ is also midpoint-convex/symmetric.

Exercise 15.4. Let E be a normed space, and let $A \subseteq E$ be a *closed* midpoint-convex subset of E. Show that then E is *convex*, i.e., whenever $f, g \in A$ and $0 < \lambda < 1$, then also $\lambda f + (1-\lambda)g \in A$. [Hint: Consider first dyadic rationals $\lambda = {}^{j}\!/_{2^n}$ via successively taking midpoints, then employ an approximation argument.]

Exercise 15.5. Let E, F be Banach spaces, and let $T : E \to F$ be a bounded linear mapping such that $\ker(T) = \{0\}$. Show that if $\operatorname{ran}(T)$ is closed, then there is $c > 0$ such that $\|f\| \leq c\,\|Tf\|$ for all $f \in E$.

Exercise 15.6. Let E, F be Banach spaces, let $T \in \mathcal{L}(E; F)$, and define $|\!|\!| f |\!|\!| := \|f\|_E + \|Tf\|_F$ for $f \in E$. Show that E is a Banach space with respect to this norm.

Exercise 15.7. Fill in the details in the proof of Theorem 15.13.

Exercise 15.8. Let $\alpha = (\alpha_n)_{n\in\mathbb{N}}$ be a scalar sequence. Suppose that

$$\sum_{j=1}^{\infty} \alpha_j x_j$$

converges for every $x = (x_j)_{j\in\mathbb{N}}$ in c_0. Show that $\alpha \in \ell^1$.

Exercise 15.9. Let $\alpha = (\alpha_n)_{n\in\mathbb{N}}$ be a scalar sequence. Suppose that

$$\sum_{j=1}^{\infty} \alpha_j x_j$$

converges for every $x = (x_j)_{j\in\mathbb{N}}$ in ℓ^2. Show that $\alpha \in \ell^2$.

Exercise 15.10. Prove the following generalization of Corollary 15.14: Let E, F, G be Banach space, let $T : E \to F$ and $S : F \to G$ be linear operators such that ST is bounded, S is bounded and injective. Then T is bounded, too.

Exercises 15B

Exercise 15.11 (Baire's Theorem, Full Version). One can strengthen the result of Exercise 15.2 by making use of it. Let, as in that exercise, (Ω, d) be a nonempty complete metric space, and let $O_n \subseteq \Omega$ be open in Ω for $n \in \mathbb{N}$. Show the implication

$$\overline{O_n} = \Omega \quad \text{for all } n \in \mathbb{N} \quad \Longrightarrow \quad \overline{\bigcap_{n\in\mathbb{N}} O_n} = \Omega.$$

[Hint: Fix $x \in \Omega, r > 0$ and apply Exercise 15.2 to the metric space $\Omega' := \mathrm{B}[x,r]$ and the subsets $O'_n := O_n \cap \Omega'$, open in Ω'.]

Exercise 15.12. For a linear operator $T : E \to F$ between normed spaces E, F, show that

$$r\,\|T\| = \sup_{f\in\mathrm{B}[0,r]} \|Tf\| \le \sup_{f\in\mathrm{B}[g,r]} \|Tf\|$$

for every $g \in E$ and $r > 0$. [Hint: Consider $K = \{f \in E \mid \|Tf\| \le \sup_{h\in\mathrm{B}[g,r]} \|Th\|\}$ and apply Lemma 15.5.]

Exercise 15.13. Let E be a Banach space and let $P : E \to E$ be a *projection*, i.e., a linear operator satisfying $P^2 = P$. Show that P is bounded if and only if $F := \operatorname{ran}(P)$ is closed.

Exercise 15.14 (Hellinger–Toeplitz Theorem[4,5]). Let H be a Hilbert space, and let $T : H \to H$ be a linear operator such that

$$\langle Tf, g\rangle = \langle f, Tg\rangle \quad \text{for all } f, g \in H.$$

Show that T is bounded.

Exercise 15.15. Let E, F, G be Banach spaces, and let $B : E \times F \to G$ be a bilinear mapping. Suppose that B is separately continuous, i.e., for each $f \in E$ and $g \in F$ the mappings

$$B(f,\cdot) : F \to G \quad \text{and} \quad B(\cdot, g) : E \to G$$

are continuous. Show that there is $c \ge 0$ such that

$$\|B(f,g)\|_G \le c\,\|f\|_E\,\|g\|_G \quad \text{for all } f \in E, g \in F.$$

Exercise 15.16. Let E, F be Banach spaces and let $T \in \mathcal{L}(E;F)$ be surjective. Show that for $S \in \mathcal{L}(E;F)$ such that $\|S - T\|$ is small enough, then S is also surjective. [Hint: Let a be as in Theorem 15.8 and $\|S - T\| < 1/a$. Then Theorem 15.11 can be applied to S with $q = \|S - T\|\, a$].

Exercise 15.17. Can one modify the proof of Tietze's theorem in such a way that one can replace the space $\mathrm{C_b}$ of bounded continuous functions everywhere by $\mathrm{UC_b}$, the space of bounded *uniformly* continuous functions? (See Exercise 5.14.)

---

[4]Ernst Hellinger (1883–1950), German mathematician.

[5]Otto Toeplitz (1881–1940), German mathematician.

Exercises 15C

Exercise 15.18. a) Consider, for $f \in \mathrm{C}_{\mathrm{per}}[0,1]$, the convolution operator

$$T : \mathrm{L}^1(0,1) \to \mathrm{L}^1(0,1), \qquad Tg = f * g.$$

Show that $\|T\| = \|f\|_1$. [Hint: Fejér's theorem.]

b) Use a) to show that there is at least one function $f \in \mathrm{L}^1(0,1)$ such that the sequence of partial Fourier sums

$$S_n f = \sum_{k=-n}^{n} \widehat{f}(k) \mathrm{e}^{2\pi \mathrm{i} k\, t}$$

does not converge with respect to $\|\cdot\|_1$.

Remark: It can be shown that $\|S_n f - f\|_p \to 0$ as $n \to \infty$ for each $f \in \mathrm{L}^p(0,1)$, $1 < p < \infty$. The case $p = 2$ is "simple" (Corollary 9.15), but for $p \neq 2$ this is rather involved; see [**Kat04**, II.1.5].

Chapter 16

Duality and the Hahn–Banach Theorem

In the 19th century it was realized that any statement of projective geometry about a certain configuration of points and lines remains true when one "dualizes" it, i.e., interchanges the words 'point' and 'line' and the verbs 'intersect in' and 'lie on'. Although this is an extreme example, the idea to study a mathematical object by associating with it a "dual" object has proved to be very powerful and pervades all areas of mathematics. To a graph there is a dual graph, each polytope has its dual polytope, each (locally compact) Abelian group has a dual group, just to name a few examples.

In linear algebra, the dual of a vector space E is defined as the set $\mathrm{Lin}(E;\mathbb{K})$ of all linear mappings (functionals) of E into the scalar field $\mathbb{K}$ (see Appendix A.7). But for functional analysis this purely algebraic notion of dual space is not of great use. Instead, for a *normed* space E one defines the **(topological) dual space** to be

$$E' := \mathcal{L}(E;\mathbb{K}) = \{\varphi \in \mathrm{Lin}(E;\mathbb{K}) \mid \varphi \text{ is bounded}\},$$

i.e., the space of bounded linear functionals on E. It turns out that with this definition a powerful duality theory can be developed.

In the literature one also finds the notation E^* for the topological dual of a normed space E. The reason why we prefer to use E' will be clear when it comes to dual operators (see below).

As a first step towards it we address an apparently simple question: Can we be sure that always $E' \neq \{0\}$ if $E \neq \{0\}$? The answer is "yes" in many cases: if E is finite-dimensional, e.g., each linear functional on E is bounded, whence $E' = \mathrm{Lin}(E;\mathbb{K})$ has the same dimension as E. If E is an inner product space, then every $0 \neq f \in E$ yields an element $\langle \cdot, f\rangle$ of E' (and by the Riesz–Fréchet theorem this is even a complete description of E'). If $E = \mathrm{C}[a,b]$, then typical elements of the dual space are point evaluation functionals and integration against L^1-functions.

In all these cases, E' is not just not trivial, but actually rich enough to **separate the points** of E. This means that one can distinguish any two different vectors in E by applying an appropriate linear functional. Or,
Ex.16.1 by linearity, for each $0 \neq f \in E$ there is $\varphi \in E'$ such that $\varphi(f) \neq 0$. Equivalently,

$$f = g \quad \Leftrightarrow \quad \varphi(f) = \varphi(g) \quad \text{for all } \varphi \in E'. \tag{16.1}$$

It will turn out in the following sections that this property, essential for any good duality theory, is shared by all normed spaces.

16.1. Extending Linear Functionals

Let E be a normed space and $0 \neq f \in E$ a nonzero vector. We want to produce $\varphi \in E'$ with $\varphi(f) \neq 0$.

Let us first look at the simplest case. If E is one-dimensional we have $E = \mathbb{K}f$ and

$$\varphi(\lambda f) := \lambda \quad \text{for} \quad \lambda \in \mathbb{K}$$

is a well-defined bounded linear functional $\varphi : E \to \mathbb{K}$ on E satisfying $\varphi(f) = 1 \neq 0$.

If E is not one-dimensional we may start with defining $E_0 := \mathbb{K}f$, find our functional φ_0, say, on E_0 as above and then try to *extend* φ_0 step by step to a bounded linear functional on all of E. In each step the functional $\varphi_n : E_n \to \mathbb{K}$ obtained so far is extended to a subspace E_{n+1} that has exactly one dimension more than E_n. In this way, sequences $(E_n)_n$ and $(\varphi_n)_n$ with

$$\begin{aligned} E_0 \subseteq E_1 \subseteq E_2 \subseteq \dots, &\qquad \varphi_n : E_n \to \mathbb{K}, \\ \dim(E_n) = 1 + \dim(E_{n-1}), &\qquad \varphi_n|_{E_{n-1}} = \varphi_{n-1}, \end{aligned}$$

are generated. If $\dim(E) < \infty$ this procedure will eventually stop, and we are done. If $\dim(E) = \infty$ the procedure will not stop, but we shall see that it is still useful. We have to make sure, however, that the (increasing!) values $\|\varphi_n\|$ stay bounded, as otherwise there cannot be a bounded common extension of all the φ_n.

In order to solve this problem we pass to a little more general situation. First of all, we from now on restrict to *real* spaces, i.e., we assume $\mathbb{K} = \mathbb{R}$. The step to complex spaces is postponed until the end, but will be relatively simple.

Let E be a real linear space. A mapping $p : E \to \mathbb{R}$ is called a **sublinear functional** on E if it is **subadditive** and **positively homogeneous**, i.e.,

$$p(f + g) \le p(f) + p(g), \quad p(\lambda f) = \lambda p(f) \quad \text{for all} \quad f, g \in E, \ \lambda > 0.$$

Certainly, a norm $p = \|\cdot\|$ is a sublinear functional.

We shall consider linear functionals $\varphi : E \to \mathbb{R}$ that are *dominated* by a sublinear functional p, i.e., satisfy $\varphi \le p$. In the case that $p = \|\cdot\|$ is a norm, such a domination is equivalent to being a contraction, as

$$\varphi(f) \le \|f\| \quad \text{for all } f \in F \quad \Leftrightarrow \quad |\varphi(f)| \le \|f\| \quad \text{for all } f \in F. \tag{16.2}$$

This is the situation we have in mind in the following result.

Lemma 16.1. *Let E be a real linear space and let $p : E \to \mathbb{R}$ be a sublinear functional. Furthermore, let $F \subseteq E$ be a linear subspace, $\varphi : F \to \mathbb{R}$ a linear mapping with $\varphi \le p$ on F. Given any $h \in E \setminus F$ there is $\alpha \in \mathbb{R}$ such that the definition*

$$F_1 := F \oplus \mathbb{R}\, h, \quad \varphi_1(f + th) := \varphi(f) + \alpha\, t \qquad (t \in \mathbb{R},\ f \in F)$$

yields a linear mapping $\varphi_1 : F_1 \to \mathbb{R}$ with $\varphi_1|_F = \varphi$ and $\varphi_1 \le p$ on F_1.

Proof. It is clear that with *any* $\alpha \in \mathbb{R}$ the mapping φ_1 is well-defined, linear, and extends φ. We need to choose α in such a way that

$$\varphi(f) + \alpha t \le p(f + th) \quad \text{for all} \quad t \in \mathbb{R},\ f \in F. \tag{16.3}$$

For $t = 0$ this is satisfied by hypothesis. Distinguishing the cases $t > 0$ and $t < 0$ and writing $s = -t$, $g = f$ in the latter, we can reformulate (16.3) in the form of the two conditions

$$\text{(i)} \quad \alpha \le \frac{p(f + th) - \varphi(f)}{t} \qquad (t > 0,\ f \in F),$$

$$\text{(ii)} \quad \alpha \ge \frac{\varphi(g) - p(g - sh)}{s} \qquad (s > 0,\ g \in F).$$

Hence, the proof will be complete when we can show that

$$\sup_{s>0, g\in F} \frac{\varphi(g) - p(g - sh)}{s} \le \inf_{t>0, f\in F} \frac{p(f + th) - \varphi(f)}{t}, \tag{16.4}$$

because then any α that lies between (or equals one of) these two terms satisfies (i) and (ii).

We proceed by noting that (16.4) is equivalent to

$$\frac{\varphi(g) - p(g - sh)}{s} \le \frac{p(f + th) - \varphi(f)}{t}$$

for all $f, g \in F$ and $s, t > 0$. If we multiply this inequality with ts, collect terms, employ the linearity of φ and the homogeneity of p we finally obtain the equivalent inequality

$$\varphi(tg + sf) \le p(tg - sth) + p(sf + sth).$$

But this, almost miraculously, follows directly from our assumptions:

$$\begin{aligned}\varphi(tg + sf) &\le p(tg + sf) = p(tg - sth + sf + sth) \\ &\le p(tg - sth) + p(sf + sth)\end{aligned}$$

for all $f, g \in F$ and $s, t > 0$. □

The lemma enables us to extend a given linear functional one dimension at a time from smaller to larger subspaces while keeping the condition $\varphi \le p$. This suffices to prove the Hahn–Banach theorem in the separable case.

A normed space E is **separable** if it contains a countable fundamental set, i.e., if there is a sequence $(f_n)_{n\in\mathbb{N}}$ such that $\overline{\operatorname{span}}\{f_n \mid n \in \mathbb{N}\} = E$. (By Theorem 4.37 this is equivalent with the metric definition of separability of Definition 4.35. See also page 141 where we encountered this notion in the context of Hilbert spaces.)

Theorem 16.2 (Hahn–Banach,[1] Separable Case). *Let E be a separable normed space over the scalar field $\mathbb{K}$. Let $E_0 \subseteq E$ be a subspace and $\varphi_0 \in E_0'$ a bounded linear functional on E_0. Then there is an extension $\varphi \in E'$ of φ_0 to all of E with $\|\varphi\| = \|\varphi_0\|$.*

Proof. For the proof we first suppose that $\mathbb{K} = \mathbb{R}$ and (without loss of generality) $\|\varphi_0\| = 1$. Then define the sublinear functional $p : E \to \mathbb{R}$ by $p(f) := \|f\|$ for $f \in E$. Then $\varphi_0 \le p$ on E_0.

Since E is separable we can find a countable fundamental set $\{f_n \mid n \in \mathbb{N}\}$. Define $E_n := E_0 + \operatorname{span}\{f_1, \dots, f_n\}$ for $n \in \mathbb{N}$ and obtain an increasing chain of subspaces

$$E_0 \subseteq E_1 \subseteq E_2 \cdots \subseteq E.$$

Passing from E_n to E_{n+1} either nothing changes or the dimension increases by one. By Lemma 16.1 we can extend φ_0 stepwise along this chain of subspaces to obtain a linear functional

$$\varphi_\infty : E_\infty \to \mathbb{R} \quad \text{on} \quad E_\infty := E_0 + \operatorname{span}\{f_1, f_2, \dots\} = \bigcup\nolimits_{n\in\mathbb{N}} E_n$$

extending φ_0 and satisfying $\varphi \le p$ on E_∞. But this means that

$$|\varphi_\infty(f)| \le \|f\| \quad \text{for all } f \in E_\infty,$$

[1] Hans Hahn (1879–1934), Austrian mathematician.

i.e., φ_∞ is bounded with norm ≤ 1. The extension theorem (Theorem 9.28) then yields an extension of φ_∞ to a bounded linear functional $\varphi \in E'$ with the same norm, and that is what we were aiming at.

The real case being settled, let us turn to the case of complex scalars, i.e., $\mathbb{K} = \mathbb{C}$. The main point is to realize two things: 1) every complex linear space is trivially a real linear space and 2) every complex linear functional $\varphi : E \to \mathbb{C}$ can be reconstructed from the real-linear functional $\operatorname{Re}\varphi$ by means of the identity

$$\varphi(f) = \operatorname{Re}\varphi(f) - \mathrm{i}\operatorname{Re}\varphi(\mathrm{i}f) \qquad (f \in E).$$

So the plan is clear: start with $\varphi_0 \in E_0'$ (complex-linear), pass to $\psi_0 := \operatorname{Re}\varphi_0$ (real-linear and clearly bounded), extend that to a real-linear functional $\psi : E \to \mathbb{R}$ with $\|\psi\| = \|\psi_0\|$ and then *define*

$$\varphi(f) := \psi(f) - \mathrm{i}\psi(\mathrm{i}f) \qquad (f \in E).$$

Finally, check that φ has the desired properties (complex-linear, extension of φ_0 and $\|\varphi\| = \|\varphi_0\|$). We leave the details as Exercise 16.2. □ Ex.16.2

Remarks 16.3. 1) In general, the extension guaranteed by the Hahn–Banach theorem is *far from being unique.* (Indeed, it depends on many choices, one in each step of the extension procedure.) Also, in general, no other properties of φ_0 are preserved in the extension. The Hahn–Banach theorem is therefore mainly a theoretical tool; in "concrete" situations one often also has a "concrete" way of constructing the extension.

2) As an example for the assertion just given, consider the case of a Hilbert space H and a bounded linear functional φ_0 defined on the closed subspace H_0. Then $\varphi := \varphi_0 \circ P$ is an equinormed extension, where $P : E \to E_0$ is the orthogonal projection onto E_0. Alternatively, one can first represent φ_0 via the Riesz–Fréchet theorem as $\varphi_0 = \langle \cdot, g \rangle$ where $g \in E_0$, and then define the extension φ by the same expression. Ex.16.3

3) The property of a normed space expressed in Theorem 16.2 is called the **Hahn–Banach property** (HB-property) in [**dPvR13**]. So separable spaces have the HB-property, and there are some examples of nonseparable spaces where one can prove the HB-property from the separable case [**dPvR13**, Exercise 9.5]. Also, it is a simple exercise to see that subspaces of spaces with the HB-property also do have the HB-property.

However, when one accepts the Axiom of Choice as a means of proof, it turns out that *every* normed space has the Hahn–Banach property;

cf. Corollary 16.5 below. This is the reason for us to not develop this notion any further.

A List of Separable and Nonseparable Spaces. In light of Theorem 16.2, but also for other reasons, it is helpful to have a list of concrete examples of separable and non-separable spaces. Note that, by Lemma 4.36, subspaces of separable spaces are again separable.

1) All the sequence spaces ℓ^p with $1 \le p < \infty$ are separable. (The standard unit vectors form a countable fundamental set.) The same is true for the space c_0 of scalar null sequences.
2) The sequence space ℓ^∞ is *not* separable (cf. Exercise 4.36). The reason is that the set of $\{0,1\}$-sequences is an uncountable discrete metric space, and hence not separable.
3) The space $C[a,b]$ is separable with respect to $\|\cdot\|_\infty$ since by Weierstrass' theorem the sequence of monomials has dense linear span.
4) More generally, each space $C(K)$ is separable when K a compact metric space (Exercise 4.39).
5) The spaces $L^p(a,b)$ with $1 \le p < \infty$ are separable, by 3) and the fact that they have $C[a,b]$ as a dense subspace. The space $L^\infty(a,b)$ is not
Ex.16.4 separable (Exercise 16.4).
6) The spaces $C_0(\mathbb{R})$ and $L^p(\mathbb{R})$ for $1 \le p < \infty$ are separable (Exercises 16.19 and 16.20).
7) The Sobolev spaces $H^k(a,b)$ are separable for any $k \ge 1$ (Exercise 16.21).
8) The space $C_b(\mathbb{R})$ is not separable (Exercise 16.22).

***Nonseparable Spaces and General Sublinear Majorants.** What happens when our space E is not separable? And does a Hahn–Banach extension theorem hold for general sublinear functionals in place of norms? The answer is positive if one accepts the Axiom of Choice or one of its equivalents. (If E is not separable one would face the need to make uncountably many steps (choices) in the proof of Theorem 16.2. This is a clear indicator that the Axiom of Choice has to enter the stage.) We present the usual proof employing Zorn's lemma.

Theorem 16.4 (Hahn–Banach). *Let E be a real linear space and let $p : E \to \mathbb{R}$ be a sublinear functional. Let $E_0 \subseteq E$ be a subspace and let $\varphi_0 : E_0 \to \mathbb{R}$*

be a linear functional such that $\varphi_0 \le p$ on E_0. Then there is an extension of φ_0 to a linear functional φ on E with $\varphi \le p$ on E.

Proof. We consider the set $\mathcal{M}$ of ordered pairs (φ_1, E_1) such that E_1 is a subspace of E containing E_0, $\varphi_1 : E_1 \to \mathbb{R}$ is linear and extends φ_0 and satisfies $\varphi_1 \le p$ on E_1.

There is a natural partial ordering on $\mathcal{M}$ given by

$$(E_1, \varphi_1) \le (E_2, \varphi_2) \quad :\Leftrightarrow \quad E_1 \subseteq E_2, \quad \varphi_2|_{E_1} = \varphi_1.$$

Let $\mathcal{K} \subseteq \mathcal{M}$ be a totally ordered subset of $\mathcal{M}$ (a "chain"). Let $E_\infty := \bigcup\{E_* \mid (E_*, \varphi_*) \in \mathcal{K}\}$. Then, by the total orderedness of $\mathcal{K}$, E_∞ is a subspace of E. Moreover, letting

$$\varphi_\infty(f) := \varphi_*(f) \quad \text{whenever } (E_*, \varphi_*) \in \mathcal{K} \text{ and } f \in E_*$$

yields a (well-defined!) linear functional φ_∞ on E_∞. It is easy to see that $(E_\infty, \varphi_\infty)$ is a member of $\mathcal{M}$ and it majorizes every member of $\mathcal{K}$.

So we have shown that every chain in $\mathcal{M}$ has an upper bound. Zorn's lemma (Theorem A.2) tells that $\mathcal{M}$ has a *maximal element* (E_1, φ_1), say, and it remains to show that $E_1 = E$. Suppose that this is not the case. Then there is a vector $h \in E \setminus E_1$, and Lemma 16.1 yields a proper extension of φ_1. But this amounts to an element of $\mathcal{M}$ strictly greater than (E_1, φ_1), contradicting the maximality. □

We can now drop the separability assumption from the hypotheses of Theorem 16.2.

Corollary 16.5. *Let E be an arbitrary normed space, and let $E_0 \subseteq E$ be a linear subspace and $\varphi_0 \in E_0'$ a bounded linear functional on E_0. Then there is a bounded linear functional $\varphi \in E'$ such that $\varphi|_{E_0} = \varphi_0$ and $\|\varphi\| = \|\varphi_0\|$.*

Proof. For $\mathbb{K} = \mathbb{R}$ the theorem follows directly from Theorem 16.4 with $p : E \to \mathbb{R}$ defined by $p(f) := \|\varphi_0\| \, \|f\|$. For $\mathbb{K} = \mathbb{C}$ we employ the same trick as in the proof of Theorem 16.2. □

***Separation of Convex Sets.** Let E be a real vector space E, $\varphi : E \to \mathbb{R}$ a linear functional and $c \in \mathbb{R}$. If $\varphi \neq 0$, then the set

$$[\varphi = c] := \{f \in E \mid \varphi(f) = c\}$$

is called an (affine) **hyperplane** in E, and each of the sets

$$[\varphi \le c] := \{f \in E \mid \varphi(f) \le c\} \quad \text{and} \quad [\varphi \ge c] := \{f \in E \mid \varphi(f) \ge c\}$$

are called (closed) **halfspaces**. A set $A \subseteq E$ is **separated** by the hyperplane $[\varphi = c]$ from a vector $f \in E \setminus A$ if $\varphi(A) \subseteq (-\infty, c]$ and $\varphi(f) \ge c$. Ex.16.5

In Exercise 8.20 it is shown that if H is a real Hilbert space, $A \subseteq H$ is a nonempty closed convex subset of H and $f \in H \setminus A$, then A can be separated from f by a closed hyperplane. With the help of the Hahn–Banach theorem one can show that this is actually true in every normed space.

Theorem 16.6 (Hahn–Banach Separation Theorem). *Let E be a normed real linear space, $A \subseteq E$ convex and closed and $f \in E \setminus A$. Then there is a bounded linear functional $\varphi \in E'$ and $c \in \mathbb{R}$ such that*

$$\varphi(A) \subseteq (\infty, c] \quad \text{and} \quad \varphi(f) > c.$$

Proof. For A being a subspace, see Lemma 16.7 below. In the general case a proof would lead us too far astray. We therefore refer to the standard literature, e.g., [**Con90**, Chap. 4, §3]. □

The Hahn–Banach separation theorem tells that *each closed convex subset A of a normed space is the intersection of all closed halfspaces that contain A.* Or, in other words, *a closed convex subset of a normed space is completely determined by the linear inequalities it satisfies.* This is a nontrivial result even in a finite-dimensional setting.

16.2. Elementary Duality Theory

In this section we study some of the consequences of the Hahn–Banach theorem. The first is a special case of Theorem 16.6 above.

Lemma 16.7 (Separation Lemma). *Let F be a closed subspace of a normed space E and let $f \in E \setminus F$. Then there is $\varphi \in E'$ such that $\|\varphi\| = 1$, $\varphi = 0$ on F, and $\varphi(f) = \operatorname{dist}(f, F)$.*

Proof. Let $E_0 := F \oplus \mathbb{K}f$ and $\varphi_0 : E_0 \to \mathbb{K}$ defined by

$$\varphi(h + \lambda f) := \lambda\alpha \qquad (h \in F, \lambda \in \mathbb{K}),$$

where $0 \neq \alpha \in \mathbb{K}$ is any scalar. Because F is a subspace, we obtain

$$\|\varphi_0\| = \sup_{h\in F, \lambda\neq 0} \frac{|\alpha\lambda|}{\|h + \lambda f\|} = \sup_{h\in F} \frac{|\alpha|}{\|f - h\|} = \frac{|\alpha|}{\operatorname{dist}(f, F)}.$$

Note that since F is closed and $f \notin F$ we have $\operatorname{dist}(f, F) > 0$, and hence the choice $\alpha := \operatorname{dist}(f, F) > 0$ yields $\|\varphi_0\| = 1$. Finally, extend φ_0 by virtue of the Hahn–Banach theorem. □

The separation lemma has some interesting consequences.

Corollary 16.8. *Let E be a (nontrivial) normed space. Then the following assertions hold.*

a) *For every $f \in E$ there exists $\varphi \in E'$ such that $\|\varphi\| = 1$ and $\varphi(f) = \|f\|$.*

b) *For every $f \in E$ one has the variational formula*

$$\|f\| = \sup_{\varphi \in E', \|\varphi\| \le 1} |\varphi(f)|$$

for the norm. (The supremum is actually attained.)

c) *For every $A \subseteq E$ one has*

$$\overline{\operatorname{span}}(A) = E \quad \Longleftrightarrow \quad \forall \varphi \in E' : \varphi|_A = 0 \;\Rightarrow\; \varphi = 0.$$

Proof. a) Apply Lemma 16.7 with $F = \{0\}$. b) This follows from a).
c) The implication "$\Rightarrow$" is simple (Lemma 9.3). For the converse, suppose that $F := \overline{\operatorname{span}}(A) \neq E$. Then by Lemma 16.7 there is $0 \neq \varphi \in E'$ with $\varphi|_F = 0$. A fortiori $\varphi|_A = 0$, which contradicts the right-hand side. □

Remarks 16.9. 1) For a subset $S \subseteq E$ one defines its **annihilator** by

$$S^\perp := \{\varphi \in E' \mid \varphi|_S = 0\}.$$

Then c) of Corollary 16.8 just tells that S is fundamental in E if and only if its annihilator is trivial.

2) It is a simple but useful exercise to give a direct proof of Lemma 16.7 and Corollary 16.8 for Hilbert spaces without making use of the Hahn–Banach theorem. Ex.16.6

Dual Operators. If $T : E \to F$ is a bounded linear operator, then for every $\varphi \in F'$ the linear mapping $\varphi \circ T : E \to \mathbb{K}$ is bounded, hence in E'. This amounts to a new operator

$$T' : F' \to E' \qquad T'\varphi := \varphi \circ T,$$

called the **dual operator** or **Banach space adjoint** of T. By definition, the dual operator satisfies

$$(T'\varphi)f = \varphi(Tf) \quad \text{for all} \quad f \in E,\ \varphi \in F'. \tag{16.5}$$

Theorem 16.10. *For normed spaces E, F and a bounded linear operator $T : E \to F$ the dual operator $T' : F' \to E'$ is bounded with $\|T\| = \|T'\|$. Moreover, one has*

$$(T + S)' = T' + S' \quad \text{and} \quad (\lambda T)' = \lambda T'$$

for T, S in $\mathcal{L}(E; F)$ and $\lambda \in \mathbb{K}$. If G is another normed space and $S : F \to G$ is a bounded linear mapping, then $(ST)' = T'S'$.

Proof. The proof is simple and is left as Exercise 16.7. □ Ex.16.7

The Canonical Duality. Let E be a normed space with dual space E'. Then we can consider the *bilinear* mapping

$$E \times E' \longrightarrow \mathbb{K}, \qquad (f, \varphi) \longmapsto \varphi(f),$$

called the **canonical duality**. Observe the symmetry of the variational formula in Corollary 16.8, b) and the definition of the norm on E':

$$\|f\| = \sup\nolimits_{\varphi \in E', \|\varphi\| \le 1} |\varphi(f)| \qquad (f \in E), \tag{16.6}$$

$$\|\varphi\| = \sup\nolimits_{f \in E, \|f\| \le 1} |\varphi(f)| \qquad (\varphi \in E'). \tag{16.7}$$

In many functional analysis texts you will even find the notation

$$\langle f, \varphi \rangle := \varphi(f) \quad \text{for} \quad f \in E,\ \varphi \in E'.$$

Then the Banach space adjoint T' of a bounded linear operator $T \in \mathcal{L}(E; F)$ satisfies

$$\langle f, T'\varphi \rangle = \langle Tf, \varphi \rangle \quad \text{for all} \quad f \in E,\ \varphi \in F'.$$

Of course, all this can be done only if there is no danger of confusion with a Hilbert space inner product. (Recall that an inner product is sesquilinear, whereas the canonical duality is bilinear.)

If we had written E^* for the dual space, the natural notation for the dual operator would have been T^*. In the case of complex Hilbert spaces, however, this notation has already been used for the Hilbert space adjoint. That is the reason why we prefer to use E' and not E^* for the topological dual.

Fix $f \in E$ and consider the expression $\varphi(f)$ as a function of $\varphi \in E'$. We obtain a bounded linear functional on E', i.e., an element of the **double dual** E''. This gives rise to a linear mapping

$$\mathrm{j} : E \longrightarrow E'', \qquad (\mathrm{j}f)\varphi := \varphi(f) \qquad (f \in E,\ \varphi \in E').$$

By (16.6), *the mapping* j *is isometric.* Consequently, the original space E can be identified with the subspace $\mathrm{j}(E)$ of E''.

Corollary 16.11. *Every normed space is isometrically isomorphic to a dense subspace of a Banach space.*

Proof. By Theorem 11.8 each dual space is complete, i.e., a Banach space. The linear isometry $\mathrm{j} : E \to E''$ then maps E to the subspace $\mathrm{j}(E)$ of the Banach space E'', and hence $\overline{\mathrm{j}(E)}$ is a Banach space which has $\mathrm{j}(E)$ as a dense subspace. □

Corollary 16.11 yields a "concrete" construction of "the" completion of a normed space. For more on this topic see Appendix B.

***Reflexivity.** A normed space E is called **reflexive** if its associated isometric embedding $\mathrm{j} : E \to E''$ is surjective, i.e., an isometric isomorphism. In particular, a reflexive space is complete, i.e., a Banach space. Ex.16.8

Lemma 16.12. a) *Each finite-dimensional space is reflexive.*

b) *Each Hilbert space is reflexive.*

c) *A closed subspace of a reflexive space is reflexive.*

d) *A Banach space is reflexive if and only if its dual space is reflexive.*

Proof. a) and b) are left as Exercise 16.9. Ex.16.9

c) Let E be a reflexive space and $F \subseteq E$ a closed subspace. Furthermore, let $\Phi \in F''$ be an element of the double dual. That is, $\Phi : F' \to \mathbb{K}$ is linear and bounded. The restriction mapping

$$E' \to F' \qquad \varphi \mapsto \varphi|_F$$

is linear and bounded. Hence, the mapping

$$E' \to \mathbb{K}, \qquad \varphi \mapsto \Phi(\varphi|_F)$$

is an element of E''. Since E is reflexive, there is $f \in E$ that induces this mapping, i.e.,

$$\varphi(f) = \Phi(\varphi|_F) \quad \text{for all } \varphi \in E'.$$

By Lemma 16.7, it follows that $f \in F$. So if $\psi \in F'$ is arbitrary we can take $\varphi \in E'$ to be any extension of ψ to E (by the Hahn–Banach theorem) and obtain

$$\psi(f) = \varphi(f) = \Phi(\varphi|_F) = \Phi(\psi).$$

But that means that $\mathrm{j}f = \Phi$.

d) Let E be a Banach space with canonical isometry $\mathrm{j} : E \to E''$, and suppose that E' is reflexive. Then $\mathrm{j}(E)$ is a closed subspace of E''. (Here we use that E is complete). To show that $\mathrm{j}(E) = E''$ we employ the Hahn–Banach theorem. Let $\tilde{\varphi} \in E'''$ with $\tilde{\varphi}|_{\mathrm{j}(E)} = 0$. By the reflexivity of E' there is $\varphi \in E'$ that induces $\tilde{\varphi}$, which means that

$$\tilde{\varphi}(\Phi) = \Phi(\varphi) \quad \text{for all } \Phi \in E''.$$

By specializing $\Phi = \mathrm{j}f$ for $f \in E$ we obtain

$$0 = \tilde{\varphi}(\mathrm{j}f) = (\mathrm{j}f)\varphi = \varphi(f) \quad \text{for all } f \in E.$$

But that means that $\varphi = 0$ and hence $\tilde{\varphi} = 0$. It follows by Corollary 16.8.c) that $\mathrm{j}(E) = E''$, whence E is reflexive.

For the converse suppose that E is reflexive. Then $\mathrm{j} : E \to E''$ is an isometric isomorphism, hence E'' is reflexive (Exercise 16.8). But E' is a Banach space, so it follows that E' is reflexive by the implication we already proved. □

Reflexive spaces share a couple of nice properties that a general Banach space may not have; cf. Theorem 16.15. We shall not pursue the study of reflexivity any further, but refer the interested reader to Exercises 16.33–16.40

***Weak Convergence.** A sequence $(f_n)_{n\in\mathbb{N}}$ in a normed space E is called **weakly convergent** to a vector $f \in E$, if

$$\varphi(f_n) \to \varphi(f) \qquad \text{as } n \to \infty \qquad \text{for all } \varphi \in E'.$$

In this case f is called a **weak limit** of the sequence and one writes

$$f_n \rightharpoonup f \quad (n \to \infty).$$

Ex.16.10 *A weak limit is unique*: if $f_n \rightharpoonup f$ and $f_n \rightharpoonup g$ and $\varphi \in E'$ is any element of the dual space, then it follows that $\varphi(f) = \varphi(g)$. Since E' separates the points of E (Hahn–Banach theorem), it follows that $f = g$.

Remark 16.13. Because

$$|\varphi(f) - \varphi(f_n)| = |\varphi(f - f_n)| \le \|\varphi\| \, \|f_n - f\| \quad \text{for each } \varphi \in E',$$

each norm convergent sequence is weakly convergent (to the same limit). The converse is false: let e_n be the n-th standard unit vector in ℓ^2. Then
Ex.16.11 $e_n \rightharpoonup 0$ but $e_n \not\to 0$ as $n \to \infty$ (Exercise 16.11).

The following theorem collects all important elementary facts about weakly convergent sequences.

Theorem 16.14. a) *Let E and F be normed spaces and let $T : E \to F$ be linear and bounded. Then T maps weakly convergent sequences in E to weakly convergent sequences in F. More precisely,*

$$f_n \rightharpoonup f \quad \text{in } E \quad \Rightarrow \quad Tf_n \rightharpoonup Tf \quad \text{in } F.$$

b) *Let F be a closed subspace of a Banach space E, let $(f_n)_{n\in\mathbb{N}}$ be a sequence in F and $f \in E$. Then*

$$f_n \rightharpoonup f \quad \text{in } E \quad \Leftrightarrow \quad f \in F \quad \text{and} \quad f_n \rightharpoonup f \quad \text{in } F.$$

c) *A weakly convergent sequence is norm bounded.*

Proof. a) Suppose that $f_n \rightharpoonup f$ in E and let $\varphi \in F'$. Then $T'\varphi \in E'$ and hence

$$\varphi(Tf_n) = (T'\varphi)(f_n) \to (T'\varphi)(f) = \varphi(Tf) \quad (n \to \infty).$$

b) The implication "$\Leftarrow$" follows from a) by specializing $T : F \to F$ to be the inclusion mapping. For the converse, suppose that $f \in E \setminus F$. Then by Lemma 16.7 there is $\varphi \in E'$ such that φ vanishes on F but $\varphi(f) \neq 0$. As all $f_n \in F$, it follows that $\varphi(f_n) = 0$ for all $n \in \mathbb{N}$, and hence $\varphi(f) = 0$, a contradiction.

In order to show that $f_n \rightharpoonup f$ in F, we take $\psi \in F'$. By the Hahn–Banach theorem, there is $\varphi \in E'$ such that $\varphi|_F = \psi$. Hence, by hypothesis,

$$\psi(f_n) = \varphi(f_n) \to \varphi(f) = \psi(f) \quad \text{as} \quad n \to \infty.$$

This proves the claim.

c) Let $(f_n)_{n\in\mathbb{N}}$ be a weakly convergent sequence in the normed space E. Then the set $\{\mathrm{j}f_n \mid n \in \mathbb{N}\} \subseteq \mathcal{L}(E';\mathbb{K})$ is pointwise convergent, and hence pointwise bounded. Since E' is a Banach space, the uniform boundedness principle yields $\sup_{n\in\mathbb{N}} \|f_n\| = \sup_{n\in\mathbb{N}} \|\mathrm{j}f_n\| < \infty$. □

Weak convergence is often the best one can expect from approximation procedures in infinite-dimensional situations. In particular, this is true for many standard variational methods to find solutions to partial differential equations; see [**Sho77**, Chapter VII]. In this context, reflexivity plays a key role.

Theorem 16.15. *In a reflexive Banach space every norm bounded sequence has a weakly convergent subsequence.* Ex.16.12

Proof. This is left as Exercise 16.40. □

Theorem 16.15 can be rephrased as: *The closed unit ball of a reflexive Banach space is weakly (sequentially) compact.* This terminology becomes fully clear only in the framework of general *topological vector spaces*, a concept more general than normed and even than metric spaces. See [**Rud91**, Chapters 1 and 3] for more about topological vector spaces and weak topologies.

16.3. Identification of Dual Spaces

Even if a concrete normed space E is given, its dual space E' is a quite abstract object. On the other hand, there may be rather concrete ways of constructing bounded linear functionals on E. On $(\mathrm{C}[a,b], \|\cdot\|_\infty)$, for instance, we can consider the point evaluation functionals δ_x for $x \in [a, b]$ or integration functionals $f \mapsto \int_a^b f(t)m(t)\mathrm{d}t$ where $m \in \mathrm{L}^1(a, b)$.

Since one often can work better with such concrete functionals, it is of interest to prove "representation theorems" which (in the best case) tell that *each* bounded functional has a particular concrete form. The protoype of such a result is the Riesz–Fréchet theorem: each bounded linear functional on a Hilbert space H can be written in the form $\langle \cdot, f\rangle$ for some $f \in H$. In order to formulate representation theorems for the duals of other Banach spaces, one needs a replacement for the inner product. This is what we treat next.

A **pairing** for a pair (E, F) of normed spaces E, F is any *bilinear* mapping

$$\gamma : E \times F \to \mathbb{K}$$

that satisfies

$$|\gamma(f, g)| \le \|f\| \, \|g\| \quad \text{for all} \quad f \in E,\ g \in F. \tag{16.8}$$

A pairing is a **duality** if

$$\|g\| = \sup\left\{ |\gamma(f, g)| \mid f \in E, \|f\| \le 1\right\}. \tag{16.9}$$

The standard example of a duality is, of course, the canonical duality

$$E \times E' \to \mathbb{K} \qquad \gamma(f, \varphi) := \varphi(f).$$

Here, (16.9) is simply (16.6), true by definition of the norm on E'.

Given a duality $\gamma : E \times F \to \mathbb{K}$ one can consider the linear mapping

$$\Gamma : F \to E' \qquad g \mapsto \Gamma g := \gamma(\cdot, g),$$

which is an isometry, by (16.9). In this way, F is "identified" with a certain
Ex.16.13 subspace of E', closed if F is a Banach space. A representation theorem now would state that this subspace is actually all of E', i.e., the map Γ is surjective. If this is the case, one says that "$E' \cong F$ via γ".

It is time to treat some concrete examples.

The Dual of c_0. Let us consider the **canonical pairing**

$$\gamma : \mathrm{c}_0 \times \ell^1 \longrightarrow \mathbb{K}, \qquad \beta((x_j)_j, (y_j)_j) := \sum\nolimits_{j=1}^{\infty} x_j y_j$$

between $E := \mathrm{c}_0$ the space of scalar null sequences and $F := \ell^1$ the space of absolutely summable sequences. By (the simple case of) Hölder's inequality, this is a well-defined bilinear mapping satisfying

$$\left|\sum\nolimits_{j=1}^{\infty} x_j y_j\right| \le \|x\|_\infty \, \|y\|_1 \qquad \text{for all } x \in \mathrm{c}_0,\ y \in \ell^1,$$

which is (16.8). In order to prove the stronger statement (16.9), fix $y \in \ell^1$ and define $x^{(n)}$ for $n \in \mathbb{N}$ by

$$(16.10) \qquad x_j^{(n)} := \begin{cases} \frac{\overline{y_j}}{|y_j|} & \text{if } j \le n \text{ and } y_j \neq 0, \\ 0 & \text{else,} \end{cases} \qquad \text{for } j \in \mathbb{N}.$$

Then $x^{(n)} \in \mathrm{c}_{00}$, $\left\|x^{(n)}\right\|_\infty \le 1$ and

$$\gamma(x^{(n)}, y) = \sum\nolimits_{j=1}^{\infty} x_j^{(n)} y_j = \sum\nolimits_{j=1}^{n} |y_j| .$$

Hence

$$\sup \left\{ |\gamma(x,y)| \mid x \in \mathrm{c}_0,\ \|x\|_\infty \le 1 \right\} \ge \sup_{n\in\mathbb{N}} \sum\nolimits_{j=1}^{n} |y_j| = \|y\|_1 .$$

This proves (16.9), whence γ is a duality for the pair (c_0, ℓ^1).

Theorem 16.16. *Via the canonical pairing, $\mathrm{c}_0' \cong \ell^1$.*

Proof. It remains to show that each bounded linear functional on c_0 can be represented by an element of ℓ^1 via the canonical pairing, so let us fix $\varphi \in \mathrm{c}_0'$. We want to find $y \in \ell^1$ such that $\gamma(\cdot, y) = \varphi$. *If* such a vector y exists, it certainly satisfies $y_n = \gamma(e_n, y) = \varphi(e_n)$, where e_n is the n-th canonical unit vector. Hence it is reasonable to *define* y as

$$y := (y_n)_n, \quad y_n := \varphi(e_n) \quad (n \in \mathbb{N}).$$

To show that $y \in \ell^1$, we use the same construction of the vectors $x^{(n)}$ as before; see (16.10). Then, for each $n \in \mathbb{N}$,

$$\begin{aligned} \sum\nolimits_{j=1}^{n} |y_j| &= \left|\sum\nolimits_{j=1}^{n} x_j^{(n)} y_j\right| = \left|\sum\nolimits_{j=1}^{n} x_j^{(n)} \varphi(e_j)\right| = \left|\varphi\left(\sum\nolimits_{j=1}^{n} x_j^{(n)} e_j\right)\right| \\ &= \left|\varphi(x^{(n)})\right| \le \|\varphi\| \left\|x^{(n)}\right\|_\infty \le \|\varphi\| . \end{aligned}$$

It follows that $y \in \ell^1$ with $\|y\|_1 \le \|\varphi\|$.
Finally, $\varphi = \gamma(\cdot, y)$, since both functionals are bounded and they agree on the fundamental set $\{e_j \mid j \in \mathbb{N}\}$ of canonical unit vectors (Lemma 9.3.b). □

The Dual of ℓ^1. A similar approach as for the space c_0 works for ℓ^1. As above, we consider the canonical pairing

$$\gamma : \ell^1 \times \ell^\infty \longrightarrow \mathbb{K}, \qquad \gamma((x_j)_j, (y_j)_j) := \sum\nolimits_{j=1}^{\infty} x_j y_j,$$

which is a well-defined bilinear mapping satisfying

$$\left|\sum\nolimits_{j=1}^{\infty} x_j y_j\right| \le \|x\|_1 \ \|y\|_\infty \qquad \text{for all } x \in \ell^1,\ y \in \ell^\infty,$$

by Hölder's inequality.

Theorem 16.17. *Via the canonical pairing, $\ell^{1\prime} \cong \ell^\infty$.*

Ex.16.14 **Proof.** We leave the proof as Exercise 16.14. □

The Dual of $\mathrm{L}^1(a,b)$**.** We turn to the "continuous" analogue of the previous example. Let $[a,b] \subseteq \mathbb{R}$ be any finite interval and consider the canonical pairing

$$\gamma : \mathrm{L}^1(a,b) \times \mathrm{L}^\infty(a,b) \longrightarrow \mathbb{K}, \qquad \gamma(f,g) := \int_a^b fg \,\mathrm{d}\lambda.$$

As in the discrete case, this is a well-defined bilinear mapping by virtue of Hölder's inequality. In order to establish (16.9) fix $0 \neq g \in \mathrm{L}^\infty(a,b)$. For $0 < c < \|g\|_{\mathrm{L}^\infty}$ consider the set $A_c := \{|g| \ge c\}$ and the function

$$f = \lambda(A_c)^{-1} \frac{\overline{g}}{|g|} \mathbf{1}_{A_c}. \tag{16.11}$$

(Note that $\lambda(A_c) \neq 0$ by the choice of c.) Then $\|f\|_1 = 1$ and

$$|\gamma(f,g)| = \Big|\int_a^b fg \,\mathrm{d}\lambda\Big| = \lambda(A_c)^{-1} \int_{A_c} |g| \,\mathrm{d}\lambda \ge \lambda(A_c)^{-1} \int_{A_c} c \,\mathrm{d}\lambda = c.$$

By varying c it follows that

$$\sup\{\, |\gamma(f,g)| \mid f \in \mathrm{L}^1(a,b),\ \|f\|_1 \le 1\} \ge \|g\|_{\mathrm{L}^\infty},$$

and hence (16.9). (Confer also Exercise 11.26.)

Theorem 16.18. *Via the canonical pairing,* $\mathrm{L}^1(a,b)' \cong \mathrm{L}^\infty(a,b)$.

Proof. Let $\varphi : \mathrm{L}^1(a,b) \to \mathbb{K}$ be a bounded linear functional. Since the inclusion mapping $\mathrm{L}^2(a,b) \to \mathrm{L}^1(a,b)$ is bounded, $\varphi|_{\mathrm{L}^2}$ is a bounded linear functional on $\mathrm{L}^2(a,b)$. By the Riesz–Fréchet theorem there is $g \in \mathrm{L}^2(a,b)$ such that

$$\varphi(f) = \langle f, \overline{g}\rangle_{\mathrm{L}^2} = \int_a^b fg \qquad \text{for all } f \in \mathrm{L}^2(a,b).$$

Because $\mathrm{L}^2(a,b)$ is dense in $\mathrm{L}^1(a,b)$, it remains to show that $g \in \mathrm{L}^\infty(a,b)$. To this end, fix $c > 0$, consider as above $A_c := \{|g| \ge c\}$, and suppose that $\lambda(A_c) \neq 0$. Let f be defined as above in (16.11). Then $|f| \le 1$ almost everywhere and hence $f \in \mathrm{L}^2(a,b)$. Therefore,

$$c \le \lambda(A_c)^{-1} \int_{A_c} |g| \,\mathrm{d}\lambda = \Big|\int_a^b fg \,\mathrm{d}\lambda\Big| = |\varphi(f)| \le \|\varphi\| \,\|f\|_1 = \|\varphi\|.$$

It follows that $|g| \le \|\varphi\|$ almost everywhere. □

It is true that $\mathrm{L}^1(X)' \cong \mathrm{L}^\infty(X)$ via the canonical pairing for much more general cases of X; cf. Exercise 16.27.

The Dual of $\mathrm{C}[a,b]$**?** We now examine whether there is also a continuous analogue of Theorem 16.16. For a finite interval $[a,b]$ we consider the natural pairing

$$\gamma : \mathrm{C}[a,b] \times \mathrm{L}^1(a,b) \to \mathbb{C}, \qquad \gamma(f,g) := \int_a^b fg \,\mathrm{d}\lambda.$$

By Lemma 9.17 this is indeed a duality. However, the point evaluation functionals cannot be represented by integration against L^1-functions. Ex.16.15

But even considering $\mathrm{L}^1(a,b)$ *plus* the point evaluation functionals (as in Exercise 16.29) is not sufficient to cover the whole dual of $\mathrm{C}[a,b]$. The identification of $\mathrm{C}[a,b]'$ is actually quite laborious, and it was one of the big successes in the early history of functional analysis, achieved by F. Riesz in 1909. We devote Section 16.4 to it.

***The Dual of** ℓ^p **and** L^p **for** $1 < p < \infty$**.** Let again $[a,b] \subseteq \mathbb{R}$ be any finite interval. Furthermore, let $1 < p < \infty$ and let $q \in (1,\infty)$ be the dual exponent, i.e., q satisfies

$$\frac{1}{p} + \frac{1}{q} = 1.$$

By the general form of Hölder's inequality (Theorem 7.22), the canonical pairing

$$\gamma(f,g) := \int_a^b fg \,\mathrm{d}\lambda$$

is well-defined for the pair of spaces $\mathrm{L}^p(a,b)$ and $\mathrm{L}^q(a,b)$, where $1 < p,q < \infty$ and $1/p + 1/q = 1$.

In order show that this pairing is a duality, fix $0 \neq g \in \mathrm{L}^q(a,b)$ and define

$$f := \frac{\overline{g}}{|g|} \, |g|^{q-1}$$

in the sense that $f = 0$ on the set $\{g = 0\}$. Then, since $q/p = q-1$,

$$fg = |g|^q, \quad |f| = |g|^{q-1}, \quad |f|^p = |g|^q.$$

Hence

$$f \in \mathrm{L}^p(a,b), \quad \|f\|_p = \|g\|_q^{q-1} \quad \text{and} \quad \int_a^b fg \,\mathrm{d}\lambda = \|g\|_q^q.$$

Replacing f by $f/\|f\|_p$ we obtain a unit vector f such that $|\gamma(f,g)| = \|g\|_q$, and that establishes (16.9).

Theorem 16.19. *For* $1 < p < \infty$ *and* $1/p + 1/q = 1$ *is* $\mathrm{L}^p(a,b)' \cong \mathrm{L}^q(a,b)$ *via the canonical pairing.*

Proof. We restrict to the case $1 < p \le 2$ (cf. Remark 16.20 below). Let $\varphi : \mathrm{L}^p(a,b) \to \mathbb{K}$ be any bounded linear functional. Since the inclusion mapping $\mathrm{L}^2(a,b) \to \mathrm{L}^p(a,b)$ is bounded, $\varphi|_{\mathrm{L}^2}$ is a bounded linear functional on $\mathrm{L}^2(a,b)$. By the Riesz–Fréchet theorem there is $g \in \mathrm{L}^2(a,b)$ such that

$$\varphi(f) = \langle f, \overline{g}\rangle_{\mathrm{L}^2} = \int_a^b fg \qquad \text{for all } f \in \mathrm{L}^2(a,b).$$

Because $\mathrm{L}^2(a,b)$ is dense in $\mathrm{L}^p(a,b)$, it remains to show that $g \in \mathrm{L}^q(a,b)$. For $n \in \mathbb{N}$ let $A_n := \{0 < |g| \le n\}$ and

$$f_n := \frac{\overline{g}}{|g|}\,|g|^{q-1}\,\mathbf{1}_{A_n}.$$

Then $f_n \in \mathrm{L}^\infty(a,b)$, whence in particular $f_n \in \mathrm{L}^2(a,b)$. We obtain

$$\int_{A_n} |g|^q = \Bigl|\int f_n g\Bigr| = |\varphi(f_n)| \le \|\varphi\|\,\|f\|_p = \|\varphi\|\Bigl(\int_{A_n} |g|^q\Bigr)^{1/p},$$

from which it follows that

$$\int_{A_n} |g|^q \le \|\varphi\|^q.$$

By the monotone convergence theorem, $\int_a^b |g|^q \le \|\varphi\|^q < \infty$, and this concludes the proof. □

As before, it is true that $\mathrm{L}^p(X)' \cong \mathrm{L}^q(X)$ via the canonical pairing for much more general cases of (measure spaces) X; cf. Exercise 16.31.

Remark 16.20. For the case $p > 2$ the proof of Theorem 16.19 is more complicated. The standard approach builds on the Radon-Nikodym theorem [**Rud87**, Chap. 6]. Alternatively, one can use Clarkson's inequalities [**Cla36**] to show that L^p is uniformly convex for $2 \le p < \infty$, then conclude by Milman's theorem [**Kak39**, Thm. 3] that L^p is reflexive for those p and then by general theory that L^p is reflexive for $1 < p < \infty$. From this it is immediate to show that $\mathrm{L}^{p\prime} \cong \mathrm{L}^q$ for all $1 < p < \infty$; see also [**Lax02**, Thm. 11].

We turn to the discrete case. The canonical pairing

$$\gamma : \ell^p \times \ell^q \longrightarrow \mathbb{K}, \qquad \gamma((x_j)_j, (y_j)_j) := \sum\nolimits_{j=1}^\infty x_j y_j,$$

is a well-defined pairing by Hölder's inequality; see Section 2.5. Using the same ideas as for the case $\mathrm{L}^p(a,b)$, one can show readily that this pairing is a duality. Moreover, combining the ideas of the proofs of Theorems 16.16 and 16.19 one arrives at the following result.

Theorem 16.21. *For $1 < p < \infty$ and $1/p + 1/q = 1$ is $\ell^{p\prime} \cong \ell^q$ via the canonical pairing.*

The proof is left as Exercise 16.32. Note that one neither needs to employ the Riesz–Fréchet theorem nor any restriction on the exponent p.

16.4. *The Riesz Representation Theorem

In this section we identify the dual of $\mathrm{C}[a,b]$, endowed with the uniform norm. We start by constructing a suitable duality.

A function $f : [a,b] \to \mathbb{C}$ is a **step function** if there is a partition $a = t_0 < t_1 < \cdots < t_n = b$ of the interval $[a,b]$ and scalars $c_j \in \mathbb{C}$ such that

$$f(t) = c_j \quad \text{if} \quad t_{j-1} < t < t_j.$$

(The values of f at the points t_j are inessential for the following.)

The set of step functions is denoted by $\mathrm{St}[a,b]$ (Exercise 2.36). The $\|\cdot\|_\infty$-closure of $\mathrm{St}[a,b]$ within the space of bounded functions is $\mathrm{Reg}[a,b]$, the space of **regulated functions**. For a regulated function f the left and right limits

$$f(c-) := \lim_{t \nearrow c} f(t), \qquad f(c+) := \lim_{t \searrow c} f(t)$$

exist for each $c \in [a,b]$ whenever this is meaningful, and f has only a countable number of discontinuities, see Exercise 5.27 and Exercise 16.41.

A function $f : [a,b] \to \mathbb{C}$ is called **of bounded variation** if

$$\|f\|_v := \sup \sum\nolimits_{j=1}^{n} |f(t_j) - f(t_{j-1})| < \infty,$$

where the supremum is taken over all decompositions

$$a = t_0 < t_1 < \cdots < t_n = b \qquad (n \in \mathbb{N} \text{ arbitrary}).$$

Denote by

$$\mathrm{BV}_0[a,b] := \{f : [a,b] \to \mathbb{C} \mid f(a) = 0,\ \|f\|_v < \infty\},$$

the space of all functions of bounded variation that vanish at the left endpoint; cf. Exercise 2.34. By Exercise 9.23 we can define an integral

$$\int_a^b f\,\mathrm{d}g := \int_a^b f(t)\,\mathrm{d}g(t) \quad \text{for } f \in \mathrm{Reg}[a,b],\ g \in \mathrm{BV}_0[a,b]$$

in such a way that

$$\int_a^b \mathbf{1}_{[c,d]}\,\mathrm{d}g = g(d) - g(c) \quad (a \le c \le d \le b).$$

Changing g by a constant function does not affect the integral $\int_a^b f\,\mathrm{d}g$, whence our restriction to functions satisfying $g(0) = 0$. Ex.16.16

Let us emphasize that $\int_a^b f \, dg$ — despite the notation — is *not* a classical Riemann–Stieltjes integral as developed in [**Lan93**, Chap. X]. In fact, if f is a step function that has a jump at a point of discontinuity of g, then f is not Riemann–Stieltjes integrable with respect to g. One can show that the two integral concepts coincide for pairs $f \in \mathrm{Reg}[a,b]$, $g \in \mathrm{BV}_0[a,b]$ such that f and g have no common points of discontinuity.

By the definition of the integral, step functions of the form $\mathbf{1}_{\{c\}}$ with $c \in [a,b]$ have zero integral no matter what g is. Hence the corresponding point evaluation functionals δ_c cannot be generated by means of such an integral. We shall resolve this problem by restricting to a suitable subspace of $\mathrm{Reg}[a,b]$.

Let us call a regulated function $f \in \mathrm{Reg}[a,b]$ **special** if it is left continuous at each $t \in (a,b]$ and right continuous at $t = a$. Further, let

$$\mathrm{Reg}_s[a,b] := \overline{\{f \in \mathrm{St}[a,b] \mid f \text{ special}\}}$$

be the $\|\cdot\|_\infty$-closure of the space of special step functions. Then each $f \in$
Ex.16.17 $\mathrm{Reg}_s[a,b]$ is again special (Exercise 16.17). (Conversely, it can be shown that each special regulated function is contained in $\mathrm{Reg}_s[a,b]$ (see Exercise 16.42), but we shall not use this fact.) The following lemma is a simple consequence of uniform continuity, its proof is left as an exercise.

Ex.16.18 **Lemma 16.22.** *The space* $\mathrm{C}[a,b]$ *is a closed subspace of* $\mathrm{Reg}_s[a,b]$.

Consider now the bilinear mapping

$$\gamma : \mathrm{Reg}_s[a,b] \times \mathrm{BV}_0[a,b] \to \mathbb{C}, \qquad \gamma(f,g) := \int_a^b f(t)\, \mathrm{d}g(t).$$

Note that $|\gamma(f,g)| \le \|f\|_\infty \|g\|_v$ by part b) of Exercise 9.23, whence γ is a pairing.

Lemma 16.23. *The pairing* $\gamma : \mathrm{Reg}_s[a,b] \times \mathrm{BV}_0[a,b] \to \mathbb{C}$ *is a duality.*

Proof. Fix $g \in \mathrm{BV}_0[a,b]$ and pick a partition $a = t_0 < \cdots < t_n = b$. Find, for each $j = 1, \dots, m$ a number $c_j \in \mathbb{C}$ such that

$$|c_j| = 1 \quad \text{and} \quad c_j(g(t_j) - g(t_{j-1})) = |g(t_j) - g(t_{j-1})|\,.$$

Define the step function f in such a way that it is left continuous on $(a,b]$, right continuous at a and satisfies $f = c_j$ on (t_{j-1}, t_j). Then $\|f\|_\infty = 1$ and $\int_a^b f(t)\mathrm{d}g(t) = \sum_{j=1}^n |g(t_j) - g(t_{j-1})|$. By the definition of $\|g\|_v$, the claim follows. □

Theorem 16.24. *One has* $\mathrm{Reg}_s[a,b]' \cong \mathrm{BV}_0[a,b]$ *isometrically via the duality* $\gamma(f,g) = \int_a^b f\,\mathrm{d}g$.

Proof. Fix $\varphi \in \mathrm{Reg}_s[a,b]'$ and define

$$g(t) := \begin{cases} 0, & t = a, \\ \varphi(\mathbf{1}_{[a,t]}), & a < t \le b. \end{cases}$$

Then

$$g(t) - g(s) = \begin{cases} \varphi(\mathbf{1}_{[a,t]}), & a = s < t \le b, \\ \varphi(\mathbf{1}_{(s,t]}), & a < s < t \le b. \end{cases} \tag{16.12}$$

We first prove that $g \in \mathrm{BV}[a,b]$. To this end, let $a = t_0 < t_1 < \cdots < t_m = b$ be any partition of $[a,b]$. Find $c_j \in \mathbb{C}, |c_j| = 1$ with

$$c_j(g(t_j) - g(t_{j-1})) = |g(t_j) - g(t_{j-1})|$$

for all $1 \le j \le m$. The special step function f is defined by

$$f = c_1 \mathbf{1}_{[a,t_1]} + \sum\nolimits_{j=2}^{m} c_j \mathbf{1}_{(t_{j-1},t_j]}.$$

Then $\|f\|_\infty \le 1$ and

$$\sum\nolimits_{j=1}^{m} |g(t_j) - g(t_{j-1})| = \sum\nolimits_{j=1}^{m} c_j(g(t_j) - g(t_{j-1})) = \varphi(f),$$

whence $\sum_{j=1}^{m} |g(t_j) - g(t_{j-1})| \le \|\varphi\|$. It follows that $\|g\|_v \le \|\varphi\|$.

Next, we prove that $\varphi(f) = \int_a^b f\,\mathrm{d}g$ whenever f is a special step function. Such functions are linear combinations of functions of the form $\mathbf{1}_{(s,t]}$ and $\mathbf{1}_{[a,t]}$ with $a < s < t \le b$. But for $a \le s < t \le b$,

$$g(t) - g(s) = \int_a^b \mathbf{1}_{(s,t]}\,\mathrm{d}g = \int_a^b \mathbf{1}_{[s,t]}\mathrm{d}g,$$

hence the claim follows from (16.12) above.

Finally, $\varphi(f) = \int_a^b f\,\mathrm{d}g$ holds for all $f \in \mathrm{Reg}_s[a,b]$ by approximation. □

Corollary 16.25 (Riesz Representation Theorem). *Let* $\varphi \in \mathrm{C}[a,b]'$. *Then there is* $g \in \mathrm{BV}_0[a,b]$ *such that* $\|g\|_v = \|\varphi\|$ *and*

$$\varphi(f) = \int_a^b f(t)\,\mathrm{d}g(t) \quad \text{for all} \quad f \in \mathrm{C}[a,b].$$

Proof. By the Hahn–Banach theorem there is an extension of φ to a bounded linear functional ψ on $\mathrm{Reg}_s[a,b]$ with $\|\psi\| = \|\varphi\|$. By Theorem 16.24 we can represent ψ on $\mathrm{Reg}_s[a,b]$ by integration against a function $g \in \mathrm{BV}_0[a,b]$ with $\|g\|_v = \|\psi\|$. The claim follows. □

The Riesz representation theorem yields a representing function g for a given bounded functional φ on $\mathrm{C}[a,b]$. However, in order to determine g uniquely, we need an additional condition. Define by

$$\mathrm{BV}_0^r[a,b] := \{g \in \mathrm{BV}_0[a,b] \mid g(a) = 0 \text{ and } f(c) = f(c+) \; \forall\, c \in (a,b)\}$$

the space of function of bounded variation that are continuous from the right on (a,b) and vanish at $t = a$. Then we have the following result.

Theorem 16.26. *Via the canonical pairing,* $\mathrm{C}[a,b]' \cong \mathrm{BV}_0^r[a,b]$.

Proof. The proof is contained in Exercises 16.43–16.47. □

The Riesz representation theorem dates back to [**Rie10**] and is one of the major results from the early period of functional analysis. Later, with the help of measure-theoretic integration theory, it was considerably extended to a characterization of the dual space of $\mathrm{C}(K)$, where K is any compact topological space. This is the famous

Riesz–Kakutani[2] Theorem: *Let K be a compact topological space, and let $\mathrm{M}(K)$ be the space of complex regular Borel measures on K, endowed with the total variation norm. Then $\mathrm{C}(K)' \cong \mathrm{M}(K)$ via the duality*

$$(f, \mu) \mapsto \int_K f \,\mathrm{d}\mu.$$

Under this isomorphism positive measures correspond to positive functionals.

Proofs can be found in many textbooks on functional analysis; see, e.g., [**Lan93**, Chap. IX]. Lax's proof from [**Lax02**, Appendix A] is particularly interesting from a functional analytic perspective.

Remark 16.27. Let $g \in \mathrm{BV}_0^r[a,b]$ and $\mu \in \mathrm{M}[a,b]$ be the measure from the Riesz–Kakutani theorem satisfying

$$\int_a^b f \mathrm{d}g = \int_{[a,b]} f \,\mathrm{d}\mu \tag{16.13}$$

for all $f \in \mathrm{C}[a,b]$. Then $g(x) = \mu([a,x])$ for all $a < x \le b$, and identity (16.13) holds for all $f \in \mathrm{Reg}_s[a,b]$.

Exercises 16.48 and 16.49 describe how to retrieve integration against L^1-functions and point evaluation functionals from the description of $\mathrm{C}[a,b]'$ in Theorem 16.26. But on page 293 we claimed that these do not suffice to

[2]Shizuo Kakutani (1911–2004), Japanese-American mathematician.

describe all of $C[a,b]'$. In order to see this, for simplicity let $[a,b] = [0,1]$. We need to find a function $g \in BV_0^r[0,1]$ that cannot be written as

$$g = Jh + \sum_{0<c\le 1} \lambda_c \mathbf{1}_{[c,1]} + \lambda_0 \mathbf{1}_{(0,1]} \tag{16.14}$$

with $\sum_{c\in[0,1]} |\lambda_c| < \infty$ and $h \in L^1(0,1)$. An example is the so-called **Lebesgue singular function** [**Ran02**, Example 6.2.4 (iv)] or **Cantor function.** This function — let us denote it by g in the following — is continuous and monotonically increasing from 0 to 1 on $[0,1]$. In particular, it is contained in $BV_0^r[0,1]$. Moreover, since it is continuous, any representation (16.14) would imply that all the $\lambda_c = 0$. Hence it remains to show that $g = Jh$ cannot be true for any $h \in L^1(0,1)$, and this can be done by realizing that whenever $h \in L^1(0,1)$, then Jh is **absolutely continuous** on $[0,1]$; see Exercise 16.50.

Exercises 16A

Exercise 16.1. Show that for $E = C[a,b]$ with the supremum norm its dual space E' separates the points of E.

Exercise 16.2. Let E be a complex linear space.

a) Show that if $\varphi : E \to \mathbb{C}$ is $\mathbb{C}$-linear and $\psi := \operatorname{Re}\varphi$, then

$$\varphi(f) := \psi(f) - \mathrm{i}\psi(\mathrm{i}f) \qquad (f \in E)$$

and $\|\varphi\| = \|\psi\|$. [Hint: (A.1) is helpful.]

b) Show that if $\psi : E \to \mathbb{R}$ is $\mathbb{R}$-linear, then $\varphi : E \to \mathbb{C}$ defined by

$$\varphi(f) := \psi(f) - \mathrm{i}\psi(\mathrm{i}f) \qquad (f \in E)$$

is $\mathbb{C}$-linear with $\operatorname{Re}\varphi = \psi$.

Combine these facts to complete the proof of Theorem 16.2 in the case of complex scalars.

Exercise 16.3. Show that if E is a Hilbert space and $E_0 \subseteq E$ a closed subspace of E, then every $\varphi_0 \in E'$ has *precisely one* extension of equal norm to an element of E'. (Consequently, the two ways of extending a bounded linear functional from a closed subspace to the whole of a Hilbert space mentioned in 2) of Remark 16.3 amount to the same.)

Exercise 16.4. Consider the map

$$T : \ell^\infty \longrightarrow L^\infty(0,\infty) \qquad Tx := \sum\nolimits_{n\in\mathbb{N}} x_n \mathbf{1}_{(n,n+1)}.$$

Show that T is a linear isometry. Conclude that $L^\infty(0,\infty)$ is not separable. Prove the same for $L^\infty(a,b)$ for arbitrary $-\infty \le a < b \le +\infty$.

Exercise 16.5. Let E be a linear space and $\varphi : E \to \mathbb{K}$ a linear functional. Show that φ is surjective if and only if $\varphi \neq 0$. Show that in this case $\ker(\varphi)$ has codimension 1 in E, and that for any $c \in \mathbb{K}$ one has

$$[\varphi = c] = f + \ker(\varphi)$$

for every $f \in E$ such that $\varphi(f) = c$.

Exercise 16.6. Prove Lemma 16.7 and Corollary 16.8 under the assumption that $E = H$ is a Hilbert space and without using the Hahn–Banach theorem.

Exercise 16.7. Provide a proof of Theorem 16.10.

Exercise 16.8. Let E, F be normed spaces such that there exists an invertible mapping $T \in \mathcal{L}(E; F)$. Show that E is reflexive if and only if F is reflexive.

Exercise 16.9. Show that each Hilbert space and each finite dimensional normed space is reflexive.

Exercise 16.10. Let $(f_n)_{n\in\mathbb{N}}$ and $(g_n)_{n\in\mathbb{N}}$ be sequences in a normed space E and $\lambda \in \mathbb{K}$. Show: if $f, g \in E$ with $f_n \rightharpoonup f$ and $g_n \rightharpoonup g$, then $f_n + \lambda g_n \rightharpoonup f + \lambda g$.

Exercise 16.11. Let e_n, $n \in \mathbb{N}$, be the n-th standard unit vector in ℓ^2. Show that $e_n \rightharpoonup 0$ but $e_n \not\to 0$ as $n \to \infty$.

Exercise 16.12. Let $f_n := e_1 + \cdots + e_n$ where e_n is the n-th standard unit vector in c_0. Show that $(f_n)_{n\in\mathbb{N}}$ is a bounded sequence that does not have any weakly convergent subsequence.

Exercise 16.13. Let $\gamma : E \times F \to \mathbb{K}$ be a duality of the pair of *Banach* spaces (E, F) with induced map $\Gamma : F \to E'$. Show that $\Gamma(F)$ is a closed subspace of E'.

Exercise 16.14. Show that $\ell^{1'} \cong \ell^\infty$ via the canonical pairing (Theorem 16.17).

Exercise 16.15. Let $x \in [a, b]$. Show that there cannot be any function $g \in \mathrm{L}^1(a, b)$ such that

$$\int_a^b f(s)g(s)\,\mathrm{d}s = f(x) \quad \text{for all } f \in \mathrm{C}[a, b].$$

[Hint: Try functions f with $f(x) = 1$ and small support.]

Exercise 16.16. For $c \in (a, b)$ let $g \in \mathrm{BV}[a, b]$ be defined as

$$g(t) := \begin{cases} 0 & a \le t < c, \\ 1 & c < t \le b, \end{cases}$$

and $g(c) \in \mathbb{C}$ is arbitrary. Show that

$$\int_a^b f(t)\,\mathrm{d}g(t) = g(c)f(c-) + (1 - g(c))f(c+)$$

for all $f \in \mathrm{Reg}[a, b]$; cf. Exercise 9.23. What happens for $c = a$ or $c = b$?

Exercise 16.17. Show that each $f \in \mathrm{Reg}_s[a, b]$ is special (left continuous on $(a, b]$ and right continuous at a).

Exercise 16.18. Show that each continuous function $f \in \mathrm{C}[a, b]$ can be uniformly approximated by special step functions.

Exercises 16B

Exercise 16.19. Show that the space $\mathrm{C}_0(\mathbb{R})$ is separable.

Exercise 16.20. Show that the spaces $\mathrm{L}^p(\mathbb{R})$, $1 \le p < \infty$ are separable.

Exercise 16.21. Show that the spaces $\mathrm{H}^k(a,b)$, $k \in \mathbb{N}$, are separable.

Exercise 16.22. Show that the space $\mathrm{C}_\mathrm{b}(\mathbb{R})$ is not separable.

Exercise 16.23. Let E be a normed space such that its dual space E' is separable. Show that E is itself separable. [Hint: let $\{\varphi_n \mid n \in \mathbb{N}\}$ be fundamental in E'; pick for every $n \in \mathbb{N}$ a sequence $(f_{n,m})_m$ in E with $\|f_m\| \le 1$ and $|\varphi_n(f_{n,m})| \to \|\varphi_n\|$ as $m \to \infty$; then show that $A := \{f_{n,m} \mid n,m \in \mathbb{N}\}$ is fundamental in E.]

Exercise 16.24. For a subset $B \subseteq E'$, E any Banach space, the **pre-annihilator** is defined by

$$B^\top := \{f \in E \mid \varphi(f) = 0 \text{ for all } \varphi \in B\}.$$

Show that for $A \subseteq E$ one has

$$(A^\perp)^\top = \overline{\operatorname{span}}(A).$$

Exercise 16.25. Let E, F be normed spaces, and let $T : E \to F$ be a bounded linear operator. Show that

$$\overline{\operatorname{ran}}(T) = (\ker T')^\top;$$

cf. Exercise 16.24.

Exercise 16.26. Let $(f_n)_{n\in\mathbb{N}}$ be sequence in a Banach space E, and let $A \subseteq E'$ be a fundamental subset of E'. Show that for $f \in E$ the following two assertions are equivalent:

(i) $f_n \rightharpoonup f$ as $n \to \infty$.

(ii) $\sup_{n\in\mathbb{N}} \|f_n\| < \infty$ and $\varphi(f_n) \to \varphi(f)$ as $n \to \infty$ for each $\varphi \in A$.

Exercise 16.27. Show that $\mathrm{L}^1(X)' \cong \mathrm{L}^\infty(X)$ via the canonical pairing if X is an unbounded interval. Depending on your knowledge of measure theory, prove the same statement if X is an arbitrary measurable subset of $\mathbb{R}^d$, or even an arbitrary σ-finite measure space.

Exercise 16.28 (Moment Operator). Consider the operator

$$T : \mathrm{L}^1(0,1) \longrightarrow c_0, \qquad Tf(n) = \int_0^1 f(s)s^n \,\mathrm{d}s \quad (n \in \mathbb{N}_0);$$

cf. Exercise 9.21. Compute the dual operator $T' : \ell^1 \to \mathrm{L}^\infty(0,1)$.

Exercise 16.29. Show that $\gamma : \mathrm{C}[a,b] \times \big(\mathrm{L}^1(a,b) \oplus \ell^1[a,b]\big) \to \mathbb{C}$, defined by

$$\gamma(f, g \oplus \alpha) := \int_a^b fg \,\mathrm{d}\lambda + \sum_{x\in[a,b]} \alpha_x f(x)$$

is a duality, where on the right factor we take the norm

$$\|g \oplus \alpha\| := \|g\|_{\mathrm{L}^1} + \|\alpha\|_{\ell^1} := \|g\|_1 + \sum_{x\in[a,b]} |\alpha_x|\,.$$

Exercises 16C

Exercise 16.30. Prove the following complex version of Theorem 16.4: *Let E be a complex linear space and let $p : E \to \mathbb{R}$ be subadditive with*

$$p(\lambda f) = |\lambda|\, p(f) \quad \textit{for all } \lambda \in \mathbb{C},\ f \in E.$$

Let $E_0 \subseteq E$ be a subspace and let $\varphi_0 : E_0 \to \mathbb{R}$ be a complex-linear functional such that $|\varphi_0| \le p$ on E_0. Then there is an extension of φ_0 to a complex-linear functional φ on E with $|\varphi| \le p$ on E.

Exercise 16.31. Show that $\mathrm{L}^p(X)' \cong \mathrm{L}^q(X)$ via the canonical pairing if X is an unbounded interval, $1 < p \le 2$ and q is the dual exponent. Depending on your knowledge of measure theory, prove the same statement if X is an arbitrary measurable subset of $\mathbb{R}^d$, or even an arbitrary (σ-finite) measure space.

Exercise 16.32. Let $1 < p < \infty$ and $1/p + 1/q = 1$. Show that $\ell^{p\prime} \cong \ell^q$ via the canonical pairing.

Exercise 16.33. Suppose E, F are Banach spaces, and $E' \cong F$ via a duality $\gamma : E \times F \to \mathbb{K}$. Show that then also

$$\gamma^r : F \times E \longrightarrow \mathbb{K}, \qquad \gamma^r(g, f) := \gamma(f, g)$$

is a duality, and E is reflexive if and only if $F' \cong E$ via γ^r.

Exercise 16.34. Show that the spaces c_0, ℓ^1 and ℓ^∞ are not reflexive.

Exercise 16.35. Show that for any interval $X \subseteq \mathbb{R}$ the space $\mathrm{L}^\infty(X)$ is not reflexive. [Hint: Exercises 16.4 and 16.34.]

Exercise 16.36. Show that for any interval $X \subseteq \mathbb{R}$ the space $\mathrm{L}^1(X)$ is not reflexive. [Hint: Exercise 16.35.]

Exercise 16.37. Show that each space ℓ^p with $1 < p < \infty$ is reflexive. [Hint: Exercise 16.33 and Theorem 16.21.]

Exercise 16.38. Show, using Theorem 16.19 that $\mathrm{L}^p(a, b)$ is reflexive if $1 < p < \infty$. (This is actually true for general measure spaces.)

Exercise 16.39. Show that the space $\mathrm{C}[a, b]$ is not reflexive. [Hint: Use Exercise 16.36 and Lemma 16.12.]

Exercise 16.40. Prove Theorem 16.15, i.e., that *every norm-bounded sequence in a reflexive Banach space is weakly convergent.* [Plan: Start with a sequence $(f_n)_{n\in\mathbb{N}}$ in the unit ball of a reflexive space E; use Theorem 16.14.b) and Lemma 16.12.a) to reduce to the case that $(f_n)_{n\in\mathbb{N}}$ is fundamental in E. Use the reflexivity and Exercise 16.23 to conclude that E' is separable and let $(\varphi_m)_{m\in\mathbb{N}}$ be a fundamental sequence in E'. Now employ a *diagonal argument* to find a subsequence $(g_n)_{n\in\mathbb{N}}$ of $(f_n)_{n\in\mathbb{N}}$ such that $l(\varphi_m) := \lim_{n\to\infty} \varphi_m(g_n)$ exists for each $m \in \mathbb{N}$. Conclude that

$$l(\varphi) := \lim_{n\to\infty} \varphi(g_n)$$

exists for all $\varphi \in E'$ and that $l \in E''$. Finally employ the reflexivity of E again.]

Exercise 16.41. Let $f : [a, b] \to \mathbb{C}$ be a function such that the left and right limits $f(c+)$ and $f(c-)$ exist for all $c \in [a, b]$ whenever this is meaningful. Show that f is regulated; cf. [**Die69**, Sec. 7.6]

Exercise 16.42. Show that each special regulated function (left continuous on $(a, b]$ and right continuous at a) can be uniformly approximated by special step functions, i.e., is contained in $\mathrm{Reg}_s[a, b]$. [Hint: By Exercise 16.41 obtain an approximating sequence of step functions in the first place. Then modify these functions to make them special.]

Exercise 16.43. Let $g \in \mathrm{BV}[a, b]$ and define $V_g(t) := \left\| g|_{[a,t]} \right\|_v$, the variation of g on $[a, t]$, for $t \in [a, b]$.

a) Show that V_g is an increasing function on $[a, b]$.
b) Show that if g is right (left) continuous at $t_0 \in [a, b)$ ($t_0 \in (a, b]$) then also V_g is right (left) continuous at t_0.

Exercise 16.44. Let $g \in \mathrm{BV}_0[a, b]$ and suppose that g is right continuous on (a, b). Show that

$$\|g\|_v = \sup\Big\{ \Big| \int_a^b f \mathrm{d}g \Big| \mid f \in \mathrm{C}[a, b],\ \|f\|_\infty \le 1 \Big\}.$$

[Hint: Start with a partition $a = t_0 < \cdots < t_m = b$, find for each $j = 1, \dots, m$ a number $c_j \in \mathbb{C}$ such that $|c_j| = 1$ and $c_j(g(t_j) - g(t_{j-1})) = |g(t_j) - g(t_{j-1})|$. Choose $s_j \in (t_j, t_{j+1})$ close to t_j and define a piecewise linear function $f \in \mathrm{PL}[a, b]$ such that $\|f\|_\infty \le 1$ and

$$f = \begin{cases} c_1 & \text{on} \quad [a, t_1], \\ c_j & \text{on} \quad [s_j, t_{j+1}] \quad (t = 1, \dots, m-1). \end{cases}$$

Use Exercise 16.43 to show that the integrals $\int_{t_j}^{s_j} f \, \mathrm{d}g$ can be made arbitrarily small as s_j is sufficently close to t_j. Conclude that the integral $\int_a^b f \, \mathrm{d}g$ can be made arbitrarily close to $\sum_{j=1}^m |g(t_j) - g(t_{j-1})|$.]

Exercise 16.45. Let $g \in \mathrm{BV}_0[a, b]$.

a) Show that the right limit $g(c+)$ and the left limit $g(c-)$ exist for $c \in [a, b]$ whenever they are meaningful.
b) Show that

$$\sum_{c \in [a,b)} |g(c+) - g(c)| + \sum_{c \in (a,b]} |g(c) - g(c-)| \le \|g\|_v \,.$$

c) Show that

$$\tilde{g} := g + \sum_{c \in (a,b)} (g(c+) - g(c)) \mathbf{1}_{\{c\}}$$

is a well-defined function in $\mathrm{BV}_0[a, b]$, right continuous on (a, b) and such that

$$\int_a^b f \, \mathrm{d}\tilde{g} = \int_a^b f \, \mathrm{d}g \quad \text{for all } f \in \mathrm{C}[a, b].$$

Exercise 16.46 (The Dual of $\mathrm{C}[a, b]$)**.** Prove Theorem 16.26, i.e., that

$$\mathrm{C}[a, b]' \cong \{ g \in \mathrm{BV}_0[a, b] \mid g \text{ right continuous on } (a, b) \}$$

via the pairing $\gamma(f, g) = \int_a^b f \, \mathrm{d}g$.

Exercise 16.47. A linear functional $\varphi \in \mathrm{C}[a,b]'$ is called **positive**, if $f \geq 0$ implies that $\varphi(f) \geq 0$.

a) Show that if φ is positive, then $|\varphi(f)| \leq \varphi(|f|)$ for all $f \in \mathrm{C}[a,b]$. [Hint: Imitate the proof of the triangle inequality for integrals, see Lemma 7.8.]

b) Show that if φ is positive, then $\|\varphi\| = \varphi(\mathbf{1})$.

c) Suppose that φ is positive and $g \in \mathrm{BV}_0[a,b]$ is a representing function. Show that g is increasing. [Hint: Think about in which cases one can have equality in the triangle inequality for complex scalars.]

Exercise 16.48. Let $h \in \mathrm{L}^1(a,b)$ and let $J : \mathrm{L}^1(a,b) \to \mathrm{C}[a,b]$ be the integration operator $Jh(t) := \int_a^b h(s)\,\mathrm{d}s$. Show that

$$\int_a^b f\,\mathrm{d}(Jh) = \int_a^b f\,h\,\mathrm{d}\lambda$$

for all $f \in \mathrm{Reg}[a,b]$.

Exercise 16.49. Given $c \in [a,b]$, for which function $g \in \mathrm{BV}_0[a,b]$ do we have

$$\int_a^b f\,\mathrm{d}g = g(c+) \quad \text{for all } f \in \mathrm{Reg}_s[a,b]?$$

For which function $g \in \mathrm{BV}_0[a,b]$ do we have

$$\int_a^b f\,\mathrm{d}g = g(c) \quad \text{for all } f \in \mathrm{Reg}_s[a,b]?$$

Exercise 16.50. A function $f : [a,b] \to \mathbb{C}$ is **absolutely continuous** if for each $\epsilon > 0$ there is $\delta > 0$ such that for all choices of $n \in \mathbb{N}$ and $a \leq a_1 < b_1 < a_2 < b_2 < \cdots < a_n < b_n \leq b$ one has

$$\sum_{j=1}^n (b_j - a_j) < \delta \quad \Longrightarrow \quad \sum_{j=1}^n |f(b_j) - f(a_j)| < \epsilon.$$

Show that each absolutely continuous function is continuous, and that every function of the form $f(t) = \int_a^t h(s)\,\mathrm{d}s$ with $h \in \mathrm{L}^1(a,b)$, is absolutely continuous. Then look up, e.g., in [**Ran02**, pp. 181–184], the construction of the Lebesgue singular function and the proof that it is continuous and increasing but not absolutely continuous.

Historical Remarks

The history of functional analysis is, contrary to its subject matter, highly nonlinear. We give a short account based on Siegmund-Schultze's excellent article [**SS03a**] and, occasionally, on the more detailed books [**Mon73**] and [**Die81**]. References for the original sources can be found in these works.

Surprisingly, what comes first in the systematic development of the theory — abstract linear algebra — was historically the last ingredient to be developed. The axioms of an abstract vector space were for the first time described in Giuseppe Peano's[1] book *Calcolo Geometrico* (1888). Peano was influenced by ideas from the work *Ausdehnungslehre* (1844) of the genial but for a long time neglected German mathematician-philosopher-linguist Hermann Grassmann.[2] Interesting enough, also Peano's axiomatization remained without much influence for about 30 years until it was rediscovered and incorporated into the modern concept of a normed linear space around 1920 by Stefan Banach.[3] So it came to be that a purely algebraic concept received wide acceptance only after having proved valuable in an analytic context.

In the year 1932, three books appeared marking the end of what could be termed the childhood and adolescence of functional analysis: *Linear Transformations on Hilbert Space* by Marshall H. Stone,[4] *Mathematische Grundlagen der Quantenmechanik* by John von Neumann,[5] and *Théorie des*

[1] Giuseppe Peano (1858–1932), Italian mathematician.

[2] Hermann Günther Grassmann (1809–1877), German mathematician, philosopher and linguist.

[3] Stefan Banach (1892–1945), Polish mathematician.

[4] Marshall H. Stone (1903–1989), American mathematician.

[5] John von Neumann, orig. János Lajos Neumann (1903–1957), Hungarian-American Mathematician.

Opérations Linéaires by Banach. Although von Neumann's succesful Hilbert space operator formulation of quantum mechanics helped much to convince the mathematical community of the importance and usefulness of functional analysis, it was Banach's book that established functional analysis as an abstract theory and as an independent mathematical discipline.

Banach's book was the result of a period of intense research, starting with Banach's doctoral dissertation (completed 1920 and published in 1922), where for the first time the abstract definition of a normed linear space and of what would later be called a Banach space appears. This reseach was mainly carried out by Banach and his colleages (Hugo Steinhaus,[6] Stanisław Mazur,[7] Stanisław Ulam[8] and others) in the eastern Polish town Lwów (now Lviv, Ukraine). Much of the fundamental corpus of functional analysis was created or put into a final form in these years between 1920 and 1932. However, this was only the last step of a coming-of-age process that has its roots back in the final decades of the 19th century.

As Siegmund-Schultze writes, functional analysis has mainly two origins. "On the one hand, there were concrete applied problems, especially the Dirichlet problem of potential theory, which gave birth to various notions of generalized analysis via the calculus of variations and the theory of integral equations. On the other hand, there was the search for a unifying, generalizing point of view which is inherent in mathematics ..." [**SS03a**, p. 386].

This strive for generalization was indeed particularly strong in the period after 1880 and changed the face of mathematics profoundly. Set theory, originated by Georg Cantor[9] and Richard Dedekind[10] in the 1870s, facilitated the "construction" of higher order mathematical objects, like functions of functions. A new, permissive, attitude towards actually infinite collections of objects was about to replace the traditional restriction to the potentially infinite, one of the remnants of Aristotelian epistemology (cf. Section A.5). In his little booklet *Was sind und was sollen die Zahlen?* (What are numbers and what are they good for?) from 1888, Dedekind called the numbers a "free creation of the human mind". This marks a decisive emancipation of mathematics from physics, equally present at the same time in many other areas of mathematics, like in the discussions about non-Euclidean geometries.

---

[6]Władysław Hugo Steinhaus (1887–1972), Polish mathematician.
[7]Stanisław Mazur (1905–1981), Polish mathematician.
[8]Stanisław Ulam (1909–1984), Polish-American mathematician.
[9]Georg Cantor (1845–1918), German mathematician.
[10]Richard Dedekind (1831–1916), German mathematician.

David Hilbert's[11] revival and re-interpretation of the axiomatic method beginning with his groundbreaking work *Grundlagen der Geometrie* (1899) was as much a consequence as it was a catalyst for this development towards generalization and abstraction. It led in particular to exploring more and more abstract and general notions of "space". In his doctoral dissertation from 1906 the Frenchman Maurice Fréchet[12] introduced the concept of a metric space (the term itself is due to Felix Hausdorff[13]). Hausdorff, in his famous book *Grundzüge der Mengenlehre* (1914) coined the even more abstract notion of a "topological space".

Functional analysis, as we see it nowadays, emerges when one applies these abstract, but essentially geometric, notions of space to the objects of analysis, i.e., to *functions*. To be able to view a function as a single mathematical object is already one of the fruits of the paradigm change associated with the advent of set theory. However, to view it as a "point" in a "space" is a different thing and maybe even more daring. Fréchet in 1906 clearly had adopted this view — one of his prime examples is the space $\mathrm{C}[a,b]$ of continuous functions on an interval, with metric $d(f,g) = \sup_{x\in[a,b]} |f(x) - g(x)|$ — but it would still need years in order to experience wide acceptance.

One of the most important impulses for functional analysis came from the field of "integral equations" (the name was coined by Paul du Bois-Reymond[14]). Vito Volterra,[15] in a series of publications starting in 1887, considers "functions that depend on other functions". He observes that the value of the expression

$$I = \int_a^b \varphi(x)\,\mathrm{d}x$$

depends on the whole "line" (=graph) $y = \varphi(x)$, and hence on infinitely many parameters. Volterra studied integral equations of the form

$$x(s) + \int_a^b K(s,t)x(t)\,\mathrm{d}t = f(s) \tag{1}$$

where x, f and K are continuous. His main method consisted of approximating the integral by finite sums and hence considering the integral equations as a limit of finite systems of linear equations. In a paper of 1896 he uses the "Neumann series" to solve these equations in the case that $b = s$ is variable. (This is exactly what we did in Corollary 11.15 on "Volterra integral equations".)

[11]David Hilbert (1862–1943), German mathematician.
[12]Maurice René Fréchet (1878–1973).
[13]Felix Hausdorff (1868–1942), German mathematician.
[14]Paul Du Bois-Reymond (1831–1889), German mathematician.
[15]Vito Volterra (1860–1940), Italian mathematician.

Paul Lévy[16] — who seems to be responsible for the name "functional analysis" through the title of his book *Leçons d'analyse fonctionelle* (1922) — continued Volterra's work. It also had a strong influence on Ivar Fredholm,[17] who in 1903 developed a theory of integral equations culminating in his famous "Fredholm alternative". (We only touched upon a special case in Theorem 13.13.) Fredholm views (1) as a "transformation" of a function, whence his work can be regarded as the beginning of operator theory in function spaces [**SS03a**, p. 393].

The study of integral equations was continued by Hilbert in six famous *Mitteilungen* (communications) in a local journal of the University of Göttingen in the years 1904–1910. In the opinion of Siegmund-Schultze, Hilbert was "perhaps the most important early functional analyst" [**SS03a**, p. 394]. Hilbert employs the analogy with eigenvalue theory of matrices and speaks of eigenfunctions and eigenvalues of an integral equation. He gives conditions that a function can be expanded into an infinite linear combination of eigenfunctions, notes their orthogonality (modern terminology) and the analogy to Fourier series. In effect, Hilbert establishes the spectral theory of certain symmetric operators in analogy with the matrix case (principal axes theorem). It is here, where real sequences $(x_n)_{n\in\mathbb{N}}$ with the property $\sum_{n=1}^{\infty} x_n^2 < \infty$ feature prominently.

However, Hilbert did not consider the set ℓ^2 (modern notation) as a "space". Also, like Volterra, Hilbert approached integral equations by virtue of discretization. It was Hilbert's doctoral student Erhard Schmidt[18] who in his dissertation (1905) promoted an entirely different approach. Schmidt put forward the *structural* features of the situation, in particular the inner product and the linear structure, and developed the "Hilbert–Schmidt theory" in a quasi-axiomatic way. In a paper from 1908 he "introduced geometric language (projection, decomposition, orthogonality, scalar product) into the "Hilbert space" ℓ^2 [**SS03a**, p. 399]. It is through Schmidt's work that the abstract structure of a "Hilbert space" appeared, although a fully abstract axiomatic description of it would only be given by von Neumann in 1927. In that framework, the Hilbert–Schmidt spectral theory then quickly received the form it has today, the spectral theorem for compact self-adjoint operators (Theorem 13.11).

It was the Hungarian Frigyes Riesz,[19] travelling to and fro between Paris and Göttingen, who took the next step. The French mathematician Henri Lebesgue[20] had just created his notion of integral in his doctoral dissertation

[16]Paul Lévy (1886–1971), French mathematician.
[17]Erik Ivar Fredholm (1866–1927), Swedish mathematician.
[18]Erhardt Schmidt (1876–1959), German mathematician.
[19]Frigyes Riesz (1880–1956), Hungarian mathematician.
[20]Henri Léon Lebesgue (1875–1941), French mathematician.

from 1902, and it was a straightforward problem to ask whether the Hilbert–Schmidt results could be extended by means of the new integral and the new classes of functions associated with it. Riesz noted in 1906 that the square mean distance $\left(\int_a^b (f-g)^2\right)^{1/2}$ meets the abstract axioms of a metric in the sense of Fréchet, and using this notion of convergence he extended the Hilbert–Schmidt theory to bounded measurable functions. In 1907 he and the Austrian Ernst Fischer[21] proved what is now called the "Riesz–Fischer theorem", namely the isometric isomorphism of $\ell^2 \cong \mathrm{L}^2(a,b)$ and hence the completeness of the space L^2.

Riesz also made decisive contributions to the representation theory of *functionals*, i.e., Volterra's "functions of functions". The name "functional" in its modern meaning appears for the first time in a paper (1906) of Jacques Hadamard.[22] He proved a first representation theorem (cf. Section 16.3) for continuous linear functionals on $\mathrm{C}[a,b]$. This was motivated by the so-called "moment problem" from probability theory, where one asks whether a probability distribution is characterized completely by its sequence of moments. Subsequently, Fréchet (who was Hadamard's doctoral student) considered functionals on other "function spaces". In 1907, he and Riesz independently proved the "Riesz–Fréchet theorem" for $\mathrm{L}^2(a,b)$ (Theorem 8.12). In 1909, Riesz improved Hadamard's result and established his representation theorem for functionals on $\mathrm{C}[a,b]$ (Theorem 16.25) employing Riemann–Stieltjes[23] integrals. In 1910 he defined L^p-spaces for p different from 1, 2, ∞ and extended the Riesz–Fréchet theorem to them (Theorem 16.19).

Up to about 1910 and with only a few exceptions (e.g., Fredholm), functionals and matrices were the only kind of operators under consideration. This changed with the work of Riesz in the years 1910–1913. He studied "completely continuous" transformations — nowadays called compact operators (Definition 12.5) — on L^p-spaces, and developed a spectral theory for them. It seems that in this study, Riesz came very close to the abstract notion of a complete normed space [**SS03a**, p. 402].

However, it was Banach in close cooperation with Steinhaus — who had been a doctoral student of Hilbert in the years 1906–1911 — to take this final step. In the years to come, operators (transformations, linear operations) became a central object of study. The contraction principle (Theorem 6.1), motivated by the Picard iteration method for solving differential equations, is due to Banach in his dissertation (1922). A first version of the uniform boundedness principle (Theorem 15.4) is also contained there, the final version was established by Banach and Steinhaus (1927) by employing

[21]Ernst Sigismund Fischer (1875–1954), Austrian mathematician.

[22]Jacques-Salomon Hadamard (1865–1963), French mathematician.

[23]Thomas Joannes Stieltjes (1856–1894), Dutch mathematician.

methods from René Baire's[24] dissertation (1899). The open mapping theorem (Theorem 15.8) appears for the first time in Banach's book (1932), and the same is true for the general Hahn–Banach theorem (Theorem 16.4), although previous versions were proved by Hans Hahn (1927) and Banach (1929).

Therefore, one may say that much of the abstract theoretical body of functional analysis (and certainly that part of it that is treated in our main text) was well established by 1932 when the three books, mentioned above, appeared. The next step in the development of the theory is marked by the epoch-making three volume set *Theory of Linear Operators* (1956) by Nelson Dunford[25] and Jacob T. Schwartz.[26] For the younger history of Banach space theory and linear operators we refer to Pietsch's book [**Pie07**].

The term "Sobolev space" derives from Sergei Sobolev,[27] who studied these spaces extensively during the 1930s. However, as Naumann points out in [**Nau02**], functions with weak derivatives in L^2 were already considered by Beppo Levi[28] in his paper *Sul principio di Dirichlet* from 1906. The elegant application of Hilbert space methods (orthogonal projection, Riesz–Fréchet theorem) to the Dirichlet and the Poisson problem goes back to Hermann Weyl[29] [**Wey40**]. In 1954 Peter Lax[30] and Arthur Milgram[31] generalized the Riesz–Fréchet theorem to the "Lax–Milgram theorem" (Section 12.3) and applied it to more general partial differential equations [**LM54**].

[24]René-Louis Baire (1874–1932), French mathematician.
[25]Nelson James Dunford (1906–1986), American mathematician.
[26]Jacob Theodore Schwartz (1930–2009), American mathematician.
[27]Sergei Lvovich Sobolev (1908–1989), Russian mathematician.
[28]Beppo Levi (1875–1961), Italian mathematician.
[29]Hermann Klaus Hugo Weyl (1885–1955), German mathematician.
[30]Peter David Lax (1926–), Hungarian-American mathematician, Abel Prize 2005.
[31]Arthur Norton Milgram (1912–1961), American mathematician.

Appendix A

Background

We denote by $\mathbb{N} := \{1, 2, 3, \dots\}$ the set of natural numbers and by $\mathbb{N}_0 := \mathbb{N} \cup \{0\} = \{0, 1, 2, \dots\}$ the natural numbers with 0 adjoined. The set of integers is denoted by $\mathbb{Z} := \{\dots, -1, 0, 1, 2, \dots\}$ and the field of rational numbers by $\mathbb{Q} := \{n/m \mid n, m \in \mathbb{Z}, m \neq 0\}$. The symbols $\mathbb{R}$ and $\mathbb{C}$ denote the fields of real and complex numbers, respectively (see Sections A.5 and A.6 below).

A.1. Sequences and Subsequences

Let X be an arbitrary nonempty set. A **sequence in** X is a mapping

$$x : \mathbb{N} \longrightarrow X.$$

If x is a sequence in X then one often writes

$$x_n \quad \text{instead of} \quad x(n)$$

for the n-th member and $x = (x_n)_{n\in\mathbb{N}}$ to denote the whole sequence.

Note the difference between

$$(x_n)_{n\in\mathbb{N}} \quad \text{and} \quad \{x_n \mid n \in \mathbb{N}\}.$$

The first denotes the sequence, the second denotes the *range* of the sequence. For example, the range of the sequence $x_n := (-1)^n$, $n \in \mathbb{N}$, is

$$\{(-1)^n \mid n \in \mathbb{N}\} = \{1, -1\}.$$

By abuse of notation one sometimes writes "$(x_n)_{n\in\mathbb{N}} \subseteq \mathbb{N}$" in place of "$\{x_n \mid n \in \mathbb{N}\} \subseteq X$".

Note also the difference between "x_n" and "$(x_n)_{n\in\mathbb{N}}$". The first denotes the n-th member of the sequence and is an element of X, the second denotes the sequence as a whole, and is an element of the set of functions from $\mathbb{N}$ to X.

If $x = (x_n)_{n\in\mathbb{N}}$ is a sequence in X, then a **subsequence** of x is a sequence of the form

$$y_n := x_{\pi(n)} \qquad (n \in \mathbb{N})$$

where $\pi : \mathbb{N} \to \mathbb{N}$ is a *strictly increasing* map. Intuitively, π selects certain members of the original sequence with increasing indices. One sometimes writes n_k instead of $n_{\pi(k)}$.

A.2. Equivalence Relations

An equivalence relation is the mathematical model of a fundamental operation of the human mind: neglecting differences between objects and identifying them when they share certain properties. For example, the box office of your local theatre will probably charge different prices for children and adults, with a strict definition separating the two classes for this very purpose.

Mathematically, an **equivalence relation** on a set X is a binary relation $\sim$ on X satisfying the following three axioms for all x, y, $z \in X$:

1) $x \sim x$ **(reflexivity)**,
2) $x \sim y \;\;\rightarrow\;\; y \sim x$ **(symmetry)**,
3) $x \sim y,\ y \sim z \;\;\rightarrow\;\; x \sim z$ **(transitivity)**.

If $\sim$ is an equivalence relation on a set X, then to each $x \in X$ one can form its **equivalence class**

$$[x] := [x]_\sim := \{y \in X \mid x \sim y\},$$

the set of all elements of X that are equivalent to x. Two such classes are either equal or disjoint: $[x] = [a]$ if and only if $x \sim a$. All equivalence classes are collected in a new set

$$X/\!\sim\; := \{[x] \mid x \in X\}.$$

One speaks of **dividing** the set X by the equivalence relation.

Suppose one has a binary operation $*$ on X such that if $x \sim a$ and $y \sim b$, then $x * y \sim a * b$. (One says that the operation $*$ is **compatible** with the equivalence relation $\sim$.) Then one can **induce** this operation on $X/\sim$ by defining

$$[x] * [y] := [x * y].$$

By hypothesis, this definition does not depend on the choice of representatives, hence is a good definition. One says that the operation $*$ on $X/\sim$ is **well-defined**.

This mechanism works with functions of more variables and with relations in general. In this way structural elements (relations, operations) are transported ("induced") on the new set $X/\sim$.

The standard examples are found in algebra. For instance, if U is a subgroup of an Abelian group G, then one defines

$$x \sim_U y \quad :\Longleftrightarrow \quad x - y \in U$$

for $x, y \in G$. This is an equivalence relation on G (check it!). Let us denote the equivalence class containing $x \in G$ by $[x]$, as above. Then $[x] = [y]$ if and only if $x - y \in G$ and $[x] = x + U$ as sets. The set of equivalence classes

$$G/U := \{[x] \mid x \in G\}$$

is called the **factor group** or **quotient group**. It is itself an Abelian group with respect to the operation

$$[x] + [y] := [x + y] \qquad (x, y \in G).$$

(Of course one has to check that these are well-defined, i.e., that the sum is compatible with the equivalence relation.) The mapping

$$s : G \longrightarrow G/U, \quad sx := [x] \quad (x \in G)$$

is then a group homomorphism is called the **canonical surjection**.

A typical example occurs when $G = \mathbb{Z}$ and $U = n\mathbb{Z}$ for some natural number $n \in \mathbb{N}$. Other examples are quotient spaces of vector space (see below).

System of Representatives. Let $\sim$ be an equivalence relation on a nonempty set X. Each member of an equivalence class is called a **representative** for it. So each $y \in [x]$ is a representative for $[x]$.

A mapping $r : X/\sim \to X$ assigning to each equivalence class a representative for it, i.e., such that $[r(t)] = t$ for all $t \in X/\sim$, is called a **system of representatives**. For example, if one fixes a natural number $n \in \mathbb{N}$, then one can consider the quotient group $\mathbb{Z}/n\mathbb{Z}$ and $\{0, \ldots, n-1\}$ would be a system of representatives for it.

An important theoretical tool in mathematics is the following set-theoretic axiom.

Axiom of Choice. *For each equivalence relation on a nonempty set X there exists a system of representatives.*

The Axiom of Choice is a theoretical tool that helps in proofs and the "creation" of mathematical objects. Its strength lies in the fact that it does not presuppose a "constructive" way in which the equivalence relation is given; its weakness lies in the fact that it does not give a "constructive" description of the new object either.

Because of this weakness, the use of the Axiom of Choice is considered as doubtful by some people. This debate takes us right into the center of the philosophy of mathematics with its speculations about the meaning of 'existence' of mathematical objects and the 'foundations' of mathematical reasoning. My own position here is pragmatic: the Axiom of Choice is very useful since a mathematical formalization of the intuitive notion of "constructive" is clumsy and awkward, if not impossible. And in a "concrete" or "constructively given" situation, the Axiom of Choice is superfluous. It is a theoretical tool that helps us to avoid making precise what "concrete" or "constructively given" should mean in the first place.

A.3. Ordered Sets

A **partial ordering** of a set X is a binary relation $\leq$ on X such that the following three axioms are satisfied for all $x, y, z \in X$:

1) $x \leq x$ **(reflexivity)**,

2) $x \leq y,\ y \leq x \quad \to \quad x = y$ **(antisymmetry)**,

3) $x \leq y,\ y \leq z \quad \to \quad x \leq z$ **(transitivity)**.

A **(partially) ordered set** is a pair $(X, \leq)$ where X is a set and $\leq$ is a partial ordering of X. If one has in addition that

4) $x \leq y \quad \text{or} \quad y \leq x$

for all $x, y \in X$, then $(X, \leq)$ is called a **totally ordered set**.

We use $\leq$ as a generic symbol for an ordering, but in concrete situations other symbols may be used.

Examples A.1. a) If Ω is a set, then $X := \mathcal{P}(\Omega)$ (the power set of Ω) is ordered either by set inclusion $\subseteq$ or by set containment $\supseteq$. These orderings are not total if Ω has at least two points.

b) $X := \mathbb{N}$ is ordered by $n|m$ (meaning that m is divisible by n).

c) $\mathbb{R}$ is totally ordered by the usual $\leq$.

d) $\mathbb{R}^d$ is ordered by $x \leq y$ being defined as $x_j \leq y_j$ for all $j = 1, \ldots, d$.

e) If $(X, \leq)$ is an ordered set, then $x \leq_r y$ defined by $y \leq x$ and usually written $x \geq y$, defines a new ordering on X, called the **reverse ordering**.

f) If $(X, \leq)$ is an ordered set and $A \subseteq X$, then A is also ordered by restriction of the ordering of X to A. For example, $\mathbb{N}$ is ordered by the ordering coming from $\mathbb{R}$.

Let $(X, \leq)$ be an ordered set, let $A \subseteq X$ and $x \in X$. We write

$$A \leq x \quad \text{to abbreviate} \quad \forall\, a \in A : a \leq x.$$

In this case x is called an **upper bound** for A. And $x \in X$ is called a **lower bound** for A if $x \leq a$ for all $a \in A$, abbreviated by $x \leq A$. A **greatest** element of A is an upper bound for A that, moreover, *belongs to* A. Analogously, a **least** element of A is a lower bound for A that belongs to A.

Greatest (or least) elements may not exist. However, if they exist, they are unique: if $a, b \in A$ are both greatest elements of A, then $a \leq b$ and $b \leq a$, so $a = b$ by antisymmetry.

The **supremum**, or the **least upper bound**, of A is the least element of the set of all upper bounds for A, provided such an element exists. In the case that it exists, it is denoted by $\sup A$ and it is characterized uniquely by the properties:

$$A \leq \sup A \quad \text{and} \quad \forall x \in X : A \leq x \to \sup A \leq x.$$

Analogously, the greatest element of the set of all lower bounds of A, if it exists, is called an **infimum** of A.

An element $a \in A$ is called **maximal** in A if no other element of A is strictly greater, i.e., if it satisfies

$$\forall x \in A : (\, a \leq x \implies a = x\,).$$

Analogously, $a \in A$ is called **minimal** in A if no other element of A is strictly smaller. Maximal (minimal) elements may not exist, and when they exist they may not necessarily be unique. We give an important criterion for an ordered set to have a maximal element.

Theorem A.2 (Zorn's Lemma[1]). *Let $(X, \leq)$ be an ordered set such that every totally ordered subset $A \subseteq X$ of X (a so-called "chain") has an upper bound. Then X has a maximal element.*

[1]Max August Zorn (1906–1993), German-American mathematician.

Proof. Zorn's lemma is equivalent to the Axiom of Choice. See [**Rud87**, Appendix] for a proof of Hausdorff's maximality theorem,[2] which directly implies Zorn's lemma. □

A.4. Countable and Uncountable Sets

A set X is called **countable** if it is the range of a sequence in X: $X = \{x_n \mid n \in \mathbb{N}\}$. In this case one can even find a *bijective* mapping between $\mathbb{N}$ or a finite segment $\{1, \ldots, N\}$ of $\mathbb{N}$ and X, by discarding double occurrences in the sequence $(x_n)_{n\in\mathbb{N}}$.

Clearly, finite sets are countable. Subsets of countable sets are countable. If X is a set and $A_n \subseteq X$ is countable for every $n \in \mathbb{N}$, then

$$A := \bigcup_{n\in\mathbb{N}} A_n \subseteq X$$

is countable, too. If X, Y are countable, then so is their Cartesian product

$$X \times Y = \{(x, y) \mid x \in X,\ y \in Y\}.$$

In particular, the sets $\mathbb{N}^2, \mathbb{N}^3, \ldots$ are all countable.

A set that is not countable is called **uncountable**. Any nonempty open interval of the real numbers is uncountable, by Cantor's famous diagonal argument.

A.5. Real Numbers

There is no science without presuppositions. One has to start somewhere, take something for granted. A proof is an argument that convinces us of a statement, but the argument has to recur on our pre-established knowledge, things that we call "facts". However, these "facts" are only conditional to more basic "facts". In mathematics, the most basic "facts" are called *axioms*. Ideally, all other mathematical statements should rest logically on those — and only on those — axioms.

The most well-known axioms are those from elementary (synthetic) geometry. In the beginning of Euclid's *Elements*[3] the basic concepts of *point, line, and plane* are introduced together with some very basic statements about them, and all other statements are derived from these basic ones by means of logical deduction. This has been a role model for over 2000 years, until in the 19th century people became dissatisfied with the old approach. The discussion centered around the problem of what points and lines "really" are. (Euclid had given 'definitions', which were at best obscure, and

[2]Felix Hausdorff (1868–1942), German mathematician.

[3]Euclid (around 280 BC), Greek mathematician.

never used as such in mathematical arguments.) Also, the invention (or discovery?) of non-Euclidean geometries shattered the old dogma that mathematics is about the "real" (=physical) world. Finally, Hilbert[4] in his book "Grundlagen der Geometrie" (1899) gave the first modern axiomatization of geometry, cleared away the defects of Euclid's text and, moreover, paved the way for a more flexible, non(meta)physical view of mathematics. According to this new approach, axioms can be chosen freely depending on what you want to model with them. The actual *meaning* of the words is inessential, what matters are the relations described by the axioms.

Although axioms are fundamental from a logical point of view, they are not "more true" than any mathematical statement derived from them. Axioms help us to structure and organize our knowlege. In fact, according to the modern approach advocated by Hilbert there is no *intrinsic* criterion of truth other than the mere logical consistency. We cannot ask whether an axiom (or better: axiom system) is true or not, unless we mean by it whether it appropriately describes what it is supposed to describe.

Let us turn to what many consider to be the most central object in mathematics, the *real number system*. What seems so natural to us was actually unknown for a long time. The ancient Greeks had a very different concept of a "number" than we do: they considered only the positive natural numbers as numbers (a concept associated with the process of counting) and objects as the rational numbers or the real numbers were not known. However, they had the related concepts of a *quantity* (e.g. length of a line segment, area of inclosed regions, etc.) and of a *ratio* of quantities. Originally it was believed that the intuitive notion of ratio could be reduced to ratios of natural numbers. But already within the school of Pythagoras[5] it was discovered that the length of the diagonal of a square and its side length are "incommensurable", i.e., cannot be expressed as a ratio of whole numbers. Nowadays this is expressed by saying that $\sqrt{2}$ is an irrational number.

The Greek fixation on geometry hindered very much the development of the number concept; for example, an algebraic equation like $x^2 + x = 3$ was inconceivable for the Greeks since x would represent a length of a line, and x^2 the area of a square; and how in the world could one add a length and an area? So a big step towards the modern concept of number was by stripping off geometric interpretations and focusing on algebraic properties, a step that is already present in the work of Diophantus of Alexandria around 250 AD, but was properly established only after the important work of Arab mathematicians after 800 AD. Another big step was made by Descartes[6]:

[4]David Hilbert (1862–1943), German mathematician.

[5]Pythagoras (around 520 BC), Greek philosopher and religious leader.

[6]René Descartes (1596–1650), French mathematcian and philosopher.

through his "analytic geometry" geometry was designed to become a subdiscipline of arithmetic.

In the 17th century the invention of the calculus by Newton[7] and Leibniz[8] revolutionized the world. However, despite the obvious successes of the new mathematics, many things remained unclear and were subject to harsh critique, in particular the use of "infinitesimal (= infinitely small) quantities". Eventually, these vanished from the mathematical discourse during the 19th century and were replaced by the new but yet imprecise concept of a *limit*. At the same time, mainly due to Fourier's[9] theory of heat, the hitherto quite restrictive notion of a "function" was questioned towards greater generality.

In a development of about 50 years it became clear that these notions could only be made mathematically rigorous by allowing *actually infinite* totalities in mathematics. This was a real revolution, as since the times of Aristotle[10] only *potentially infinite* collections were allowed in scientific discourse. For example, one could not speak of the "set" of natural numbers as a complete (actually) infinite totality, but only of the natural numbers as being capable of (potentially) infinite augmentation. (Analogously, a line was always given by a finite line segment, capable of indefinite prolongation.)

This paradigm change towards actually infinite collections was put forward, initially, by only a few mathematicians, mainly Bolzano,[11] Dedekind[12] and Cantor.[13] It was Dedekind who for the first time introduced real numbers in a mathematically rigorous way thereby eliminating the use of "quantities" in mathematics. He also gave the first rigorous definition of an "infinite collection of objects". Finally, in close correspondence with Dedekind, Cantor coined the notion of a "set" and made it the central object in mathematics. Real numbers could now be considered as special sets "constructed" from the set of rational numbers. Shortly after, the concept of "function" was given a precise set-theoretic definition by Dedekind, and Weierstrass[14] gave the first proper definition of "limit" at around the same time. The "arithmetization" of analysis had been successful.

Axioms for the Real Numbers. Instead of constructing the real numbers from more elementary objects, one might as well give axioms for real numbers directly instead of for sets. These axioms usually consist of three

---

[7] Isaac Newton (1643–1727), English mathematician and physicist.
[8] Gottfried Wilhelm Leibniz (1646–1716), German philosopher and mathematician.
[9] Joseph Fourier (1768–1830), French mathematician.
[10] Aristotle (384–322 BC), Greek philosopher.
[11] Bernard Bolzano (1781–1848), Bohemian mathematician, theologian and philosopher.
[12] Richard Dedekind (1831–1916), German mathematician.
[13] Georg Cantor (1845–1918), German mathematician.
[14] Karl Weierstrass (1815–1897), German mathematician.

groups which describe that the real numbers form a "complete, totally ordered, Archimedean field". The different structural elements are

1) the algebraic structure: $\mathbb{R}$ is a *field*, i.e., one may add, subtract, multiply and divide according to the usual rules;
2) the order structure: $\mathbb{R}$ is *totally ordered*, and the order is compatible with the algebraic structure (e.g., if $x, y \geq 0$, then also $x + y \geq 0 \ldots$);
3) the order is *Archimedean*, i.e., if $0 < x, y$ then there is a natural number n such that $y < x + \cdots + x$ (n-times);
4) $\mathbb{R}$ is *complete.*

The completeness axiom (4) comes in different forms, all equivalent in the presence of the other axioms. It is a matter of personal taste which one considers more fundamental, but the common spirit is the geometric intuition of the continuum: the real numbers have no holes.

Theorem A.3. *In the presence of the axioms* 1)–3) *the following statements are equivalent:*

(i) *Every bounded sequence of real numbers has a convergent subsequence.*
(ii) *Every Cauchy sequence of real numbers converges.*
(iii) *Every monotonic and bounded sequence of real numbers converges.*
(iv) *Every nonempty set of real numbers, bounded from above (below), has a supremum (infimum).*

Property (ii) is metric completeness, (iii) is called **order completeness** or the **Dedekind axiom**, and (i) is called the **Bolzano–Weierstrass property**.

Proof. (i)⇒(ii): This follows from the fact that a Cauchy sequence is bounded, and from the fact that a Cauchy sequence having a convergent subsequence must converge, see Lemma 5.2.c).

(ii)⇒(iii): If a monotonic sequence is not Cauchy, it has a subsequence with every two points having distance larger than a certain $\epsilon > 0$. The Archimedean axiom prevents the sequence from being bounded. This shows that a monotonic and bounded sequence is Cauchy, and hence converges, by hypothesis.

(iii)⇒(iv): Suppose that A is a set of real numbers, bounded by b_0. Choose any $a_0 \in A$. (This works since A is assumed to be not empty.) Consider $x := (a_0 + b_0)/2$. Either x is also an upper bound for A, then set $a_1 := a_0$, $b_1 := x$; or x is not an upper bounded for A, and then there must exist $x < a_1 \in A$. In this case we set $b_1 := b_0$. In either situation, b_1 is an upper bound for A, $a_1 \in A$, and $b_1 - a_1 \le (b_0 - a_0)/2$. Continuing inductively we construct an increasing sequence $a_n \le a_{n+1}$ of elements in A and a decreasing sequence $b_{n+1} \le b_n$ of upper bounds of A such that $b_n - a_n \to 0$. By hypothesis, both sequences converge to a and b, say. Clearly b is an upper bound for A, but since a_n comes arbitrarily close to b, there cannot be any strictly smaller upper bound of A. So $b = \sup A$.

(iv)$\Rightarrow$(i): Suppose that $(x_n)_{n\in\mathbb{N}}$ is a Cauchy sequence. Then one finds $a < b$ such that $x_n \in [a, b]$ for all n. Consider the set $A_n := \{x_k \mid k \ge n\}$. As $A_n \subseteq [a, b]$ there exists $a_n := \sup A_n \in [a, b]$, by hypothesis. The set $A := \{a_n \mid n \in \mathbb{N}\}$ is contained in $[a, b]$ and hence $x := \inf_n a_n$ exists. Using this, it is easy to find a subsequence of $(x_n)_{n\in\mathbb{N}}$ that converges to x. But since $(x_n)_{n\in\mathbb{N}}$ is Cauchy, it itself must converge to x; see Lemma 5.2c). □

Corollary A.4. *The Euclidean space $\mathbb{R}^d$ is complete.*

Proof. Suppose that $(x_n)_{n\in\mathbb{N}}$ is a Cauchy sequence in the Euclidean metric. Then, since $|x_{n,j} - x_{m,j}| \le \|x_n - x_m\|_2$, each coordinate sequence $(x_{n,j})_{n\in\mathbb{N}}$, $j = 1, \dots, d$, is a Cauchy sequence in $\mathbb{R}$. By the completeness of $\mathbb{R}$, it must have a limit $x_{\infty,j}$. It is now easy to see that $x_\infty := (x_{\infty,1}, \dots, x_{\infty,d})$ is the limit of $(x_n)_{n\in\mathbb{N}}$. □

Corollary A.5 (Bolzano–Weierstrass). *A subset $A \subseteq \mathbb{R}^d$ is (sequentially) compact if and only if it is closed and bounded.*

Proof. Suppose that A is sequentially compact; then A must be bounded, since a sequence which wanders off to "infinity" cannot have a convergent subsequence. It also must be closed, since if $(x_n)_{n\in\mathbb{N}} \subseteq A$ and $x_n \to x \in \mathbb{R}^d$ then by hypothesis there is a subsequence $(x_{n_k})_k$ which converges to some $a \in A$. But clearly $x_{n_k} \to x$ as well, and since limits are unique, $x = a \in A$. Hence A is closed, by Definition 4.1.

Conversely, suppose that A is closed and bounded, take any sequence $(x_n)_{n\in\mathbb{N}}$ in A and write $x_n = (x_{n1}, \dots, x_{nd})$. Then every coordinate sequence $(x_{nj})_{n\in\mathbb{N}}$ in $\mathbb{R}$ is bounded. By the Bolzano–Weierstrass axiom/theorem, one finds a strictly increasing map $\pi_1 : \mathbb{N} \to \mathbb{N}$ such that $x_{\infty,1} := \lim_n x_{\pi_1(n),1}$ exists. The same argument yields a subsequence $\pi_2 : \mathbb{N} \to \mathbb{N}$ such that $x_{\infty,2} := \lim_n x_{\pi_1(\pi_2(n)),2}$ also converges. Continuing in this manner one finds strictly increasing maps $\pi_j : \mathbb{N} \to \mathbb{N}$ such that

$$x_{\infty,j} := \lim_{n\to\infty} x_{\pi_1 \dots \pi_j(n),j}$$

exists for every $j = 1, \dots j$. Setting $\pi := \pi_1 \circ \dots \circ \pi_d$, we have that $x_{\pi(n),j} \to x_{\infty,j}$ for all j, and hence

$$x_{\pi(n)} \to x_\infty := (x_{\infty,1}, \dots, x_{\infty,d})$$

in $\mathbb{R}^d$. Since A is closed, $x_\infty \in A$, and we are done. $\square$

Note that one may exchange $\mathbb{R}$ for $\mathbb{C}$ in the previous two statements: also $\mathbb{C}^d$ is complete and a subset of $\mathbb{C}^d$ is compact if and only if it is closed and bounded. This is true since, metrically, $\mathbb{C}^d = \mathbb{R}^{2d}$.

A.6. Complex Numbers

Complex numbers can be constructed from the real numbers in many different ways. (Recall from the previous section that it is not important what complex numbers "really are", but which properties they satisfy; as different constructions lead to objects with the same properties, we may consider them equivalent and freely choose each one of them.) The easiest, probably, of these constructions goes back to Gauss and is based on the geometric picture of complex numbers as points in the Euclidean plane. It was Hamilton[15] who first presented the modern approach, which we adopt.

The set of **complex numbers** $\mathbb{C}$ is defined as

$$\mathbb{C} := \mathbb{R}^2 = \{(a, b) \mid a, b \in \mathbb{R}\}$$

the set of pairs of real numbers. If $z = (a, b)$ is a complex number, then $a := \operatorname{Re} z$ is the **real part** and $b := \operatorname{Im} z$ is the **imaginary part** of z. Clearly a complex number is uniquely determined by its real and imaginary part.

The algebraic operations on $\mathbb{C}$ are given by

$$\begin{aligned}(x, y) + (a, b) &:= (x + a, y + b), \\ (x, y) \cdot (a, b) &:= (xa - yb, xb + ya),\end{aligned}$$

for all $(x, y), (a, b) \in \mathbb{C}$. One then checks that all usual rules of computation (associativity and commutativity of addition and multiplication, distributivity) are valid. Then one realises that

$$(x, 0) + (y, 0) = (x + y, 0) \quad \text{and} \quad (x, 0) \cdot (y, 0) = (xy, 0)$$

for all real numbers $x \in \mathbb{R}$. So the complex numbers of the form $(x, 0)$ behave the "same" as the corresponding real numbers, whence there is no confusion writing x instead of $(x, 0)$. Defining the **imaginary unit** i by

$$\mathrm{i} := (0, 1)$$

[15]Sir William Rowan Hamilton (1805–1865).

and using the writing convention $x = (x, 0)$ for real numbers, we see that every complex number $z = (a, b)$ can be uniquely written as

$$z = a + ib = \operatorname{Re} z + \operatorname{Im} z \cdot \mathrm{i}.$$

Moreover, $\mathrm{i}^2 = -1$, and hence $1/\mathrm{i} = -\mathrm{i}$.

Two more operations play an imporant role for complex numbers. The first is **conjugation** defined by

$$\overline{z} := a - ib \quad \text{if} \quad z = a + ib,\ a, b \in \mathbb{R},$$

or, equivalently, by $\operatorname{Re} \overline{z} := \operatorname{Re} z, \operatorname{Im} \overline{z} := -\operatorname{Im} z$. Then

$$\operatorname{Re} z = 1/2(z + \overline{z}) \quad \text{and} \quad \operatorname{Im} z = 1/2\mathrm{i}(z - \overline{z}),$$

and $\overline{z} = z$ if and only if $z \in \mathbb{R}$. The following computation rules are easy to check:

$$\overline{z + w} = \overline{z} + \overline{w}, \quad \overline{zw} = \overline{z}\,\overline{w}, \quad z\overline{z} = (\operatorname{Re} z)^2 + (\operatorname{Im} z)^2.$$

The **modulus** or **absolute value** of a complex number $z = a + ib$ is defined by

$$|z| = \sqrt{a^2 + b^2} = \sqrt{(\operatorname{Re} z)^2 + (\operatorname{Im} z)^2}.$$

If one pictures complex numbers geometrically as points in the plane, $|z|$ just gives the usual Euclidean distance of z from the origin. It is then clear that we have

$$|\operatorname{Re} z| \le |z|, \quad |\operatorname{Im} z| \le |z|, \quad |\overline{z}| = |z| \quad \text{and} \quad z\overline{z} = |z|^2$$

for all complex numbers $z \in \mathbb{C}$. The last formula gives us a clue how to compute a *multiplicative inverse* for a nonzero complex number z. Indeed, $z \neq 0$ if and only if $|z| \neq 0$ and so the last formula becomes $z\left(|z|^{-2}\,\overline{z}\right) = 1$, and this amounts to

$$z^{-1} = |z|^{-2}\,\overline{z}.$$

Finally we note the useful identity

$$|z| = \sup_{|c|=1} \operatorname{Re}(cz) \qquad (z \in \mathbb{C}), \tag{A.1}$$

which follows from $\operatorname{Re}(cz) \le |cz| = |c|\,|z| = |z|$ and taking $c := \overline{z}/\,|z|$ in case $z \neq 0$. (So the supremum is even attained, i.e., a maximum.)

A.7. Linear Algebra

A **vector space** over the field $\mathbb{K} \in \{\mathbb{R}, \mathbb{C}\}$ is a set E together with an operation

$$E \times E \longrightarrow E, \qquad (x, y) \longmapsto x + y$$

called **addition**, an operation

$$\mathbb{K} \times E \longrightarrow E, \qquad (\lambda, x) \longmapsto \lambda x$$

called **scalar multiplication**, and a distinguished element $0 \in E$ called the **zero vector**, such that for $x, y, z \in E$ and $\lambda, \mu \in \mathbb{K}$ the following statements hold:

1) $x + (y + z) = (x + y) + z$;
2) $x + y = y + x$;
3) $x + 0 = x$;
4) $x + (-1)x = 0$;
5) $(\lambda + \mu)x = \lambda x + \mu x$;
6) $\lambda(x + y) = \lambda x + \lambda y$;
7) $(\lambda\mu)x = \lambda(\mu x)$;
8) $1 \cdot x = x$.

(These are called the **vector space axioms**). Vector spaces over $\mathbb{K} \in \{\mathbb{R}, \mathbb{C}\}$ are also called **linear spaces**. We use both expressions synonymously.

If E is a vector space, $\{v_1, \dots, v_n\} \subseteq E$ is a collection of vectors and $\lambda_1, \dots, \lambda_n \in \mathbb{K}$ are scalars, the expression

$$\sum\nolimits_{j=1}^{n} \lambda_j v_j = \lambda_1 v_1 + \cdots + \lambda_n v_n$$

is called a (finite) **linear combination** of the vectors v_j.

If E is a vector space and $F \subseteq E$ is a subset, then F is called a **(linear) subspace** of E if

$$0 \in F \quad \text{and} \quad x, y \in F,\, \lambda \in \mathbb{K} \;\Rightarrow\; x + y \in F,\, \lambda x \in F.$$

It is then clear that F is a vector space in its own right (with respect to the induced operations and the same zero vector).

If $A \subseteq E$ is any subset there is a "smallest" linear subspace, called the **linear span**, $\operatorname{span} A$ of A, consisting of all finite linear combinations of vectors in A, i.e.,

$$\operatorname{span} A := \Bigl\{ \sum\nolimits_{a \in A} \lambda_a a \mid \lambda_a \in \mathbb{K},\ \#\{a \in A \mid \lambda_a \neq 0\} < \infty \Bigr\}.$$

(Note that we do not assume A to be a finite set here, so we have to set all coefficients $\lambda_a = 0$ except for finitely many.) If U is a subspace and $A \subseteq U$ is such that $\operatorname{span} A = U$, then A is said to **generate** U.

Let E be any $\mathbb{K}$-vector space. A family $(x_j)_{j \in J}$ of vectors in E (J an arbitrary nonempty index set) is called **linearly independent** if there is only the trivial way to write the zero vector 0 as a finite linear combination of the x_j. Formally:

$$\text{If } \sum_{j \in J} \lambda_j x_j = 0, \quad \text{where} \quad \lambda_j \in \mathbb{K},\ \#\{j \in J \mid \lambda_j \neq 0\} < \infty,$$
$$\text{then} \quad \lambda_j = 0 \quad \forall j \in J.$$

An **(algebraic) basis** of E is a linearly independent family $(x_j)_{j \in J}$ such that $E = \text{span}\{x_j \mid j \in J\}$. Equivalently, every element from E can be written in a *unique way* as a finite linear combination of the x_j.

A vector space E is called **finite-dimensional** if it has a finite basis, otherwise **infinite-dimensional**. A famous theorem of Steinitz[16] says that each two bases of a finite-dimensional vector space E have the same number of elements. This number is then called the **dimension** of E, $\dim E$.

Theorem A.6. *Let E be a vector space over $\mathbb{K}$ and let $A \subseteq B \subseteq E$ be subsets. If A is linearly independent and B generates E then there is a basis C of E such that $A \subseteq C \subseteq B$.*

Proof. Apply the following procedure: discard from B all vectors v such that $A \cup \{v\}$ is linearly dependent, and call this the new B. Then, if $A = B$, we are done. If not, then there is a vector $v \in B \setminus A$ such that $A \cup \{v\}$ is linearly independent. Call this set the new A and repeat the procedure.

If B is finite, this procedure must terminate and one ends up with a basis. In the case of infinite B one may have to make an infinite (and even uncountable) number of steps. Such a "transfinite" procedure is unwieldy and may be replaced by an application on Zorn's lemma. To this end, let

$$X := \{C \mid A \subseteq C \subseteq B,\ B \text{ is linearly independent}\}$$

and order X by ordinary set inclusion. If $C \in X$ is maximal, then each vector from B is linearly dependent on C (otherwise we could enlarge C). But since B is generating, C is then generating as well, and so C is the desired basis.

Hence a maximal element in X would solve the problem. To find it, we apply Zorn's Lemma A.2. Let $K \subseteq X$ be any totally ordered subset. We have to show that K has an upper bound in the ordered set X. Define

$$C_K := \bigcup_{C \in K} C.$$

Clearly, $A \subseteq C_K \subseteq B$. To show that C_K is linearly independent, take $x_1, \dots, x_n \in C_K$. For any $1 \leq j \leq n$ we find a $C_j \in K$ such that $x_j \in C_j$. But since K is totally ordered, one of the C_j, say C_{j_0}, contains all the others, hence $x_1, \dots, x_n \in C_{j_0}$. But C_{j_0} is linearly independent, and so

[16]Ernst Steinitz (1871–1928), German mathematician.

must be $\{x_1, \dots, x_n\}$. This shows that $C_K \in X$ and thus is an upper bound for $K \subseteq X$ in X. □

Linear Mappings. Let E, F be $\mathbb{K}$-vector spaces. A mapping $T : E \to F$ is called **linear** if

$$T(x + y) = Tx + Ty \quad \text{and} \quad T(\lambda x) = \lambda Tx$$

for all $x, y \in E$, $\lambda \in \mathbb{K}$. (Note that we often write Tx instead of $T(x)$.) Linear mappings are also called **linear operators**, especially when $E = F$. When $F = \mathbb{K}$, i.e., when $T : E \to \mathbb{K}$ then T is called a **(linear) functional**.

If E, F, G are K vector spaces and $T : E \to F$, $S : F \to G$ are linear, then

$$ST := S \circ T : E \longrightarrow G$$

is linear as well. If $T : E \to T$ is linear we write $T^0 := \mathrm{I}$ and $T^n = T \circ \ldots \circ T$ n-times, for $n \in \mathbb{N}$.

For a linear mapping $T : E \to F$, its **kernel**

$$\ker(T) := \{x \in E \mid Tx = 0\}$$

is a linear subspace of E and its **range**

$$\operatorname{ran}(T) := \{Tx \mid x \in E\}$$

is a linear subspace of F. The kernel is sometimes also called the **null space** and is denoted by $N(T)$, the range is sometimes denoted by $R(T)$. The linear mapping $T : E \to F$ is injective if and only if $\ker(T) = 0$ and surjective if and only if $\operatorname{ran}(T) = F$. If T is bijective, then T^{-1} is also a linear mapping, and T is called an **isomorphism**. Two spaces E, F are **isomorphic**, in symbols: $E \cong F$, if there is an isomorphism $T : E \to F$.

Coordinatization. Let E be a *finite-dimensional* $\mathbb{K}$-vector space. Each ordered basis $B = \{b_1, \dots, b_d\}$ defines an isomorphism $\Phi : \mathbb{K}^d \to E$ by

$$\Phi(x) := \sum\nolimits_{j=1}^{d} x_j b_j \qquad (x = (x_1, \dots, x_d) \in \mathbb{K}^d).$$

The injectivity of T is due to the linear independence of B, the surjectivity due to the fact that B is generating. If $v = \Phi(x)$, then the tuple

$$\Phi^{-1}(v) = x = (x_1, \dots, x_d)$$

is called the **coordinate vector** of v with respect to the (ordered) basis B.

Direct Sums and Projections. Let E be a vector space and U, V linear subspaces. If

$$E = U + V := \{u + v \mid u \in U,\, v \in V\} \quad \text{and} \quad U \cap V = \{0\},$$

then we call E the **algebraic direct sum** of U and V and write

$$E = U \oplus V.$$

If $x \in E$, we can write $x = u + v$ with $u \in U$ and $v \in V$. The condition $U \cap V = \{0\}$ implies that this representation of x is *unique*. Hence we may write

$$P_U x := u, \quad P_V x := v.$$

Then $P_U : E \to U$, $P_V : E \to V$ are linear, $P_U + P_V = \mathrm{I}$, $P_U^2 = P_U$, $P_V^2 = P_V$. The operators P_U, P_V are called the **canonical projections** associated with the **direct sum decomposition** $E = U \oplus V$.

Conversely, let $P : E \to E$ satisfy $P^2 = P$. Then also $(\mathrm{I} - P)^2 = \mathrm{I} - P$ and $\operatorname{ran} P = \ker(\mathrm{I} - P)$. Defining $U := \operatorname{ran} P$ and $V := \ker P$ then

$$U + V = E, \quad \text{and} \quad U \cap V = \{0\}$$

hence $E = U \oplus V$ and $P = P_U, \mathrm{I} - P = P_V$.

Vector Spaces of Functions. Clearly, $E = \mathbb{K}$ itself is a vector space over $\mathbb{K}$. If E is any $\mathbb{K}$-vector space and X is any nonempty set then the set of E-valued functions on X,

$$\mathcal{F}(X; E) := \{f \mid f : X \longrightarrow E\},$$

is also a $\mathbb{K}$-vector space, with respect to the **pointwise operations**:

$$(f + g)(x) := f(x) + g(x),$$
$$(\lambda f)(x) := \lambda f(x),$$

whenever $f, g \in \mathcal{F}(X; E)$, $\lambda \in \mathbb{K}$. Note that for $X = \mathbb{N}$, an element $f \in \mathcal{F}(\mathbb{N}; E)$ is nothing other than a **sequence** $(f(n))_{n \in \mathbb{N}}$ in E. If $X = \{1, \dots, d\}$ then $\mathcal{F}(X; \mathbb{K}) = \mathbb{K}^d$.

If E is a $\mathbb{K}$-vector space and X is a nonempty set, then each $a \in X$ defines a linear mapping

$$\delta_a : \mathcal{F}(X; E) \longrightarrow E$$

by **point evaluation** $\delta_a(f) := f(a)$, $f \in \mathcal{F}(X; E)$.

The Space of all Linear Mappings. Let E, F be vector spaces over $\mathbb{K}$. We write

$$\operatorname{Lin}(E;F) := \{T : E \longrightarrow F \mid T \text{ is linear}\}$$

the set of all linear mappings from E to F. This is clearly a subset of $\mathcal{F}(E;F)$, the vector space of *all* mappings from E to F.

Lemma A.7. *The set* $\operatorname{Lin}(E;F)$ *is a linear subspace of* $\mathcal{F}(E;F)$ *and hence a vector space. If* $E = F$, *then* $\operatorname{Lin}(E) := \operatorname{Lin}(E;E)$ *is an algebra, i.e., the multiplication (= composition) satisfies*

$$\begin{aligned} R(ST) &= (RS)T, \\ R(S+T) &= RS + RT, \\ (R+S)T &= RT + ST, \\ \lambda(ST) &= (\lambda S)T = S(\lambda T), \end{aligned}$$

for all $R, S, T \in \operatorname{Lin}(E;F), \lambda \in \mathbb{K}$.

Quotient Spaces. If F is a linear subspace of a vector space E, then an equivalence relation on E is defined by

$$x \sim_F y \quad :\Longleftrightarrow \quad x - y \in F$$

for $x, y \in E$. Let us, as above, denote the equivalence class containing $x \in E$ by $[x]$. Then $[x] = [y]$ if and only if $x - y \in F$ and $[x] = x + F$ as sets. The set of equivalence classes

$$E/F := \{[x] \mid x \in E\}$$

is called the **factor space** or **quotient space**. It is itself a vector space with respect to the operations

$$[x] + [y] := [x + y] \quad \text{and} \quad \lambda[x] := [\lambda x] \qquad (x, y \in E, \lambda \in \mathbb{K}).$$

(Of course one has to check that these are well-defined.) The mapping

$$s : E \longrightarrow E/F, \quad sx := [x] \quad (x \in E)$$

is linear (by definition of addition and scalar multiplication on E/F) and is called the **canonical surjection**.

Dual Spaces and Mappings. The **dual space** of a vector space E over $\mathbb{K}$ is the set

$$E^* := \operatorname{Lin}(E;\mathbb{K}),$$

of all linear functionals on E. It is a vector space over $\mathbb{K}$ again, by Lemma A.7. If $(b_j)_{j\in J}$ is an algebraic basis for E, then one can consider the *coordinate functionals* $b_k^* \in E^*$ defined by

$$b_k^*\Big(\sum\nolimits_{j\in J} \lambda_j b_j\Big) := \lambda_k \qquad (k \in J).$$

In this case

$$E \cong \{f \in \mathcal{F}(J;\mathbb{K}) \mid f(j) = 0 \text{ for all but finitely many } j\}$$

and $E^* \cong \mathcal{F}(J;\mathbb{K})$. If E is finite-dimensional, i.e., $d := \dim(E) = \operatorname{card}(J) < \infty$, then $(b_k^*)_{k\in J}$ is a basis for E^*, the so-called **dual basis**.

Each linear mapping $T : E \to F$ induces a linear mapping $T^* : F^* \to E^*$, the **dual mapping**, by $T^*(x^*) := x^* \circ T$ for $x^* \in E^*$. One has

$$(T+S)^* = T^* + S^*, \quad (\lambda T)^* = \lambda T^* \quad \text{and} \quad (T \circ S)^* = S^* \circ T^*$$

whenever these expressions make sense.

Sesquilinear Forms. Let E be a vector space over $\mathbb{K}$. A mapping $a : E \times E \to \mathbb{K}$ is called a **sesquilinear form** on E, if

1) $a(\alpha f + \beta g, h) = \alpha a(f,h) + \beta a(g,h)$,
2) $a(h, \alpha f + \beta g) = \overline{\alpha} a(h,f) + \overline{\beta} a(h,g)$,

for all $f, g, h \in E$ and $\alpha, \beta \in \mathbb{K}$. By

$$q_a(f) := a(f,f) \qquad (f \in E)$$

we denote the associated **quadratic form**.

In short, 1) says that $a(\cdot, h)$ is linear and 2) says that $a(h, \cdot)$ is "anti-linear" for each $h \in E$. In particular, in the case that $\mathbb{K} = \mathbb{R}$ a sesquilinear form is simply bilinear, i.e., linear in each component.

A sesquilinear form a on E is called **hermitian** or **symmetric** if

$$a(g,f) = \overline{a(f,g)} \qquad (f, g \in E).$$

Symmetric sesquilinear forms have special properties.

Lemma A.8. *Let $a : E \times E \to \mathbb{K}$ be a symmetric sesquilinear form on the $\mathbb{K}$-vector space E. Then the following assertions hold:*

a) $q_a(f) \in \mathbb{R}$,
b) $q_a(f+g) = q_a(f) + 2\operatorname{Re} a(f,g) + q_a(g)$,
c) $q_a(f+g) - q_a(f-g) = 4\operatorname{Re} a(f,g)$ **(polarization identity)**,
d) $q_a(f+g) + q_a(f-g) = 2q_a(f) + 2q_a(g)$ **(parallelogram law)**,

for all $f, g \in E$.

Proof. By symmetry, $q_a(f) = \overline{q_a(f)}$; but only a real number is equal to its own conjugate. The sesquilinearity and symmetry of a yields

$$\begin{aligned} q_a(f+g) &= a(f+g, f+g) = a(f,f) + a(f,g) + a(g,f) + a(g,g) \\ &= q_a(f) + a(f,g) + \overline{a(f,g)} + q_a(g) = q_a(f) + 2\operatorname{Re} a(f,g) + q_a(g) \end{aligned}$$

since $z + \overline{z} = 2\operatorname{Re} z$ for every complex number $z \in \mathbb{C}$. This is b). Replacing g by $-g$ yields

$$q_a(f-g) = q_a(f) - 2\operatorname{Re} a(f,g) + q_a(g)$$

and addding this to b) yields d). Subtracting it leads to c). □

Suppose that $\mathbb{K} = \mathbb{R}$. Then the polarization identity reads

$$a(f,g) = \frac{1}{4}(q_a(f+g) - q_a(f-g)) \qquad (f, g \in E).$$

This means that the values of the form a are uniquely determined by the values of its associated quadratic form. The same is true in the case $\mathbb{K} = \mathbb{C}$, since in this case

$$a(f,g) = \operatorname{Re} a(f,g) + \mathrm{i} \operatorname{Re} a(f, \mathrm{i}g)$$

for all $f, g \in E$.

A.8. Set-theoretic Notions

According to Cantor, a set is a "collection of well-distinguished objects of our mind". This is not a mathematical definition, but helps to set up the decisive properties of sets. The basic relation in the universe of sets is the $\in$-relation: $a \in A$ indicates that some object a (which may be a set itself) is an **element** of (= is contained in) the set A.

One writes $A \subseteq B$ and calls A a **subset** of B if every element of A is also an element of B:

$$A \subseteq B \quad :\Longleftrightarrow \quad \forall x : x \in A \implies x \in B.$$

Two sets are **equal** if they have the same elements:

$$A = B \quad :\Longleftrightarrow \quad A \subseteq B \quad \text{and} \quad B \subseteq A.$$

The **emtpy set** is the set $\emptyset$ that has *no* elements. If $a, b, c \dots$ are objects, then we denote by

$$\{a, b, c \dots\}$$

the set that contains them. In particular, $\{a\}$ is the **singleton** whose only element is a. If X is a set and P is a *property* that an element of X may or may not have, then

$$\{x \mid x \in X, \ x \text{ has P}\}$$

denotes the set whose elements are precisely those elements of X that have P. The **power set** of a set X is the unique set $\mathcal{P}(X)$ whose elements are precisely the subsets of X:

$$\mathcal{P}(X) := \{A \mid A \subseteq X\}.$$

Note that always $\emptyset, X \in \mathcal{P}(X)$. For subsets $A, B \subseteq X$ we define their **union, intersection, difference** as

$$\begin{aligned} A \cup B &:= \{x \in X \mid x \in A \text{ or } x \in B\}, \\ A \cap B &:= \{x \in X \mid x \in A \text{ and } x \in B\}, \\ A \setminus B &:= \{x \in X \mid x \in A \text{ but } \ x \notin B\}. \end{aligned}$$

The **complement** in X of a subset $A \subseteq X$ is

$$A^{\mathrm{c}} := X \setminus A = \{x \in X \mid x \notin A\}.$$

If A, B are sets we let

$$A \times B := \{(a, b) \mid a \in A,\ b \in B\}$$

be the set of all **ordered pairs** of elements in A and B. The set $A \times B$ is also called the **Cartesian product** of A and B. Each subset

$$R \subseteq A \times B$$

is called a (binary) **relation**. Instead of writing $(a, b) \in R$ one often has other notations, e.g., $a \le b$, $a \sim b$, $\dots$, depending on the context.

A **mapping** $f : X \to Y$ is an assignment that associates with each element $x \in X$ a *value* or *image* $f(x) \in Y$. Set-theoretically, we can *identify* a mapping f with its **graph**

$$\operatorname{graph}(f) = \{(x, y) \mid x \in X, y = f(x)\} \subseteq X \times Y.$$

This is a special kind of a binary relation R, satisfying

$$\begin{aligned} &(1) \quad \forall x \in X \ \exists y \in Y : (x, y) \in R; \\ &(2) \quad \forall x \in X \ \forall y, y' \in Y : (x, y), (x, y') \in R \implies y = y'. \end{aligned}$$

Properties (1) and (2) say that this relation is **functional**. Every functional relation is the graph of a mapping.

If $f : X \to Y$ is a mapping, then we call $X = \operatorname{dom}(f)$ the **domain** of f and Y the **codomain** or **target set**. We write

$$\begin{aligned} f(A) &:= \{f(x) \mid x \in A\} = \{y \in Y \mid \exists x \in A : f(x) = y\}, \\ f^{-1}(B) &:= \{x \in X \mid f(x) \in B\}, \end{aligned}$$

for subsets $A \subseteq X$, $B \subseteq Y$ and call it the **image** of A and the **inverse image** of B under f. The mapping $f : X \to Y$ is called **surjective** (onto) if $f(X) = Y$, and **injective** (one-to-one) if $f(x) = f(y)$ implies that $f(x) =$

$f(y)$, for all $x, y \in X$. A mapping f which is injective *and* surjective, is called **bijective**. In this case one can form its **inverse**, denoted by $f^{-1} : Y \to X$.

An **indexed family** of elements of a set X is simply a mapping $J \to X$, where we call $J \neq \emptyset$ the **index set**. One often writes $(x_j)_{j\in J} \subseteq X$ to denote this mapping, in particular, when one does not want to use a name for it (like "f" or so).

Given an indexed family $(A_j)_{j\in J} \subseteq \mathcal{P}(X)$ of subsets of a set X one considers their **union** and their **intersection**

$$\bigcup\nolimits_{j\in J} A_j = \{x \in X \mid \exists j \in J : x \in A_j\},$$

$$\bigcap\nolimits_{j\in J} A_j = \{x \in X \mid \forall j \in J : x \in A_j\}.$$

De Morgan's laws[17] say that

$$\Big(\bigcup\nolimits_{j\in J} A_j\Big)^c = \bigcap\nolimits_{j\in J} A_j^c \quad \text{and} \quad \Big(\bigcap\nolimits_{j\in J} A_j\Big)^c = \bigcup\nolimits_{j\in J} A_j^c.$$

If $f : X \to Y$ is a mapping, then

$$f\Big(\bigcup\nolimits_{j\in J} A_j\Big) = \bigcup\nolimits_{j\in J} f(A_j) \quad \text{and} \quad f\Big(\bigcap\nolimits_{j\in J} A_j\Big) \subseteq \bigcap\nolimits_{j\in J} f(A_j).$$

Attention: the inclusion on the right-hand side is proper in general!

If $f : X \to Y$ is a mapping and $(B_j)_{j\in J} \subseteq \mathcal{P}(Y)$ is an indexed family of subsets of Y, then

$$f^{-1}\Big(\bigcup\nolimits_{j\in J} B_j\Big) = \bigcup\nolimits_{j\in J} f^{-1}(B_j) \quad \text{and} \quad f^{-1}\Big(\bigcap\nolimits_{j\in J} B_j\Big) = \bigcap\nolimits_{j\in J} f^{-1}(B_j).$$

[17] Augustus De Morgan (1806–1871), English mathematician.

Appendix B

The Completion of a Metric Space

Recall from Chapter 5 that a metric space (Ω, d) is **complete** if every Cauchy sequence in Ω has a limit in Ω. Typical examples of incomplete metric spaces are $\mathbb{Q}$ with the standard metric and c_{00} with the ℓ^2-norm. However, the incompleteness is somehow virtual here, as both spaces appear as subsets of the complete metric spaces $\mathbb{R}$ and ℓ^2, respectively, with the induced metric. Hence, each Cauchy sequence has indeed a limit, just not necessarily in the original space.

It is the purpose of this appendix to demonstrate that this is not a coincidence, but that in a certain sense *each* metric space "lies within" a complete one.

A mapping $f : (\Omega, d) \to (\Omega', d')$ between metric spaces is called **isometric** or an **isometry** if

$$d'(f(x), f(y)) = d(x, y) \qquad \text{for all } x, y \in \Omega.$$

An isometry is uniformly continuous and injective. If in addition it is surjective, it is called an **isometric isomorphism** of the metric spaces. In this case its inverse mapping is an isometry as well (Exercise B.1).

A **completion** of a metric space (Ω, d) is any complete metric space (Ω', d') together with an isometry

$$j : (\Omega, d) \to (\Omega', d') \quad \text{such that} \quad \overline{j(\Omega)} = \Omega'.$$

Example B.1. Let (Ω', d') be a complete metric space and $A \subseteq \Omega'$ a dense subset, i.e., $\overline{A} = \Omega'$. On A we take the induced metric, i.e., the restriction of d' to $A \times A$. Then Ω' is a completion of A. Simply take $j : A \to \Omega'$ the inclusion mapping defined by $jx := x$

For example, $\mathbb{R}$ is a completion of $\mathbb{Q}$. And ℓ^1 and ℓ^2 are completions of c_{00} with respect to the different metrics given by $\|\cdot\|_1$ and $\|\cdot\|_2$, respectively.

In general, given a completion $j : (\Omega, d) \to (\Omega', d')$, the restriction of j to its image $j : \Omega \to j(\Omega)$ is an isometric isomorphism. That means that as long as we restrict to the metric structure, we may *identify* Ω with $j(\Omega)$ via j. In this sense, Ω always is "sitting inside" each completion Ω' of it as a dense metric subspace.

Example B.2. Let $j : \mathbb{Q} \to \mathbb{C}$ be defined by $jx = (x + x\mathrm{i})/\sqrt{2}$. Then j is an isometry, and $\Omega' := (1 + \mathrm{i})\mathbb{R} = \overline{j(\mathbb{Q})}$ is a completion of $\mathbb{Q}$. Via the mapping j we can regard $\mathbb{Q}$ as sitting inside Ω' as a dense subset.

We can now formulate the main result of this appendix.

Theorem B.3. *Every metric space has a completion. Moreover, if* $j_1 : (\Omega, d) \to (\Omega_1, d_1)$ *and* $j_2 : (\Omega, d) \to (\Omega_2, d_2)$ *are completions of a metric space* (Ω, d)*, then there is a unique isometric isomorphism of metric spaces* $F : (\Omega_1, d_1) \to (\Omega_2, d_2)$ *such that* $F \circ j_1 = j_2$.

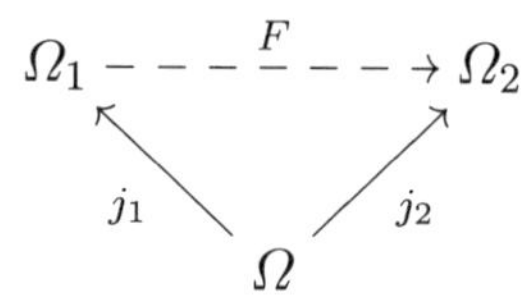

The second part of the theorem states that there is "essentially" only one completion of a metric space, i.e., up to an isometric isomorphism (and this is the strongest uniqueness statement that one can expect).

B.1. Uniqueness of a Completion

We first address the issue of uniqueness. Let $j_1 : (\Omega, d) \to (\Omega_1, d_1)$ and $j_2 : (\Omega, d) \to (\Omega_2, d_2)$ be completions of a metric space (Ω, d). Then we can define a mapping $f : j_1(\Omega_1) \to \Omega_2$ by $f(x) := j_2(j_1^{-1}x)$ for $x \in j_1(\Omega)$. Since the composition of isometries is again an isometry, f is isometric (Exercise B.2). To prove the second part of Theorem B.3 it suffices to show that f can be uniquely extended to an isometry $\Omega_1 \to \Omega_2$. This is a consequence from the following more general result.

Theorem B.4. *Let* (Ω, d) *be a metric and* (Ω', d') *a complete metric space. Let* $A \subseteq \Omega$ *with* $\overline{A} = \Omega$ *and* $f : A \to \Omega'$ *a uniformly continuous mapping.*

Then there is a unique continuous mapping $F : \Omega \to \Omega'$ *such that* $F|_A = f$.

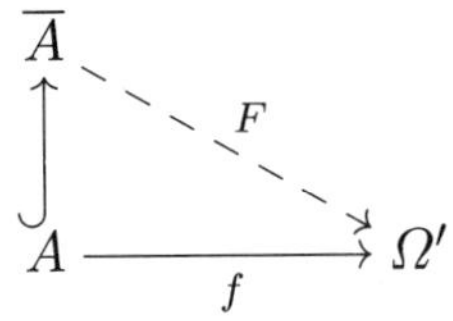

Moreover, F *is uniformly continuous, and if* f *is isometric, so is* F.

Proof. The uniqueness of the extension F follows from Exercise 4.6/Lemma 9.3.b). For the existence, take $x \in \Omega$. By hypothesis there is a sequence $(x_n)_{n\in\mathbb{N}}$ in A with $x_n \to x$. Since f is uniformly continuous (cf. Theorem 4.23), $(f(x_n))_{n\in\mathbb{N}}$ is a Cauchy sequence (Exercise B.3). But (Ω', d') is complete, and hence $x' := \lim_{n\to\infty} f(x_n)$ exists.

The limit x' is actually independent of the original choice of the sequence $(x_n)_{n\in\mathbb{N}}$. To see this, suppose that $y_n \to x$ for another sequence $(y_n)_{n\in\mathbb{N}}$ in A. Then $d(x_n, y_n) \to 0$ and hence $d'(f(x_n), f(y_n)) \to 0$ by uniform continuity (Exercise B.3 again). The triangle inquality now yields $f(y_n) \to x'$ as claimed.

We can therefore define

$$F(x) := \lim_{n\to\infty} f(x_n) \quad \text{whenever} \quad x_n \in A, \lim_{n\to\infty} x_n \to x.$$

To show that F is uniformly continuous, fix $\epsilon > 0$. By uniform continuity of f there is $\delta > 0$ such that $x, y \in A, d(x,y) < \delta$ implies that $d'(f(x), f(y)) < \epsilon$. Now take $x, y \in \Omega$ with $d(x,y) < \delta$. Then find sequences $(x_n)_{n\in\mathbb{N}}$ and $(y_n)_{n\in\mathbb{N}}$ in A with $x_n \to x$ and $y_n \to y$. By the second triangle inequality (Lemma 4.9) and the definition of F,

$$d(x_n, y_n) \to d(x,y) \quad \text{and} \quad d'(f(x_n), f(y_n)) \to d'(F(x), F(y)).$$

Consequently, $d(x_n, y_n) < \delta$ for sufficiently large $n \in \mathbb{N}$. By choice of δ and since $x_n, y_n \in A$, we obtain $d'(f(x_n), f(y_n)) < \epsilon$ for all sufficiently large $n \in \mathbb{N}$, and hence $d'(F(x), F(y)) \le \epsilon$. It follows that F is uniformly continuous. If, moreover, f is isometric, then

$$d'(F(x), F(y)) = \lim_{n\to\infty} d'(f(x_n), f(y_n)) = \lim_{n\to\infty} d(x_n, y_n) = d(x,y),$$

whence F is isometric, too. □

B.2. Existence of a Completion

Let (Ω, d) be any nonempty metric space. To obtain an idea how to construct a completion $j : (\Omega, d) \to (\Omega', d')$, let us assume for the moment that we already successfully have constructed one. Then to each point $x' \in \Omega'$ we can find a sequence $(x'_n)_{n\in\mathbb{N}}$ in $j(\Omega)$ with $x'_n \to x'$. For each $n \in \mathbb{N}$ there is

$x_n \in \Omega$ with $jx_n = x'_n$. Since j is an isometry and $(x'_n)_{n\in\mathbb{N}}$ is Cauchy, so is $(x_n)_{n\in\mathbb{N}}$. On the other hand, if $(x_n)_{n\in\mathbb{N}}$ is a Cauchy sequence in Ω then, since j is isometric, $x'_n := jx_n$ is Cauchy in Ω', and by completeness has a limit $x' := \lim_{n\to\infty} x'_n$ in Ω'.

In this way points in Ω' correspond to Cauchy sequences in Ω. This correspondence, however, is not one-to-one: the same point $x' \in \Omega'$ can be the limit of many different Cauchy sequences in $j(\Omega)$. We see an equivalence relation emerging here: two Cauchy sequences in Ω are equivalent if they correspond to the same point in Ω'. Fortunately, this can be formulated entirely without reference to the completion. Indeed, the Cauchy sequences $x'_n = jx_n$ and $y'_n = jy_n$ converge to the same limit in Ω' if and only if $d(x_n, y_n) \to 0$ as $n \to \infty$. (Check this!) Furthermore, even the metric d' can be reconstructed via the correspondence: if $x'_n = jx_n \to x'$ and $y'_n = jy_n \to y'$, then by the second triangle inequality (Lemma 4.9)

$$d'(x', y') = \lim_{n\to\infty} d'(jx_n, jy_n) = \lim_{n\to\infty} d(x_n, y_n).$$

We now employ these ideas to *construct* a completion. Given our original metric space (Ω, d) we consider the set

$$\mathrm{Cs}(\Omega) := \{(x_n)_{n\in\mathbb{N}} \mid (x_n)_{n\in\mathbb{N}} \text{ is a Cauchy sequence in } \Omega\}$$

of all Cauchy sequences in Ω. On $\mathrm{Cs}(\Omega)$ we define an equivalence relation by

$$(x_n)_{n\in\mathbb{N}} \sim (y_n)_{n\in\mathbb{N}} \quad \text{if} \quad d(x_n, y_n) \to 0 \quad (n \to \infty).$$

This relation is obviously symmetric and reflexive, and it is transitive by the triangle inequality. Now we let

$$\text{(B.1)} \qquad \Omega_c := \mathrm{Cs}(\Omega)/\sim \quad \text{and} \quad j_c : \Omega \to \Omega', \quad j_c x := [(x, x, x, \dots)]_\sim.$$

That is, Ω_c is the set of all $\sim$-equivalence classes of Cauchy sequences in Ω, and $j_c x$ is the class of the constant (hence Cauchy) sequence $x, x, x \dots$.

It follows from the considerations above that the metric d_c on Ω_c should be defined as

$$\text{(B.2)} \qquad d_c([(x_n)_n]_\sim, [(y_n)_n]_\sim) := \lim_{n\to\infty} d(x_n, y_n).$$

Of course, this will work only if we can show that the limit on the right-hand side exists and is independent of the chosen representatives of the equivalence classes. By the second triangle inequality again,

$$|d(x_n, y_n) - d(x_m, y_m)| \le d(x_n, x_m) + d(y_n, y_m) \qquad \text{for all } n, m \in \mathbb{N},$$

so $(d(x_n, y_n))_{n\in\mathbb{N}}$ is a Cauchy sequence in $\mathbb{R}$ and hence convergent. Moreover, suppose that $(x_n)_n \sim (u_n)_n$ and $(y_n)_n \sim (v_n)_n$. The usual triangle inequality yields

$$d(u_n, v_n) \le d(u_n, x_n) + d(x_n, y_n) + d(y_n, v_n),$$

and hence $\lim_{n\to\infty} d(u_n, v_n) \le \lim_{n\to\infty} d(x_n, y_n)$. Finally, by symmetry,

$$\lim_{n\to\infty} d(u_n, v_n) = \lim_{n\to\infty} d(x_n, y_n)$$

as desired.

We have thus shown that $d_c : \Omega_c \times \Omega_c \to \mathbb{R}_+$ is well-defined by (B.2) and it remains to show that $j_c : (\Omega, d) \to (\Omega_c, d_c)$ is a completion. To this end, note first that d_c is a metric and j_c is isometric (Exercise B.4). Next, let $x' = [(x_n)_n]_\sim \in \Omega_c$, let $\epsilon > 0$, and let $N \in \mathbb{N}$ be such that $d(x_n, x_m) \le \epsilon$ whenever $n, m \ge N$. Then

$$d'(x', j_c x_m) = \lim_{n\to\infty} d(x_n, x_m) \le \epsilon$$

if $m \ge N$. Hence, $j_c x_m \to x'$ as $m \to \infty$, and therefore $j_c(\Omega)$ is dense in Ω_c.

Finally, we show that (Ω_c, d_c) is complete. Let $(x'_n)_{n\in\mathbb{N}}$ be a Cauchy sequence in Ω_c. By the density of $j_c(\Omega)$ in Ω_c we can pick $x_n \in \Omega$ such that $d_c(x'_n, j_c x_n) \to 0$ as $n \to \infty$. By the triangle inequality,

$$d(x_n, x_m) = d_c(j_c x_n, j_c x_m) \le d_c(j_c x_n, x'_n) + d_c(x'_n, x'_m) + d_c(x'_m, j_c x_m),$$

and hence $(x_n)_{n\in\mathbb{N}}$ is a Cauchy sequence in Ω. Let $x' := [(x_n)_n]_\sim \in \Omega_c$. Then we have seen above that $j_c x_n \to x'$, and hence

$$d_c(x', x'_n) \le d_c(x', j_c x_n) + d(j_c x_n, x'_n) \to 0 \quad \text{as } n \to \infty,$$

i.e., $x'_n \to x'$. This concludes the proof, and the existence of a completion is established. □

B.3. The Completion of a Normed Space

Let $(E, \|\cdot\|)$ be a normed space. Then E is a metric space with respect to the metric $d(f, g) = \|f - g\|$. Hence, we can apply Theorem B.3 and find a completion $j : E \to E^\sim$.

It is now a tedious but straightforward task to turn $E^\sim$ into a normed space in such a way that the mapping j becomes linear. Actually, there is only one way to define a vector space structure on $E^\sim$, because on $j(E)$ addition and scalar multiplication are prescribed, and since these operations are continuous, there can be at most one extension to all of $E^\sim$. The norm is then determined by the relation $\|f\| = d(0, f)$. We leave the required arguments to Exercise B.5, and simply state the result.

Theorem B.5. *Every normed space has a completion, i.e, there is a Banach space $E^\sim$ and an isometric linear map $j : E \to E^\sim$ onto a dense subspace $j(E)$ of $E^\sim$.*

Moreover, such a completion is essentially unique, i.e., any two such completions $j_1 : E \to E_1^\sim$ and $j_2 : E \to E_2^\sim$ give rise to a unique isometric linear isomorphism $J : E_1^\sim \to E_2^\sim$ with $J \circ j_1 = j_2$.

Exercises

Exercise B.1. Check that an isometry $f : (\Omega, d) \to (\Omega', d')$ between metric spaces is uniformly continuous and injective, and $f^{-1} : (f(\Omega), d') \to (\Omega, d)$ is an isometry as well. Moreover, (Ω, d) is complete if and only if $(f(\Omega), d')$ is complete.

Exercise B.2. Show that the composition of isometries is again an isometry, and hence the isometries of a given metric space Ω form a group under composition.

Exercise B.3. Let (Ω, d) and $\Omega', d')$ be metric spaces. Show that for a mapping $f : \Omega \to \Omega'$ the following assertions are equivalent:

(i) f is uniformly continuous.

(ii) Whenever $(x_n)_{n\in\mathbb{N}}$ and $(y_n)_{n\in\mathbb{N}}$ are sequences in Ω such that $d(x_n, y_n) \to 0$, then $d'(f(x_n), f(y_n) \to 0$.

Moreover, show that a uniformly continuous mapping maps each Cauchy sequence to a Cauchy sequence.

Exercise B.4. Let (Ω, d) be a metric space, and let Ω_c, d_c, j_c be defined by (B.1) and (B.2). Show that $d_c : \Omega_c \times \Omega_c \to \mathbb{R}_+$ is a metric and $j_c : \Omega \to \Omega_c$ is an isometry.

Exercise B.5. Prove Theorem B.5.

Appendix C

Bernstein's Proof of Weierstrass' Theorem

The purpose of this appendix is to present Bernstein's proof[1] of the following famous result. (See Section 9.1 for Lebesgue's approach.)

Theorem C.1 (Weierstrass). *Let $[a, b]$ be a compact interval in $\mathbb{R}$. Then the space of polynomials $\mathrm{P}[a, b]$ is dense in $\mathrm{C}[a, b]$ for the sup-norm.*

Note that every complex-valued $f \in \mathrm{C}[a, b]$ can be written as $f = u + iv$ for some real-valued $u, v \in \mathrm{C}[a, b]$, and f is a polynomial if and only if u and v are polynomials. Hence the theorem for $\mathbb{C}$-valued functions follows easily from the real version. We may therefore suppose $\mathbb{K} = \mathbb{R}$ in the following.

As a second reduction step we claim that it suffices to consider the case that $[a, b] = [0, 1]$. Otherwise we employ the change of variables

$$s = \frac{t-a}{b-a},\ t \in [a, b], \quad t = a + s(b-a),\ s \in [0, 1],$$

which yields an isometric isomorphism of $\mathrm{C}[a, b]$ onto $\mathrm{C}[0, 1]$, mapping $\mathrm{P}[a, b]$ onto $\mathrm{P}[0, 1]$.

Let us introduce the special polynomials $\mathbf{1}, \mathbf{t}, \mathbf{t}^2$ by

$$\mathbf{1}(t) := 1, \quad \mathbf{t}(t) = t, \quad \mathbf{t}^2 := t^2 \qquad (t \in \mathbb{R}).$$

Moreover, we write $f \leq g$ for functions f, g on an interval $[a, b]$ as an abbreviation of $f(t) \leq g(t)$ for all $t \in [a, b]$.

[1] Sergei Natanowitsch Bernstein (1880–1968), Russian mathematician.

For an arbitrary function $f : [0,1] \to \mathbb{R}$ consider its n-th **Bernstein polynomial** $B_n(f, \cdot)$, defined by

$$B_n(f,t) := \sum_{k=0}^{n} \binom{n}{k} t^k (1-t)^{n-k} f(k/n) \qquad (t \in [0,1]).$$

The operation

$$B_n : \mathrm{C}[0,1] \longrightarrow \mathrm{P}[0,1], \quad f \longmapsto B_n(f,\cdot)$$

is obviously linear; it is also **monotonic**, by which we mean that if $f \le g$, then also $B_n f \le B_n g$. We shall show that if f is continuous, $B_n f \to f$ uniformly on $[0,1]$.

To this aim, we first compute the Bernstein polynomials of our three special functions introduced above:

$$B_n(\mathbf{1},t) = \sum_{k=0}^{n} \binom{n}{k} t^k (1-t)^{n-k} = (t + (1-t))^n = 1^n = 1$$

and hence $B_n\mathbf{1} = \mathbf{1}$ for all $n \in \mathbb{N}$.

$$\begin{aligned} B_n(\mathbf{t},t) &= \sum_{k=0}^{n} \binom{n}{k} t^k (1-t)^{n-k} (k/n) = \sum_{k=1}^{n} \binom{n-1}{k-1} t^k (1-t)^{n-k} \\ &= \sum_{k=0}^{n-1} \binom{n-1}{k} t^{k+1} (1-t)^{n-1-k} = t(t+(1-t))^{n-1} = t \end{aligned}$$

and hence $B_n\mathbf{t} = \mathbf{t}$ for all $n \in \mathbb{N}$. Finally,

$$B_n(\mathbf{t}^2,t) = \frac{t(1-t)}{n} + t^2 \qquad (t \in [0,1], n \in \mathbb{N}).$$

(We leave the computation as Exercise C.1 below). Now take $f \in \mathrm{C}[0,1]$ and fix $\epsilon > 0$. Since $[0,1]$ is compact, f is uniformly continuous. This means that we find $\delta > 0$ such that

$$|s-t| \le \delta \quad \Longrightarrow \quad |f(s) - f(t)| \le \epsilon \qquad (s,t \in [0,1]).$$

Define $\alpha := 2/\delta^2 \, \|f\|_\infty$. If $s,t \in [0,1]$ are such that $|s-t| \ge \delta$, then

$$|f(s) - f(t)| \le |f(s)| + |f(t)| \le 2\,\|f\|_\infty = \alpha\delta^2 \le \alpha(s-t)^2.$$

Combining both yields

$$|f(s) - f(t)| \le \epsilon + \alpha(s-t)^2 \qquad (s,t \in [0,1]).$$

Fix $s \in [0,1]$ and define $h_s(t) := (s-t)^2$. Then we may write

$$-\epsilon - \alpha h_s \le f(s)\mathbf{1} - f \le \epsilon + \alpha h_s.$$

We now take Bernstein polynomials of all these functions. This yields

$$B_n(-\epsilon\mathbf{1} - \alpha h_s) \le B_n(f(s)\mathbf{1} - f) \le B_n(\epsilon\mathbf{1} + \alpha h_s).$$

Since taking Bernstein polynomials is linear and $B_n(\mathbf{1}, t) = 1$, we get

$$-\epsilon - \alpha B_n(h_s, t) \le f(s) - B_n(f, t) \le \epsilon + \alpha B_n(h_s, t) \qquad (t \in [0, 1], n \in \mathbb{N}).$$

Recall that $h_s = s^2\mathbf{1} - 2s\mathbf{t} + \mathbf{t}^2$ and that we computed $B_n(g, \cdot)$ for $g = \mathbf{1}, \mathbf{t}, \mathbf{t}^2$ already above. Using these results we obtain

$$|f(s) - B_n(f, t)| \le \epsilon + \alpha B_n(h_s, t) = \epsilon + \alpha \left(s^2 - 2st + \frac{t(1-t)}{n} + t^2 \right)$$

for all $t \in [0, 1], n \in \mathbb{N}$. Equating $s = t$ here finally yields

$$|f(s) - B_n(f, s)| \le \epsilon + \alpha \frac{s(1-s)}{n} \le \epsilon + \frac{\alpha}{n}$$

for all $s \in [0, 1]$. Taking suprema yields $\|f - B_n f\|_\infty \le \epsilon + \alpha/n \le 2\epsilon$ as $n \ge \alpha/\epsilon$. □

Exercise C.1. Verify that

$$B_n(\mathbf{t}^2, t) = \frac{t(1-t)}{n} + t^2 \qquad (t \in [0, 1], n \in \mathbb{N}).$$

Show directly that $B_n(\mathbf{t}^2, \cdot) \to \mathbf{t}^2$ uniformly on $[0, 1]$.

Appendix D

Smooth Cutoff Functions

The purpose of this appendix is to give a proof of the following result, which is Theorem 9.11 in the main text.

Theorem D.1. *Given any finite interval* $[a,b] \subseteq \mathbb{R}$ *there are functions* $\rho_n \in \mathrm{C}^\infty[a,b]$ *satisfying:*

1) $0 \le \rho_n \le 1$;
2) $\rho_n = 1$ *on* $[a + 1/n, b - 1/n]$;
3) $\rho_n = 0$ *on* $[a, a + 1/2n]$ *and on* $[b - 1/2n, b]$.

In order to construct such functions ρ_n, we need an auxiliary result from real analysis.

Theorem D.2. *The function*

$$\eta(x) := \begin{cases} e^{-1/x} & \text{if } x > 0, \\ 0 & \text{if } x \le 0, \end{cases}$$

is in $\mathrm{C}^\infty(\mathbb{R})$.

For the proof we note first that clearly $\eta \in \mathrm{C}^\infty(\mathbb{R} \setminus \{0\})$, so a problem occurs only at $x = 0$. For $x \neq 0$ we obtain the following.

Lemma D.3. *For each* $n \in \mathbb{N}$ *there is a real polynomial* p_n *such that*

$$\eta^{(n)}(x) = \begin{cases} p_n(1/x) e^{-1/x} & \text{if } x > 0, \\ 0 & \text{if } x < 0. \end{cases}$$

Proof. This is a standard induction argument. (We obtain $p_0 = \mathbf{1}$ and $p_{n+1}(t) = t^2(p_n(t) - p_n'(t))$.) □

Since $\lim_{t\to\infty} p_n(t)e^{-t} = 0$, we see that

$$\eta_n(x) := \begin{cases} \eta^{(n)}(x) & \text{if } x \neq 0, \\ 0 & \text{if } x = 0, \end{cases}$$

is a continuous function on $\mathbb{R}$ for each $n \in \mathbb{N}_0$. Now we need the following lemma from elementary analysis.

Lemma D.4. *Let $f \in \mathrm{C}^1(\mathbb{R} \setminus \{0\})$ such that f is continuous at 0 and $a := \lim_{t\to 0, t\neq 0} f'(t)$ exists. Then $f \in \mathrm{C}^1(\mathbb{R})$ and $f'(0) = a$.*

Proof. By the mean value theorem, for $t \neq 0$ we have

$$\frac{f(t) - f(0)}{t - 0} = f'(\theta_t),$$

where θ_t is between 0 and t. If t is close to 0, then θ_t is so as well, and hence $f'(\theta_t)$ is close to a. It follows that $\lim_{t\to 0} {}^1\!/\!_t(f(t) - f(0)) = a$, i.e., f is differentiable at 0 with $f'(0) = a$. Since $\lim_{t\to 0, t\neq 0} f'(t) = a = f'(0)$, f' is continuous at 0, and hence on all of $\mathbb{R}$. □

Proof of Theorem D.2. If we apply Lemma D.4 to the function η_n from above we see that each η_n is differentiable with $\eta_n' = \eta_{n+1}$. Since $\eta_0 = \eta$, Theorem D.2 is proved. □

One can use the function η for a couple of useful constructions. For example, the function

$$\eta_{a,b}(x) := \eta\big((x - a)(b - x)\big) \qquad (x \in \mathbb{R})$$

is in $\mathrm{C}^\infty(\mathbb{R})$, equal to 0 outside (a, b) and strictly positive on (a, b). Hence

$$c_{a,b} := \int_a^b \eta_{a,b}(t)\, \mathrm{d}t > 0.$$

Then the function

$$f_{a,b}(x) := \frac{1}{c_{a,b}} \int_a^x \eta_{a,b}(t)\, \mathrm{d}t$$

is in $\mathrm{C}^\infty(\mathbb{R})$, equal to 0 on $(-\infty, a]$, strictly increasing on $[a, b]$, and equal to 1 on $[b, \infty)$. Finally, given real numbers $a < b < c < d$ the function

$$h := f_{a,b} - f_{c,d}$$

is in $\mathrm{C}^\infty(\mathbb{R})$, equal to 0 outside (a, d), strictly increasing on $[a, b]$, equal to 1 on $[b, c]$ and strictly decreasing on $[c, d]$. Hence by choosing the parameters appropriately, we find our functions ρ_n as required.

Appendix E

Some Topics from Fourier Analysis

Recall that the Fourier transform $\mathcal{F}f$ of a function $f \in \mathrm{L}^1(\mathbb{R})$ is defined by

$$\mathcal{F}f(t) = \int_{\mathbb{R}} \mathrm{e}^{-\mathrm{i}st}\,\mathrm{d}s \qquad (t \in \mathbb{R}).$$

It is known from Section 9.5 that $\mathcal{F}f \in \mathrm{C}_0(\mathbb{R})$, with norm estimate

$$\|\mathcal{F}f\|_\infty \le \|f\|_1 \qquad (f \in \mathrm{L}^1(\mathbb{R})).$$

In this appendix we shall prove Plancherel's identity and the Fourier inversion theorem in elementary formulations. This means that we do not make use of Lebesgue integration theory. Fubini's theorem (cf. Section 11.1), i.e., the interchanging of integrals, will be used only in an elementary form as in [**SS03b**, Appendix 2].

For example, the important formula

$$\int_{\mathbb{R}} (\mathcal{F}f)(s)\,g(s)\,\mathrm{d}s = \int_{\mathbb{R}} f(t)\,(\mathcal{F}g)(t)\,\mathrm{d}t \tag{E.1}$$

is easily proved by interchanging integrals and it holds for $f, g \in \mathrm{L}^1(\mathbb{R})$, by Fubini's theorem. We shall, however, use it only for special cases of f, g, where the validity of Fubini's theorem can be established by elementary means.

E.1. Plancherel's Identity

Consider the special function $\varphi(r) := \mathrm{e}^{-|r|}$. Then

$$\begin{aligned}\mathcal{F}\varphi(s) &= \int_{-\infty}^{\infty} \mathrm{e}^{-|r|}\mathrm{e}^{-\mathrm{i}rs}\,\mathrm{d}r = \int_0^{\infty} \mathrm{e}^{-r}\mathrm{e}^{-\mathrm{i}rs}\,\mathrm{d}r + \int_0^{\infty} \mathrm{e}^{-r}\mathrm{e}^{\mathrm{i}rs}\,\mathrm{d}r \\ &= \frac{1}{1+\mathrm{i}s} + \frac{1}{1-\mathrm{i}s} = \frac{2}{1+s^2} \qquad (s \in \mathbb{R}).\end{aligned}$$

Let $\varphi_\epsilon(r) := \varphi(\epsilon r) = \mathrm{e}^{-\epsilon|r|}$; then by a simple change of variables

$$(\mathcal{F}\varphi_\epsilon)(s) = \frac{2}{\epsilon(1 + (s/\epsilon)^2)}.$$

We have the following auxiliary result, which states that the family of functions $(\frac{1}{2\pi}\mathcal{F}\varphi_\epsilon)_{\epsilon>0}$ is an "approximation of the identity".

Lemma E.1. *If $h \in \mathrm{C_b}(\mathbb{R})$, then*

$$\lim_{\epsilon \searrow 0} \int_{\mathbb{R}} h(s)(\mathcal{F}\varphi_\epsilon)(s)\,\mathrm{d}s = 2\pi\, h(0).$$

Proof. We first note that

$$\int_{\mathbb{R}} (\mathcal{F}\varphi_\epsilon)(s)\,\mathrm{d}s = \int_{\mathbb{R}} \frac{2/\epsilon}{1 + (s/\epsilon)^2}\,\mathrm{d}s = \int_{\mathbb{R}} \frac{2\mathrm{d}s}{1+s^2} = 2\arctan s\Big|_{-\infty}^{\infty} = 2\pi.$$

Therefore we can compute for $\epsilon, \delta > 0$,

$$\begin{aligned}\Big|\int_{\mathbb{R}} h(s)(\mathcal{F}\varphi_\epsilon)(s)\,\mathrm{d}s - 2\pi\, h(0)\Big| &= \Big|\int_{\mathbb{R}} [h(s) - h(0)](\mathcal{F}\varphi_\epsilon)(s)\,\mathrm{d}s\Big| \\ &\le \int_{\mathbb{R}} |h(\epsilon s) - h(0)|\,\frac{2\,\mathrm{d}s}{1+s^2} \\ &\le \int_{-\delta/\epsilon}^{\delta/\epsilon} |h(\epsilon s) - h(0)|\,\frac{2\,\mathrm{d}s}{1+s^2}\,\mathrm{d}s + \int_{|s|\ge\delta/\epsilon} |h(\epsilon s) - h(0)|\,\frac{2\,\mathrm{d}s}{1+s^2}\,\mathrm{d}s \\ &\le 2\pi \sup_{|s|\le\delta} |h(s) - h(0)| + 2\,\|h\|_\infty \int_{|s|\ge\delta/\epsilon} \frac{2\mathrm{d}s}{1+s^2}.\end{aligned}$$

By the continuity of h at $s = 0$ we can choose $\delta > 0$ such that the first summand is arbitrarily small; but as $\epsilon \searrow 0$ the second summand tends to zero. □

For Plancherel's identity, to be treated next, recall the definition

$$\mathrm{C_c}(\mathbb{R}) := \{f \in \mathrm{C}(\mathbb{R}) \mid \exists\, N > 0\ \forall\ |s| \ge N : f(s) = 0\}$$

of the space of continuous functions *of compact support*. We regard a function $f \in \mathrm{C}_0[a,b]$ as an element of $\mathrm{C_c}(\mathbb{R})$ by setting it equal to 0 outside of $[a,b]$.

Theorem E.2 (Plancherel's identity). *If $f \in \mathrm{C_c}(\mathbb{R})$, then $\mathcal{F}f \in \mathrm{L}^2(\mathbb{R})$ and*

$$\|\mathcal{F}f\|_2 = \sqrt{2\pi}\,\|f\|_2 .$$

Proof. Let $f \in \mathrm{C_c}(\mathbb{R})$ and define

$$h(s) := \int_{\mathbb{R}} f(s+r)\overline{f(r)}\,\mathrm{d}r \qquad (s \in \mathbb{R}).$$

Then $h \in \mathrm{C_c}(\mathbb{R})$ again, $h(0) = \|f\|_2^2$ and

$$\begin{aligned}(\mathcal{F}h)(t) &= \int_{\mathbb{R}} h(s)\mathrm{e}^{-ist}\,\mathrm{d}s = \int_{\mathbb{R}}\int_{\mathbb{R}} f(s+r)\overline{f(r)}\mathrm{e}^{-ist}\,\mathrm{d}r\,\mathrm{d}s \\ &= \int_{\mathbb{R}}\int_{\mathbb{R}} f(s+r)\mathrm{e}^{-ist}\,\mathrm{d}s\,\overline{f(r)}\,\mathrm{d}r = \int_{\mathbb{R}}\int_{\mathbb{R}} f(s)\mathrm{e}^{-i(s-r)t}\,\mathrm{d}s\,\overline{f(r)}\,\mathrm{d}r \\ &= \mathcal{F}f(t)\cdot\overline{\mathcal{F}f(t)} = |\mathcal{F}f(t)|^2\end{aligned}$$

for $t \in \mathbb{R}$. Note that since f is integrable, $\mathcal{F}f$ is bounded, and hence $|\mathcal{F}f|^2\,\varphi_\epsilon$ is integrable for each $\epsilon > 0$. Hence we can compute

$$\int_{\mathbb{R}} |\mathcal{F}f(t)|^2\,\varphi_\epsilon(t)\,\mathrm{d}t = \int_{\mathbb{R}} (\mathcal{F}h)(t)\,\varphi_\epsilon(t)\,\mathrm{d}t = \int_{\mathbb{R}} h(s)(\mathcal{F}\varphi_\epsilon)(s)\,\mathrm{d}s$$

by (E.1). By Lemma E.1 we have

$$\lim_{\epsilon\searrow 0}\int_{\mathbb{R}} h(s)\mathcal{F}\varphi_\epsilon(s)\,\mathrm{d}s = 2\pi\,h(0) = 2\pi\,\|f\|_2^2 .$$

On the other hand, as $\epsilon \searrow 0$ one has $\varphi_\epsilon(t) \nearrow 1$ for each $t \in \mathbb{R}$ and uniformly on each bounded interval. It follows that

$$\|\mathcal{F}f\|_2^2 = \int_{\mathbb{R}} |\mathcal{F}f(t)|^2\,\mathrm{d}t = \lim_{\epsilon\searrow 0}\int_{\mathbb{R}} |\mathcal{F}f(t)|^2\,\varphi_\epsilon(t)\,\mathrm{d}t = 2\pi\,\|f\|_2^2$$

as claimed. □

Remark E.3. Our choice $f \in \mathrm{C_c}(\mathbb{R})$ for Plancherel's identity is motivated solely to keep the proof elementary. However, assuming Lebesgue integration theory as known, the same proof works under the more general assumption that $f \in \mathrm{L}^1(\mathbb{R}) \cap \mathrm{L}^2(\mathbb{R})$. In this case, by Cauchy–Schwarz one can show that h is a bounded function, and approximating f with functions from $\mathrm{C_c}(\mathbb{R})$ yields that $h \in \mathrm{C_0}(\mathbb{R})$. In the computation of $\mathcal{F}h$ one has to interchange the order of integration, i.e., use Fubini's theorem.

E.2. The Fourier Inversion Formula

The Fourier inversion formula is a means to compute a function from its Fourier transform. We formulate it here under quite special assumptions in order to avoid arguments from general integration theory.

Theorem E.4 (Fourier Inversion Formula). *Let $f \in C_c(\mathbb{R})$ be such that $\mathcal{F}f \in L^1(\mathbb{R})$. Then*

$$f(s) = \frac{1}{2\pi} \int_{\mathbb{R}} (\mathcal{F}f)(t) e^{ist} \, dt$$

for every $s \in \mathbb{R}$.

Proof. For fixed $s \in \mathbb{R}$ the function $t \mapsto (\mathcal{F}f)(t)e^{ist}$ is continuous and integrable. Since $\varphi_\epsilon \nearrow 1$ uniformly on bounded intervals, it follows that

$$\text{(E.2)} \qquad \frac{1}{2\pi} \int_{\mathbb{R}} (\mathcal{F}f)(t) e^{ist} \, dt = \lim_{\epsilon \searrow 0} \frac{1}{2\pi} \int_{\mathbb{R}} (\mathcal{F}f)(t) e^{ist} \varphi_\epsilon(t) dt.$$

On the other hand, we can interchange the order of integration and obtain

$$\begin{aligned} \int_{\mathbb{R}} (\mathcal{F}f)(t) e^{ist} \varphi_\epsilon(t) dt &= \int_{\mathbb{R}} \int_{\mathbb{R}} f(r) e^{-irt} \, dr \, e^{ist} \varphi_\epsilon(t) dt \\ &= \int_{\mathbb{R}} f(r) \int_{\mathbb{R}} e^{-i(r-s)t} \varphi_\epsilon(t) dt \, dr = \int_{\mathbb{R}} f(r) (\mathcal{F}\varphi_\epsilon)(r-s) \, dr \\ &= \int_{\mathbb{R}} f(r+s) (\mathcal{F}\varphi_\epsilon)(r) \, dr \to 2\pi f(0+s) = 2\pi f(s) \end{aligned}$$

as $\epsilon \searrow 0$ by Lemma E.1. □

E.3. The Carlson–Beurling Inequality

The following result gives a criterion for a function $f \in C_c(\mathbb{R})$ to satisfy $\mathcal{F}f \in L^1(\mathbb{R})$, i.e., the hypothesis in the Fourier inversion theorem.

Theorem E.5 (Carlson–Beurling Inequality[1,2]). *Let $f \in C_0^1[a,b]$. Then $\mathcal{F}f \in L^1(\mathbb{R})$ and*

$$\|\mathcal{F}f\|_1 \le 2\pi \sqrt{\|f\|_2 \cdot \|f'\|_2}.$$

Proof. Integration by parts yields

$$(\mathcal{F}f)(t) = \int_a^b f(s) e^{-ist} \, ds = \frac{1}{it} \int_a^b f'(s) e^{ist} \, ds = \frac{(\mathcal{F}f')(t)}{it} \qquad (t \neq 0).$$

[1]Fritz David Carlson (1888–1952), Swedish mathematician.

[2]Arne Carl-August Beurling (1905–1986), Swedish mathematician.

Hence, for $a, b > 0$ we have

$$\begin{aligned}\|\mathcal{F}f\|_1^2 &= \left(\int_{\mathbb{R}} \frac{\sqrt{a^2+b^2t^2}}{\sqrt{a^2+b^2t^2}} \, |(\mathcal{F}f)(t)| \, \mathrm{d}t\right)^2 \\ &\le \int_{\mathbb{R}} \frac{\mathrm{d}t}{a^2+b^2t^2} \cdot \int_{\mathbb{R}} (a^2+b^2t^2) \, |\mathcal{F}f(t)|^2 \, \mathrm{d}t \\ &= \frac{\pi}{ab}\left(a^2 \|\mathcal{F}f\|_2^2 + b^2 \|\mathcal{F}f'\|_2^2\right) = 2\pi^2\left({}^a\!/_b \, \|f\|_2^2 + {}^b\!/_a \, \|f'\|_2^2\right)\end{aligned}$$

by the Cauchy–Schwarz inequality and Plancherel's identity. (Since $f' \notin \mathrm{C}_0[a,b]$ in general, we have to employ Lemma 9.31 to conclude that $\mathcal{F}f' \in \mathrm{L}^2(\mathbb{R})$.) Now letting $b := 1/a$ and taking the infimum over $a > 0$ yields the claim. □

Remark E.6. Also Theorem E.5 can be generalized to a wider range of functions f. The most common formulation uses the Sobolev space $\mathrm{H}^1(\mathbb{R})$ of L^2-functions on $\mathbb{R}$ that have a weak L^2-derivative.

Corollary E.7. *Every function $f \in \mathrm{C}_0^1[a,b]$ is the Fourier transform of a function $g \in \mathrm{L}^1(\mathbb{R}) \cap \mathrm{C}_0(\mathbb{R})$.*

Proof. By the Carlson–Beurling inequality we have $\mathcal{F}f \in \mathrm{L}^1(\mathbb{R})$, and by Theorem E.4,

$$f(s) = \frac{1}{2\pi} \int_{\mathbb{R}} (\mathcal{F}f)(t) \mathrm{e}^{\mathrm{i}ts} \, \mathrm{d}t = (\mathcal{F}g)(s) \qquad (s \in \mathbb{R}),$$

where $g(t) := (\mathcal{F}f)(-t)/2\pi$. Then $g \in \mathrm{L}^1(\mathbb{R})$ as well. By the Riemann-Lebesgue theorem, $\mathcal{F}f \in \mathrm{C}_0(\mathbb{R})$, hence $g \in \mathrm{C}_0(\mathbb{R})$. □

Note that by Plancherel's identity we also have $g \in \mathrm{L}^2(\mathbb{R})$. However, this also follows from the Cauchy–Schwarz inequality. Indeed,

$$\int_{\mathbb{R}} |g(t)|^2 \, \mathrm{d}t \le \int_{\mathbb{R}} \|g\|_\infty \cdot |g(t)| \, \mathrm{d}t = \|g\|_\infty \, \|g\|_1,$$

and hence $\|g\|_2 \le \sqrt{\|g\|_\infty \, \|g\|_1}$.

Exercises

Exercise E.1. Let $f, g \in \mathrm{C}_c(\mathbb{R})$ and define

$$(f * g)(s) := \int_{\mathbb{R}} f(s-r) g(r) \, \mathrm{d}r \qquad (s \in \mathbb{R}).$$

Show that $f * g \in \mathrm{C}_c(\mathbb{R})$. [Hint: Use that f is uniformly continuous. Why is this so?]

Exercise E.2. Let $\varphi_\epsilon(r) = \mathrm{e}^{-\epsilon|r|}$ for $r \in \mathbb{R}$.

a) Show that $\lim_{\epsilon \searrow 0} \varphi_\epsilon = \mathbf{1}$ uniformly on each bounded subinterval $[a, b]$ of $\mathbb{R}$.

b) Let $f \in \mathrm{C}(\mathbb{R})$ with $f(t) \geq 0$ for all $t \in \mathbb{R}$. Show without using Lebesgue integration theory that

$$\lim_{\epsilon \searrow 0} \int_{\mathbb{R}} f(t)\varphi_\epsilon(t)\, \mathrm{d}t = \int_{\mathbb{R}} f(t)\, \mathrm{d}t,$$

where the integrals are improper Riemann integrals and may take the value $+\infty$.

c) Let $f \in \mathrm{C}(\mathbb{R})$ such that $\int_{\mathbb{R}} |f(t)|\, \mathrm{d}t < \infty$. Show that

$$\lim_{\epsilon \searrow 0} \int_{\mathbb{R}} f(t)\varphi_\epsilon(t)\, \mathrm{d}t = \int_{\mathbb{R}} f(t)\, \mathrm{d}t,$$

where the integrals are improper Riemann integrals.

Exercise E.3. Show that if $f \in \mathrm{C_b}(\mathbb{R})$ is *uniformly continuous*, then

$$\int_{\mathbb{R}} f(s+t)(\mathcal{F}\varphi_\epsilon)(t)\, \mathrm{d}t \to 2\pi\, f(s)$$

uniformly in $s \in \mathbb{R}$ as $\epsilon \searrow 0$.

Appendix F

General Orthonormal Systems

F.1. Unconditional Convergence

In general, even countable orthonormal systems do not come with a canonical enumeration. For example, the trigonometric system

$$e_n(t) := e^{2\pi i n \cdot t} \qquad (t \in [0,1],\ n \in \mathbb{Z})$$

in $L^2(0,1)$ is indexed by the integers, and there are of course many ways to enumerate all integers. Fortunately, the next result shows that the convergence in Theorem 8.15 is independent of the arrangement of the summands.

Lemma F.1. *Let $(e_j)_{j\in\mathbb{N}}$ be an orthonormal system in a Hilbert space H and let $f \in H$. Then*

$$\sum_{j=1}^{\infty} \langle f, e_j \rangle\, e_j = \sum_{j=1}^{\infty} \langle f, e_{\pi(j)} \rangle\, e_{\pi(j)}$$

for every permutation (= bijective map) $\pi : \mathbb{N} \to \mathbb{N}$.

Proof. Note that of course $(e_{\pi(j)})_{j\in\mathbb{N}}$ is also an orthonormal system, and so both series converge, by Theorem 8.15. By the very same theorem,

$$\sum_{j=1}^{\infty} \langle f, e_j \rangle\, e_j = P_F f \quad \text{and} \quad \sum_{j=1}^{\infty} \langle f, e_{\pi(j)} \rangle\, e_{\pi(j)} = P_G f,$$

where

$$F = \overline{\operatorname{span}}\{e_j \mid j \in \mathbb{N}\} \quad \text{and} \quad G = \overline{\operatorname{span}}\{e_{\pi(j)} \mid j \in \mathbb{N}\}$$

Since obviously $F = G$, the lemma is proved. □

From Lemma F.1 we can conclude that if $(e_i)_{i\in I}$ is an orthonormal system in a Hilbert space H, and I is countable, one can use *any* bijection $\pi : \mathbb{N} \to I$ to define

$$\sum\nolimits_{i\in I} \langle f, e_i\rangle\, e_i := \sum\nolimits_{j=1}^{\infty} \langle f, e_{\pi(j)}\rangle\, e_{\pi(j)} \qquad (f \in H).$$

Remark F.2. The property shown in Lemma F.1 is called **unconditional summability** (of the sequence $(e_j)_{j\in\mathbb{N}}$) or **unconditional convergence** (of the series $\sum_{j=1}^{\infty} e_j$). Not every convergent series is unconditionally convergent. For example, the series

$$\sum\nolimits_{n=1}^{\infty} \frac{(-1)^n}{n}$$

converges in $\mathbb{R}$, but not unconditionally: one can rearrange the summands to get a nonconvergent series. This follows from a famous theorem of Riemann, saying that in finite-dimensional spaces unconditional summability is the same as *absolute* summability, and indeed, the series

$$\sum\nolimits_{n=1}^{\infty} \frac{1}{n} = \infty$$

is not convergent. Riemann's theorem is false in infinite dimensions, and here is an example: Let $(e_n)_{n\in\mathbb{N}}$ be an orthonormal system in a Hilbert space H and $\lambda : \mathbb{N} \to \mathbb{R}$. By Theorem 8.15 the series $\sum_{n=1}^{\infty} \lambda_n e_n$ converges unconditionally if and only if $\lambda \in \ell^2$, and it converges absolutely if and only if

$$\sum\nolimits_{n=1}^{\infty} |\lambda_n| = \sum\nolimits_{n=1}^{\infty} |\lambda_n|\, \|e_n\| = \sum\nolimits_{n=1}^{\infty} \|\lambda_n e_n\| < \infty,$$

i.e., $\lambda \in \ell^1$. Since the sequence $(1/n)_{n\in\mathbb{N}}$ is square-summable but not absolutely summable, the series

$$\sum\nolimits_{n=1}^{\infty} \frac{1}{n} e_n$$

converges unconditionally but not absolutely in H.

Remark F.3 (The Case $I = \mathbb{Z}$). In the case of the integers (as in the trigonometric system) we can also go a different way. The infinite double series

$$\sum\nolimits_{n=-\infty}^{\infty} f_n$$

can be interpreted as a double limit $\lim_{n,m\to\infty} \sum_{k=-m}^{n} f_k$, where (in a general metric space (Ω, d)) $x = \lim_{n,m\to\infty} x_{nm}$ simply means

$$\forall \epsilon > 0\ \exists N \in \mathbb{N}\ \forall n, m \geq N : d(x_{nm}, x) < \epsilon.$$

F.2. Uncountable Orthonormal Bases

We now turn to the question how to deal with general, that is, possibly uncountable orthonormal systems. So let H be a Hilbert space and let $(e_i)_{i\in I} \subseteq H$ be an arbitrary orthonormal system therein. For $f \in H$ define

$$I_f := \{j \in I \mid \langle f, e_j \rangle \neq 0\}.$$

Lemma F.4. *In the situation above, the set I_f is at most countable.*

Proof. Consider the set $I_n := \{j \in I \mid |\langle f, e_j \rangle|^2 \geq 1/n\}$. Then if $J \subseteq I_n$ is finite,

$$1/n \cdot \operatorname{card} J \leq \sum\nolimits_{j\in J} |\langle f, e_j \rangle|^2 \leq \|f\|^2,$$

by Bessel's inequality, and hence $\operatorname{card} I_n \leq n \|f\|^2$ is finite. Therefore, $I_f = \bigcup_{n\in\mathbb{N}} I_n$ is at most countable. □

For $f \in H$ we can now define the term

$$\sum\nolimits_{i\in I} \langle f, e_i \rangle e_i := \sum\nolimits_{i\in I_f} \langle f, e_i \rangle e_i$$

and for the sum on the right side we can use any enumeration of the (at most countably many) members of I_f. By what we have seen above, this defines an element of H unambiguously. Then the analogue of Theorem 8.15 holds, replacing everywhere '$\sum_{j=1}^{\infty}$' by '$\sum_{j\in I}$' and '$j \in \mathbb{N}$' by '$j \in I$'. In particular, the analogue of Corollary 8.16 is true:

Theorem F.5. *Let H be a Hilbert space, let $(e_j)_{j\in I}$ be an orthonormal system in H. Then the following assertions are equivalent:*

(i) $\{e_j \mid j \in I\}^\perp = \{0\}$.

(ii) $\operatorname{span}\{e_j \mid j \in I\}$ *is dense in H.*

(iii) $f = \sum\nolimits_{j\in I} \langle f, e_j \rangle e_j$ *for all $f \in H$.*

(iv) $\|f\|^2 = \sum\nolimits_{j\in I} |\langle f, e_j \rangle|^2$ *for all $f \in H$.*

(v) $\langle f, g \rangle_H = \sum\nolimits_{j\in I} \langle f, e_j \rangle \overline{\langle g, e_j \rangle}$ *for all $f, g \in H$.*

An orthonormal system $(e_j)_{j\in I}$ is called **maximal** (or an **orthonormal basis**), if it satisfies the equivalent conditions from the theorem.

Of course, the question arises whether a maximal orthonormal system does always exist. This is indeed the case, by an application of Zorn's lemma.

Theorem F.6. *Every Hilbert space contains a possibly uncountable maximal orthonormal system.*

Proof. Let $\mathcal{M}$ be the set of all orthonormal systems in the Hilbert space H, partially ordered by common set inclusion. Let $\mathcal{K} \subseteq \mathcal{M}$ be a totally ordered subset of $\mathcal{M}$ (a "chain"). Define $E := \bigcup \mathcal{K} = \bigcup_{B \in \mathcal{K}} B$ be the union of elements of $\mathcal{K}$. Then B is clearly an orthonormal system, and it is an upper bound for $\mathcal{K}$.

Hence, every chain in $\mathcal{M}$ has an upper bound, so by Zorn's lemma (Theorem A.2) there is a maximal element, say M, in $\mathcal{M}$. If $M^{\perp} \neq \{0\}$ there is a unit vector $e \in M^{\perp}$, and then $M \cup \{e\}$ is an orthonormal system strictly larger than M. But this contradicts the maximality of M. □

Employing some set-theoretic reasoning one can show in addition that two maximal orthonormal systems in a Hilbert space must be of the same "size" (cardinality), i.e., can be mapped bijectively onto each other. For separable spaces we leave this as an exercise.

Exercise F.1. Let H be a separable Hilbert space. Show that every orthonormal system in H is at most countable.

Bibliography

[Bar95] Robert G. Bartle, *The Elements of Integration and Lebesgue Measure*, Wiley Classics Library, John Wiley & Sons, Inc., New York, 1995.

[Bar01] ———, *A Modern Theory of Integration*, Graduate Studies in Mathematics, vol. 32, American Mathematical Society, Providence, RI, 2001.

[Bau01] Heinz Bauer, *Measure and Integration Theory*, de Gruyter Studies in Mathematics, vol. 26, Walter de Gruyter & Co., Berlin, 2001, Translated from the German by Robert B. Burckel.

[BL06] Kurt Bryan and Tanya Leise, *The* \$25,000,000,000 *eigenvector: the linear algebra behind Google*, SIAM Rev. **48** (2006), no. 3, 569–581 (electronic).

[Boa96] Ralph P. Boas, *A Primer of Real Functions*, fourth ed., Carus Mathematical Monographs, vol. 13, Mathematical Association of America, Washington, DC, 1996, Revised and with a preface by Harold P. Boas.

[Bob05] Adam Bobrowski, *Functional Analysis for Probability and Stochastic Processes*, Cambridge University Press, Cambridge, 2005.

[Car66] Lennart Carleson, *On convergence and growth of partial sums of Fourier series*, Acta Math. **116** (1966), 135–157.

[Che80] Paul R. Chernoff, *Pointwise convergence of Fourier series*, Amer. Math. Monthly **87** (1980), no. 5, 399–400.

[Che01] Ward Cheney, *Analysis for Applied Mathematics*, Graduate Texts in Mathematics, vol. 208, Springer-Verlag, New York, 2001.

[Cla36] James A. Clarkson, *Uniformly convex spaces*, Trans. Amer. Math. Soc. **40** (1936), no. 3, 396–414.

[Con90] John B. Conway, *A Course in Functional Analysis*, second ed., Graduate Texts in Mathematics, vol. 96, Springer-Verlag, New York, 1990.

[Con00] ———, *A Course in Operator Theory*, Graduate Studies in Mathematics, vol. 21, American Mathematical Society, Providence, RI, 2000.

[Die69] Jean Dieudonné, *Foundations of Modern Analysis*, Academic Press, New York, 1969, Enlarged and corrected printing, Pure and Applied Mathematics, Vol. 10-I.

[Die81] ______, *History of Functional Analysis*, North-Holland Mathematics Studies, vol. 49, North-Holland Publishing Co., Amsterdam, 1981, Notas de Matemática [Mathematical Notes], 77.

[dPvR13] Ben de Pagter and Arnoud C.M. van Rooij, *An Invitation to Functional Analysis*, Epsilon Uitgaven, vol. 75, CWI Amsterdam, 2013.

[EN06] Klaus-Jochen Engel and Rainer Nagel, *A Short Course on Operator Semigroups*, Universitext, Springer, New York, 2006.

[Enf73] Per Enflo, *A counterexample to the approximation problem in Banach spaces*, Acta Math. **130** (1973), 309–317.

[Eva98] Lawrence C. Evans, *Partial Differential Equations*, Graduate Studies in Mathematics, vol. 19, American Mathematical Society, Providence, RI, 1998.

[GGK03] Israel Gohberg, Seymour Goldberg, and Marinus A. Kaashoek, *Basic Classes of Linear Operators*, Birkhäuser Verlag, Basel, 2003.

[GL06] Sudhir R. Ghorpade and Balmohan V. Limaye, *A Course in Calculus and Real Analysis*, Undergraduate Texts in Mathematics, Springer, New York, 2006.

[Haa06] Markus Haase, *The Functional Calculus for Sectorial Operators*, Operator Theory: Advances and Applications, vol. 169, Birkhäuser Verlag, Basel, 2006.

[Haa07] ______, *Convexity inequalities for positive operators*, Positivity **11** (2007), no. 1, 57–68.

[Hal63] Paul R. Halmos, *What does the spectral theorem say?*, Amer. Math. Monthly **70** (1963), 241–247.

[Kak39] Shizuo Kakutani, *Weak topology and regularity of Banach spaces*, Proc. Imp. Acad., Tokyo **15** (1939), 169–173.

[Kat04] Yitzhak Katznelson, *An Introduction to Harmonic Analysis*, third ed., Cambridge Mathematical Library, Cambridge University Press, Cambridge, 2004.

[Kör89] Thomas W. Körner, *Fourier Analysis*, second ed., Cambridge University Press, Cambridge, 1989.

[Lan93] Serge Lang, *Real and Functional Analysis*, third ed., Graduate Texts in Mathematics, vol. 142, Springer-Verlag, New York, 1993.

[Lax02] Peter D. Lax, *Functional Analysis*, Pure and Applied Mathematics (New York), Wiley-Interscience [John Wiley & Sons], New York, 2002.

[LM54] Peter D. Lax and Arthur N. Milgram, *Parabolic equations*, Contributions to the theory of partial differential equations, Annals of Mathematics Studies, no. 33, Princeton University Press, Princeton, N. J., 1954, pp. 167–190.

[LM06] Amy N. Langville and Carl D. Meyer, *Google's PageRank and Beyond: the Science of Search Engine Rankings*, Princeton University Press, Princeton, NJ, 2006.

[Mon73] Antonie Frans Monna, *Functional Analysis in Historical Perspective*, John Wiley & Sons, New York-Toronto, Ont., 1973.

[Nau02] Joachim Naumann, *Remarks on the prehistory of Sobolev spaces*, Unpublished Preprint, 2002.

[Pie07] Albrecht Pietsch, *History of Banach Spaces and Linear Operators*, Birkhäuser Boston, Inc., Boston, MA, 2007.

[Pin00] Allan Pinkus, *Weierstrass and approximation theory*, J. Approx. Theory **107** (2000), no. 1, 1–66.

[Ran02] Inder K. Rana, *An Introduction to Measure and Integration*, second ed., Graduate Studies in Mathematics, vol. 45, American Mathematical Society, Providence, RI, 2002.

[Rie10] Frigyes Riesz, *Sur les opérations fonctionnelles linéaires.*, C. R. Acad. Sci., Paris **149** (1910), 974–977 (French).

[Rud87] Walter Rudin, *Real and Complex Analysis*, third ed., McGraw-Hill Book Co., New York, 1987.

[Rud91] ———, *Functional Analysis*, second ed., International Series in Pure and Applied Mathematics, McGraw-Hill, Inc., New York, 1991.

[Sch05] René L. Schilling, *Measures, Integrals and Martingales*, Cambridge University Press, New York, 2005.

[Sho77] Ralph E. Showalter, *Hilbert Space Methods for Partial Differential Equations*, Pitman, London, 1977, Monographs and Studies in Mathematics, Vol. 1.

[Sok11] Alan D. Sokal, *A really simple elementary proof of the uniform boundedness theorem*, Amer. Math. Monthly **118** (2011), 450–452.

[SS03a] Reinhard Siegmund-Schultze, *The origins of functional analysis*, A History of Analysis, Hist. Math., vol. 24, Amer. Math. Soc., Providence, RI, 2003, pp. 385–407.

[SS03b] Elias M. Stein and Rami Shakarchi, *Fourier Analysis*, Princeton Lectures in Analysis, vol. 1, Princeton University Press, Princeton, NJ, 2003.

[Ste73] Lynn A. Steen, *Highlights in the history of spectral theory*, Amer. Math. Monthly **80** (1973), 359–381.

[Wei85] Karl Weierstrass, *Über die analytische Darstellbarkeit sogenannter willkürlicher Functionen einer reellen Veränderlichen. (On the analytic representability of so called arbitrary functions of a real variable.)*, Sitzungsberichte der Akademie zu Berlin (1885), 633–639 and 789–806 (German).

[Wer00] Dirk Werner, *Funktionalanalysis*, extended ed., Springer-Verlag, Berlin, 2000.

[Wey40] Hermann Weyl, *The method of orthogonal projection in potential theory*, Duke Math. J. **7** (1940), 411–444.

[You88] Nicholas Young, *An Introduction to Hilbert Space*, Cambridge Mathematical Textbooks, Cambridge University Press, Cambridge, 1988.

[Zei95a] Eberhard Zeidler, *Applied Functional Analysis*, Applied Mathematical Sciences, vol. 109, Springer-Verlag, New York, 1995, Main principles and their applications.

[Zei95b] ———, *Applied Functional Analysis*, Applied Mathematical Sciences, vol. 108, Springer-Verlag, New York, 1995, Applications to mathematical physics.

Symbol Index

Subject Index

Author Index

Selected Published Titles in This Series

156 **Markus Haase,** Functional Analysis: An Elementary Introduction, 2014
155 **Emmanuel Kowalski,** An Introduction to the Representation Theory of Groups, 2014
154 **Wilhelm Schlag,** A Course in Complex Analysis and Riemann Surfaces, 2014
153 **Terence Tao,** Hilbert's Fifth Problem and Related Topics, 2014
152 **Gábor Székelyhidi,** An Introduction to Extremal Kähler Metrics, 2014
151 **Jennifer Schultens,** Introduction to 3-Manifolds, 2014
150 **Joe Diestel and Angela Spalsbury,** The Joys of Haar Measure, 2013
149 **Daniel W. Stroock,** Mathematics of Probability, 2013
148 **Luis Barreira and Yakov Pesin,** Introduction to Smooth Ergodic Theory, 2013
147 **Xingzhi Zhan,** Matrix Theory, 2013
146 **Aaron N. Siegel,** Combinatorial Game Theory, 2013
145 **Charles A. Weibel,** The K-book, 2013
144 **Shun-Jen Cheng and Weiqiang Wang,** Dualities and Representations of Lie Superalgebras, 2012
143 **Alberto Bressan,** Lecture Notes on Functional Analysis, 2013
142 **Terence Tao,** Higher Order Fourier Analysis, 2012
141 **John B. Conway,** A Course in Abstract Analysis, 2012
140 **Gerald Teschl,** Ordinary Differential Equations and Dynamical Systems, 2012
139 **John B. Walsh,** Knowing the Odds, 2012
138 **Maciej Zworski,** Semiclassical Analysis, 2012
137 **Luis Barreira and Claudia Valls,** Ordinary Differential Equations, 2012
136 **Arshak Petrosyan, Henrik Shahgholian, and Nina Uraltseva,** Regularity of Free Boundaries in Obstacle-Type Problems, 2012
135 **Pascal Cherrier and Albert Milani,** Linear and Quasi-linear Evolution Equations in Hilbert Spaces, 2012
134 **Jean-Marie De Koninck and Florian Luca,** Analytic Number Theory, 2012
133 **Jeffrey Rauch,** Hyperbolic Partial Differential Equations and Geometric Optics, 2012
132 **Terence Tao,** Topics in Random Matrix Theory, 2012
131 **Ian M. Musson,** Lie Superalgebras and Enveloping Algebras, 2012
130 **Viviana Ene and Jürgen Herzog,** Gröbner Bases in Commutative Algebra, 2011
129 **Stuart P. Hastings and J. Bryce McLeod,** Classical Methods in Ordinary Differential Equations, 2012
128 **J. M. Landsberg,** Tensors: Geometry and Applications, 2012
127 **Jeffrey Strom,** Modern Classical Homotopy Theory, 2011
126 **Terence Tao,** An Introduction to Measure Theory, 2011
125 **Dror Varolin,** Riemann Surfaces by Way of Complex Analytic Geometry, 2011
124 **David A. Cox, John B. Little, and Henry K. Schenck,** Toric Varieties, 2011
123 **Gregory Eskin,** Lectures on Linear Partial Differential Equations, 2011
122 **Teresa Crespo and Zbigniew Hajto,** Algebraic Groups and Differential Galois Theory, 2011
121 **Tobias Holck Colding and William P. Minicozzi II,** A Course in Minimal Surfaces, 2011
120 **Qing Han,** A Basic Course in Partial Differential Equations, 2011
119 **Alexander Korostelev and Olga Korosteleva,** Mathematical Statistics, 2011
118 **Hal L. Smith and Horst R. Thieme,** Dynamical Systems and Population Persistence, 2011